ISO 14001 and Beyond:
Environmental Management Systems in the Real World

edited by
Christopher Sheldon

For Jennifer, and a vision shared in Granada

ISO 14001 and Beyond

ENVIRONMENTAL MANAGEMENT SYSTEMS IN THE REAL WORLD

 *Greenleaf* **Publishing** *1997*

© 1997 Greenleaf Publishing unless otherwise stated.

Published by Greenleaf Publishing
Greenleaf Publishing is an imprint of
Interleaf Productions Limited
Broom Hall
Sheffield S10 2DR
England

Typeset by Interleaf Productions Limited and printed on environmentally friendly, acid-free paper from managed forests by The Cromwell Press, Melksham, Wiltshire.

British Library Cataloguing in Publication Data:
ISO 14001 and beyond : environmental management systems in
    the real world
    1.Environmental responsibility 2. Environmental protection -
    Standards
    I.Sheldon, Christopher
    658.4 ' 08 ' 0218

ISBN 1874719012 Pbk
ISBN 1874719063 Hbk

Contents

Part 3
Tactical Responses: Managers at the Greenface

Foreword

John Elkington
Chairman, SustainAbility
Chairman, The Environment Foundation
Member, EU Consultative Forum on the Environment

FEW FIELDS breed acronyms as prolifically as the environment and, increasingly, sustainable development. Some—such as DDT, CFCs or PCBs—are borrowed directly from other fields; some (such as TQEM, or Total Quality Environmental Management) are adapted to meet new needs; others, however, evolve directly in response to the requirements of environmental management. Just when we had all finally come to terms with CER, DfE, EIA, LCA and EMAS, for example, up pops EMS. In contrast with some acronyms, which are here today and gone tomorrow, this is one that looks set to be around for decades.

In recent years, we have all learned to talk about our environmental management systems—rather like our web-sites—as if we have had them for years, which for most of us is not the case. But the new focus on the EMS, or environmental management system, is necessary, welcome and, indeed, overdue.

Worldwide, the publication of the ISO 14001 EMS standard, alongside the European Union's EMAS approach to EMS and environmental reporting, is adding real urgency to the debate. *ISO 14001 and Beyond* provides a unique overview of progress to date and gives the reader an informed look into the future. The book pulls together EMS experts from Europe, North America and the Asia-Pacific region.

If we had approached the environmental management challenge logically, the EMS evolutionary process would have been very different. Consider. The late 1980s saw a growing interest in the concept of environmental auditing, although most progress was made in Europe in the early years of the 1990s. Some companies that had audited their operations for years warned that many of those promoting auditing saw it as a stepping stone to mandatory public reporting of environmental emissions and performance. In the event, this was the way things developed, although the main impetus in the field of corporate environmental reporting to date—with most initiatives still undertaken on a voluntary basis—has been a new sense of corporate citizenship.

As more and more companies join the ranks of the report-makers, however, reporting remains haunted with a paradox. Logically, before companies began to report externally on environmental performance, initial efforts should have been focused upon developing appropriate environmental accounting methodologies for measuring performance and then installing full management structures and systems for auditing against these. Only then would a company environmental report be produced.

So much for the ideal approach. In practice, companies have tended to kick off with auditing, followed by reporting. Only now are many of them—having already committed themselves to reporting—starting to think about whether their EMS, if it even exists, is up to the task. Very few companies have taken the next step and started to consider the implications and applications of environmental and full-cost accounting techniques.

Nonetheless, the emergence of a full-blown, international discussion of environmental accounting, and of the relevant performance indicators, is only a question of time. This is the deep-seated transition that is under way behind the fleeting fashions described by Chris Sheldon in his excellent Introduction.

Most, if not all, of the contributors would agree that environmental management systems are necessary—and that international standards and some forms of certification can add real value. Paradoxically, this is true even at a time of growing globalisation and international competition. In the old order, globalisation and competition would have been used as arguments for throttling back in the environmental area. In the emerging world order, with growing attention being focused on the real and perceived environmental performance of materials, products, processes, companies, industries, and even entire economies, the competitive challenge now is one of integrating the latest environmental management systems and tools as fast as possible— and certainly before key customers begin insisting on their existence and use as a basic condition of supply contracts.

Almost all of the contributors would accept that the ISO 14001 framework will push forward environmental management in thousands of companies worldwide. A number, particularly those in Part 3, explain what it is like to develop and operate EMS systems in companies or local authorities, and describe some of the barriers.

But there remain real questions about whether ISO 14001 will push progress fast enough, or in the right direction. The process of getting to ISO 14001 inevitably involved the dilution of many original ambitions, particularly in the face of sustained opposition from US business interests.

Longer term, we will see growing corporate interest in taking the ISO 14001, EMAS and 'Responsible Care' EMS platforms and extending them to cope with the very much more challenging requirements imposed by the sustainable development agenda. Internally, the challenge will be to integrate health, safety, environmental and quality management systems, wherever possible. Externally, stakeholder dialogue will be a key component of successful corporate environmental strategies, a point well made by Andrew Blaza and Nicky Chambers.

Whatever your business or area of activity, EMS standards and requirements will have an impact. Market approaches and instruments will increasingly replace heavy-handed command-and-control approaches to environmental protection, and self-regulation will very often be the name of the game. As a result, EMS standards such as ISO 14001 and EMAS will soon come under further evolutionary pressure. Before long, if work we did recently for European Partners for the Environment (EPE) comes to fruition, we may see a shift towards what we dub SMAS (or Sustainability Management and Auditing Scheme). Another of those ubiquitous, if useful, acronyms.

Introduction
ISO 14001 and Beyond: EMS in the Real World
Christopher Sheldon

*T*HE PUBLICATION of ISO 14001, the international standard specification for an environmental management system (EMS), is the latest imprimatur of business acceptance of this fledgling discipline. Or is it? Depending on how one looks at it, an international standard can be the embodiment of best available technique or the lowest common denominator; a global green passport or an environmental Esperanto phrasebook.

In most boardrooms, standardisation is usually considered a 'MEGO' ('my eyes glaze over') subject, so how is it that a mere technical standard can excite such polarisation and earnest breast-beating? How does one separate the movers and shakers from the ranters and ravers? How can one start a journey when the destination is shrouded in uncertainty?

Questions such as these may have obscured the issues surrounding the basic principles of environmental management systems. In the midst of such confusion, it would be easy to dismiss the whole idea as just another of the latest creations from the fashion houses of management theory. After all, the hem-lines of performance management go up and down each year, the colour co-ordination otherwise known as 'the learning organisation' seems to change on a regular basis, and even the classic cuts of management such as Total Quality get re-worked for every new 'season'. Each new business model has its turn upon the catwalk; each spawns a flurry of conferences, case studies and learned papers; each catches the headlines for a brief moment. Managers everywhere, with an almost insatiable appetite for more and better results, line up to see the new collections, year on year. Why should environmental management be any different?

Because, in a sense, it cannot afford to be fashionable.

Issues such as accelerated environmental change are no longer a matter of private belief, nor are industry's contributions to the speed of such change the starting point for an interesting philosophical discourse. In getting to grips with managing environmental impacts, we all have to start from where we are, not from where we would like to be. This book addresses some of the basic questions about the nature of EMSs, key vehicles for industry in its task: questions that managers are asking themselves, prompted by a new perspective, full of unfamiliar and vaguely expressed values: 'Where are we now?'; 'Where are we going?'; 'How long have we got?'

At the start of what promises to be a worldwide explosion of interest in standardised EMSs, this book looks at their creation, their use, and their limitations, attempting to discover the essential truth about this important management tool and where it will take industry. Collected in this one volume are some of the best leading-edge environmental management thinkers and practitioners from all over the world, sharing their experiences and thoughts. It is designed to provide the reader with enough information with which to form an opinion as to the future, and how that will influence their subsequent actions. It also provides reassurance that, although the problems are real, so are the solutions.

The first step on such a thousand-mile journey is to understand the nature of industrial self-regulation and its relationship to the law—hence the first section of this book, which is dedicated to an international overview of ISO 14001 and related standards. Here you will find in-depth analysis of the creation of the standards by those who had an active part in the development process, Dick Hortensius and Mark Barthel. Those who have been at the ringside watching with interest have different perspectives, and their differing views make interesting comparative reading. Gleckman and Krut offer their 'Uncommon Perspective' with an eye to international policy-making, which is countered by the more pragmatic approach of Christopher Bell. Even the broad future of the standards in relation to trade issues and competitiveness is open to view, thanks to Donal O'Laoire and his interesting 'chemical' equation, while Mark Smith takes a critical look at some of the issues facing the development of product labelling standards, one of the most important tools to be used in tandem with environmental management systems if improvements are to have any meaning at all.

Armed with a better knowledge of the top-level context of the 14000 series, readers can then proceed to the second section of the book, where early strategic trends have been spotted by those intimately connected with the use and development of EMSs within organisations. The implications for managers of such trans-corporate issues as training, communications and external certification are examined in detail. Andy Wells shares his experience of strategic training development for environmental work in the UK, highlighting the issues yet to be addressed worldwide. Read in conjunction with Gabriele Crognale's chapter on US-based training, the chapters give an interesting overview of this vital area. Meanwhile, the impact of 14001 on the style and content of communication strategies is assessed by Andrew Blaza and Nicky Chambers, who put forward a new approach to building a dialogue with business stakeholders that could reap benefits for many organisations. Continuing the theme of external contact with markets, Tim Sunderland calls for companies to face several questions squarely before deciding whether certification by a third party is right for them: a timely reality-check among the ISO 14001 hype.

Deeper in the second section, the focus moves from the international to the national with contributions based on national studies and work in Russia (Hutchison *et al.*), Japan (Kurasaka) and the developing economies (Davy). The reader will also see that the scope of the chapters moves from the medium to the long term. What lies beyond the systems themselves and what impacts can managers and executives expect once they have got the immediate problems under control? A key contribution to our understanding comes from Philip Sutton, who shows exactly how to get the best out of a 14001 system by maximising its business and environmental potential. This forms an interesting bridge to the proposals outlined by Andrea Spencer-Cooke, who uses her own recent work to show how environmental management can become a very different animal indeed: something she refers to as 'sustainability management'.

Finally, the third section of the book addresses the tactical responses of managers around the globe, and how they overcame the problems of implementing the ISO 14001 standard. From case studies to implementation workshops, this section contains the early-warning signals of what managers can expect to find in the way of challenges when installing such a new system. Phil Stoesser, for instance, provides a classic example of how a standard can be implemented through interpretation of the requirements to fit the needs of the business. At the same time, Alison Bird covers in-depth work on the difficulties faced by managers who set objectives and targets.

In all cases, the writing has a clarity of thinking and expression that comes with having tackled real problems on a day-to-day basis—the unmistakable mark of experience. This experience certainly informs the contribution from Professor Jacqueline Cramer, who draws from her work on product design to confirm some of the ideas expressed by both Philip Sutton and Mark Smith in their chapters. Where precision matters most, of course, is in the realm of risk assessment, a vital discipline explored in full by David Shillito, a man ideally placed to give form and meaning to its use within the context of integrated management systems of which environment is an important part.

Among those who have already chosen the route of external assessment, arguments still rage as to whether or not ISO 14001 is a Hollywood remake of an original European idea—bigger, blander and preceded by a tidal wave of publicity. Either way, the experience of those involved with EMAS may provide valuable pointers for an international audience. In so international a book, no apologies need be made for including three chapters across the book's sections devoted to different facets of that experience: its use in local government in the UK (Riglar), a case study from Germany (Gelber *et al.*) and an informed investigation of the impact of the scheme on small and medium-sized enterprises (Hillary). Whether or not EMAS is ultimately ruled to be a barrier to trade and removed, the experience of those involved will provide valuable information for whatever other initiatives are undertaken.

In all three sections, the contributors have been encouraged to outline what they think lies 'beyond ISO 14001' in terms of relationships, frameworks and value shifts. The major themes of the book are how EMSs fit in with regards to regulations, certification, previous experience and, finally, that most dimly-illuminated of goals, sustainable development.

Prior to reading the book, pragmatists may be tempted to answer the question of what lies beyond ISO 14001 by saying 'a review of the standard', due to start in 1999. This is true, but it is hardly sufficient to help frame current actions. To get the best from the book, I think it is worth pausing to look at some of the major influences that will help to shape our collective future in the field of environmental management. I would also encourage readers to keep these influences in mind throughout their contact with the ideas in this volume.

Until now, many countries have relied on the traditional command-and-control ethic of legislation in the environmental arena. There is, however, already a discernible trend in policy-making to tie regulations to market forces in such a way that it is the market itself that provides the dynamic for change. This keeps ahead of spiralling demands placed upon the law, bypasses sclerotic centralised institutions, and avoids backlogs of unrealised changes—or so the theory goes. In the particular case of ISO 14001, the standard is exciting interest in the idea that self-regulation, coupled to the motive force of consumer demand and supply-chain pressure, could eventually replace what is seen as the increasingly complex and inefficient webs of national legal requirements. Whether this is true remains to be seen.

ISO 14001 may share common principles with ISO 9000, the international standard for quality management standards, but, from the legal standpoint, the relationship between ISO 14001, legislation and public policy-making is much closer than between ISO 9000 and similar social expressions of accept-ability. The European Eco-Management and Audit Scheme (EMAS) could even be characterised as a European Regulation about self-regulation. Going further still, certification of standardised environmental management systems walks the thin line between socially-policed regulation and privately-policed self-regulation. The effect of any future blurring of that line by inconsistent application of certification practice would undermine the credibility of industry's ability to run its own affairs. Much, too, will depend on the way the standard is interpreted during the process of certification, or by individual companies *en route* to self-declaration.

There is something else that marks ISO 14001 out as unusual. International standards are normally associated with anodyne expressions of what has already become recognised as commercial and industrial common sense. Many see them as a worthy way of spreading best practice, a repository of industrial wisdom, but rarely are they seen as cutting edge. Not so with ISO 14001, where one of the reasons it has excited so much interest is because it has been published in advance of widespread industrial experience. For many, taking up environmental management is embarking on a journey to an unexplored region. Perhaps it is only natural that such travellers would question the veracity of any map under the circumstances.

Those who do take up the challenge may need reassurance that the standard can deliver against the promises made on its behalf. Such reassurance can be found, but we must first dispose of the idea that any management tool is neutral, and only as good as the manager that uses it. Even that most basic of tools beloved of engineers worldwide, the hammer, comes in a variety of shapes and sizes. It is certainly possible to choose an inappropriate version for the job in hand (e.g. the famous 'sledgehammer–nut' scenario). Not only do all tools have important characteristics that change the way they are used under particular circumstances, but some of those facets lie hidden until they are discovered in use. It is only when the standard has been applied within the context of real businesses, with real impacts and real problems, that the true nature of the system will be revealed.

Precision of thought and expression will be our only weapon to ensure that consumers and governments do not believe the wilder claims made on behalf of the standard. Certainly, there are those who are willing to exploit the ignorance and confusion that exists around EMS standards by portraying it as an answer to the problems of achieving sustainable development. This book sounds a clarion call against the damage that such woolly thinking can do—it reminds us all of the responsibility to differentiate between sustainable development (long-term/strategic/policy-making) and environmental management (short-to-medium term/tactical/problem-solving). ISO 14001 is thus not so much a 'global foundation for sustainable development', but an important stepping stone down the road that leads to that ultimate destination—only if its users maximise its potential.

In creating such potential for the global marketplace, it is also easy to react more strongly to the hyperbole of the standard's public relations campaign than it is to examine it in depth. However, the drawing-up behind arguments protecting national interests runs counter to the requirements of a response to an essentially global situation. It is not simply that the environment fails to recognise national boundaries: multinational industries left them behind years ago. The concept of the 'globalisation' of industry has now been worn smooth as business coinage to the point where smaller enterprises may feel removed from the impacts of such developments. Many companies and executives have espoused global values—now is the time to deliver against the rhetoric. As interdependence becomes acknowledged as a market force and not simply an ecological principle, this delivery may require more than the easy solution of passing the buck down the supply chain.

Let me offer this closing thought. To state the obvious, commerce and industry are an integral part of the socio-economic construct that human beings have woven in the name of co-operation and furtherance of their own species. The process of value change in that construct is akin to the movement of tectonic plates across the face of the globe: it is hard to spot anything happening on a day-to-day basis, but, when enough pressure has built up, there is an almighty earthquake that no one can ignore. The continued seismic activity in attitudes towards the environment goes deeper than mere opinion, deeper than the belief that informs them, right into the core of values that motivates individuals and organisations.

Sometimes these tremors can be spotted without the aid of complicated instrumentation (consumer buying habits and increasing legislation). Sometimes the indicators are more complex, with deeper implications for the future (changing political manifestos, the consequences of globalisation); below them lie subsonic vibrations with even more potential for explosive commercial change if they are ignored (new theories of economic growth and social equity).

One part of the response to such perceived shifts came on 1st September 1996, when ISO 14001 was published simultaneously in every country around the globe. Everything changed—because there was a new tool. Nothing changed—because a tool needs to be used to be effective. What lies beyond the standard and this book is choice. It may not be an easy choice, but it is at least ours to make.

Part 1 Laying Down the Law

How Self-Regulation Came of Age

1 Beyond 14001
An Introduction to the ISO 14000 Series

Dick Hortensius and Mark Barthel

THE IDEA of setting up an ISO Technical Committee (TC) to develop international standards on environmental management did not simply come out of thin air. During its preparations for the 1992 United Nations Conference on Environment and Development (UNCED; the 'Earth Summit'), in Rio de Janeiro, Brazil, the Business Council for Sustainable Development (BCSD) came to the conclusion that the international business community would need to develop international standards on environmental performance to ensure that companies operating around the world could do so on a level playing field. These findings complemented the prevalent thinking at that time in COPOLCO (the committee platform ISO provides for consumer interests), which believed that it was imperative that the growing number of initiatives in the field of product eco-labelling were harmonised at an international level. These two drivers, along with other forces, led ISO to conclude that the development of international standards for environmental management was essential to environmental protection and future world trade.

Accordingly, in 1991, ISO established a Strategic Advisory Group on Environment (SAGE) to investigate all the areas of environmental management and performance where the development of international standards might be beneficial to the business community. Although SAGE did not itself have the authority to develop standards, it was able to recommend to ISO that a new technical committee be established to develop international standards in environmental management.

In 1993 ISO set up this new technical committee, ISO/TC207 'Environmental Management', to develop the standards proposed by SAGE and to

---

ISO/TC207 'Environmental Management'
Secretary: Jim Dixon, CSA, Canada
Chairperson: George Connell, Canada

ISO/TC207/SC1 'Environmental Management Systems'
Secretary: Christina Senabulya, BSI, UK
Chairperson: Oswald A. Dodds MBE, NB Contract Services, UK

ISO/TC207/SC2 'Environmental Auditing and Related Environmental Investigations'
Secretary: Dick Hortensius, NNI, Netherlands
Chairperson: John. C. Stans, Det Norske Veritas, Netherlands

ISO/TC207/SC3 'Environmental Labelling'
Secretary: John Henry, SSA, Australia
Chairperson: Bill Dee, Australian Consumer and Competition Committee, Australia

ISO/TC207/SC4 'Environmental Performance Evaluation'
Secretary: Steve Cornish, ANSI, USA
Chairperson: Dorothy Bowers, Merck & Company, Inc., USA

ISO/TC207/SC5 'Life-Cycle Assessment'
Secretary: Pascal Poupet, AFNOR, France
Chairperson: Manfred Marsmann, Bayer, Germany

ISO/TC207/SC6 'Terms and Definitions'
Secretary: Einar Bache, NAS, Norway
Chairperson: Havard Hjulstad, Radet for teknisk terminologi, Norway

ISO/TC207/WG1 'Environmental Aspects in Product Standards'
Chairperson: Klaus Lehmann, DIN, Germany

ISO/TC207/WG2 'Forest Management'
Chairperson: Ken Shirley, NZ Forest Owners Association, New Zealand

---

Figure 1: *Who's Who in ISO/TC207 'Environmental Management'*

investigate the possibility of developing further supporting standards. The secretariat to this new technical committee was provided by the Canadian Standards Association (CSA). ISO/TC207 subsequently created six subcommittees and one working group,[1] and allocated work on different areas of environmental management to them. The secretariats of these subcommittees (and the working group) are held by personnel from different national standards bodies, and the chairmen of the subcommittees are appointed by the same member body that holds the Secretariat (see Fig. 1). The complete organisational structure of ISO/TC207 is given in Appendix A and some basic information on the history, organisation and working procedures of ISO in Appendix B.

The UK standards body, BSI, holds the secretariat to Subcommittee 1 (ISO/TC207/SC1), which is responsible for the development of ISO 14001, *Environmental Management Systems: Specification with Guidance for Use*, and ISO 14004, *Environmental Management Systems: General Guidelines on Principles, Systems and Supporting Techniques*.

The Dutch national standards body, NNI, holds the Secretariat to ISO/ TC207/SC2 'Environmental Auditing and Related Environmental Investigations', which has been responsible for the development of ISO 14010, *Guidelines for Environmental Auditing: General Principles*, ISO 14011, *Guidelines for Environmental Auditing: Audit Procedures: Auditing of Environmental Management Systems*, and ISO 14012, *Guidelines for Environmental Auditing: Qualification Criteria for Environmental Auditors*.

The two SC1 standards were published on 1st September 1996; while the SC2 standards were published on 1st October 1996. These five standards form the backbone of the ISO 14000 series of environmental management standards.[2] Other standards under development in the ISO 14000 series include standards for environmental performance evaluation (EPE), lifecycle assessment (LCA) and environmental labels and declarations. (See Appendix C for a summary of the current status of the ISO 14000 series standards.)

In June 1996, ISO/TC207 met for the fourth time in Rio de Janeiro four years after the Earth Summit had taken place there. This meeting was hosted by the Brazilian national standards body, ABNT, and was attended by more than 650 delegates from over 60 countries and liaison bodies, illustrating the worldwide interest in these standards.

This introductory chapter provides an overview of the development of these important environmental standards and establishes the relationships and interlinkages between the different ISO 14000 series standards. The significance of these standards to the business community, government and consumers are then discussed, along with an outline of the various stakeholder groups involved in their development. The chapter ends with some remarks on the integration of environmental management with quality and occupational health and safety management and an outlook on the future of the ISO 14000 series.

■ *The ISO 14000 Standards and their Interlinkages*

The scope of ISO/TC207, and therefore the scope of the ISO 14000 series, covers 'standards in the field of environmental management tools and systems'. Explicitly excluded from this scope are:

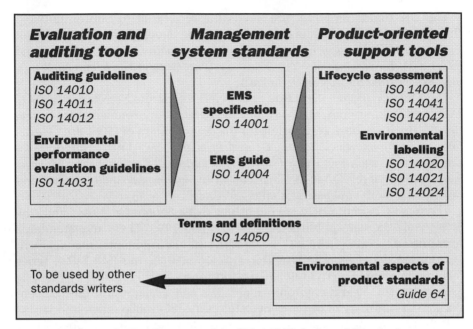

Figure 2: *The Structure of the ISO 14000 Series of Standards*

- Test methods for pollutants (because these are dealt with by other technical committees; see Appendix B)
- Setting limit values regarding pollutants and effluents
- Setting of environmental performance levels
- Standardisation of products

From this scope it becomes clear that the ISO 14000 series will not interfere with any national environmental legislation. The setting of limit values and performance levels remains the prerogative of (national) governmental authorities. The ISO 14000 series should provide management tools for organisations that aim to control their environmental aspects and improve their environmental performance. How other stakeholders, such as governments and consumers, may benefit from this will be addressed later on in this chapter.

Figure 2 shows, in schematic form, the relationship between the standards in the ISO 14000 series. The philosophy behind the whole structure is that the environmental management system of a company is of central importance and that the other standards are intended to support specific elements of the organisation's environmental policy and management.[3] In addition to the standards being produced in ISO/TC207, there are of course many other technical standards that companies can use within their environmental management system: for example, standards for the execution of environmental measurements and standards that relate to (environmental) technical requirements for installations, etc. There are many other ISO technical committees that develop these types of standards.

Environmental Management Systems

Of central importance in the ISO 14000 series are the environmental management system standards, ISO 14001 and ISO 14004. These standards allow an organisation to take a systematic approach to the evaluation of how its activities, products and services interact with the environment and to control those activities to ensure that established environmental objectives and targets are met.

The basic structure of environmental management according to ISO 14001 and ISO 14004 is shown in Figure 3.

☐ ISO 14001

ISO 14001 specifies the requirements for an environmental management system (EMS) against which an organisation may be certified by a third party.[4] It provides a specification detailing the requirements an organisation must meet in order to achieve third-party certification, including:

- The development of an environmental policy
- Identification of environmental aspects
- Establishment of relevant legal and regulatory requirements
- Development of environmental objectives and targets
- The establishment and maintenance of an environmental programme in order to achieve its objectives and targets
- Implementation of an EMS, including training, documentation, operational control and emergency preparedness and response
- Monitoring and measurement of operational activities, including record-keeping
- EMS audit procedures
- Management review of an EMS to determine its continuing suitability, adequacy and effectiveness

Annex A of ISO 14001 contains additional guidance on the use of the requirements and is intended to avoid misinterpretation of the specification;

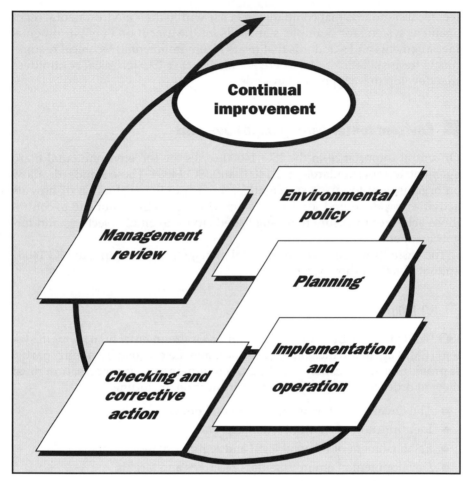

Figure 3: *Environmental Management System Model for ISO 14001 and 14004*

while Annex B contains information on the linkages and broad technical correspondences between ISO 14001 and ISO 9001, *Quality Systems: Model for Quality Assurance in Design, Development, Production and Servicing* (the equivalent Quality Management Systems standard).

☐ *ISO 14004*

ISO 14004 has been developed by ISO to provide additional guidance for organisations on the design, development and maintenance of an EMS. It is not intended to be certified against. In effect, ISO 14001 is intended to provide the specification for an organisation's EMS, while ISO 14004 will act as a 'stepping stone' to the specification for many organisations who may feel that they require some additional guidance and background information on the underlying principles, systems and supporting techniques necessary to develop an EMS.

ISO 14004 includes details of:

- Internationally-accepted principles of environmental management and how they can be applied to the design and development of all the components of an EMS
- Practical examples of the issues an organisation will need to ensure they have addressed in the design of their EMS, including guidance on how to identify the environmental aspects and impacts associated with their activities, products and services
- Practical help sections to provide an organisation with assistance in navigating the various stages of EMS design, development, implementation and maintenance

In order for the environmental management system to function properly, support tools are required. Further management tools are needed to check whether the EMS meets the requirements of the organisation, is properly implemented, and that the desired outcomes are achieved. Accordingly, several standards on environmental auditing and one standard for the evaluation of environmental performance are being developed.

■ *Environmental Auditing*

The three auditing standards, ISO 14010, 14011 and 14012, provide support tools allowing an organisation to monitor whether its EMS conforms to planned arrangements (especially the requirements of ISO 14001), to monitor its effectiveness and suitability and to suggest how the data gathered during an audit may be formatted for presentation to management in order to facilitate management review of the performance of the EMS and the organisation's overall environmental performance. The environmental audit is an independent verification of whether the environmental management system conforms to specified criteria. Such an audit can be carried out by the company itself (internal audit), but this is often done by external auditors: for example, within the framework of a certification procedure (third-party audit).

☐ *ISO 14010*

ISO 14010 is the generic environmental auditing standard that lays down guidelines on the general principles involved in environmental auditing. It contains:

- Definitions of environmental audit and related terms
- General principles of environmental auditing, e.g. objectivity, independence and competence of auditors, the application of systematic audit procedures and the reliability of audit findings and conclusions
- A framework for the structure and format of an audit report

☐ *ISO 14011*

ISO 14011 goes one step further in that it provides guidance on the audit procedures that facilitate the planning and conduct of an EMS audit. It includes guidance on:

- Audit objectives
- The roles and responsibilities of those involved in the audit, including the client
- The development of the audit scope, audit plan and working documents
- Collecting audit evidence and review of audit findings
- Preparation and documentation of the audit report

☐ *ISO 14012*

ISO 14012 lays down guidance on the minimum qualification criteria for environmental auditors and lead auditors. It provides information on the following requirements:

- Educational and professional qualifications
- Formal and on-the-job training
- Competencies and personal attributes and skills

ISO 14012 also provides pointers to employers and clients on how to evaluate the suitability of auditors according to the above criteria.

■ *Environmental Performance Evaluation*

There is a growing need for information on the environmental performance of companies. More and more companies are taking advantage of this by publishing an annual environmental report. An important question here is: What information is worth reporting? It is possible that environmental performance indicators (EPIs) can play a useful part. EPIs can be used in internal and external reporting, as well as for so-called benchmarking activities.

ISO 14031, *Environmental Performance Evaluation*, which is being developed by Subcommittee 4 to ISO/TC207, provides guidance for an organisation on how to identify suitable environmental indicators in order to measure its environmental performance against criteria set by management (e.g. the organisation's environmental policy, objectives and targets). Environmental performance evaluation (EPE) is based on the continual collection and assessment of data and information, and its conversion to appropriate environmental indicators, to provide a current evaluation, as well as trends over time, about the environmental aspects of an organisation's activities, products and services. It is this distinction that separates EPE from environmental audits, assessments, investigations and reviews which are conducted to provide information at specific points in time determined by the organisation's management.

ISO 14031 provides guidance on:

- The design and development of EPE programmes that reflect the significant environmental aspects organisations can be expected to control or have an influence over
- The establishment of environmental indicators for three core evaluation areas, namely management, operational and environmental evaluation areas
- Planning environmental performance evaluation
- Management considerations
- Selecting indicators for the three evaluation areas
- Collecting, analysing and evaluating data
- Reporting and communicating EPE information to management, employees and interested parties
- Reviewing and improving the EPE process

ISO 14031 also contains annexes that provide additional guidance on all of the above. Examples of indicator categories for each of the evaluation areas (management, operational and environmental) are given below:

- Number of employees participated in environmental training programmes, number of specific activities such as environmental audits and emergency drills, number of products undergoing lifecycle assessment in a certain time period
- The use of natural resources and energy, emissions to air, water and soil
- Greenhouse effect (e.g. global warming potential of air emissions), ozone layer depletion (e.g. increased frequency of skin cancer), and acidification (e.g. ambient air concentrations of SO_2 and NO_x)

The standard also provides basic criteria for the formulation of EPIs, such as practical usability and cost-effectiveness, objectivity, verifiability, reproducibility and comparability.

In addition to these systems-oriented support tools, standards have also been developed to support product-oriented environmental management (an essential element of environmental management, according to ISO 14001). These standards address two subject areas, namely lifecycle assessments and environmental labelling and declarations.

■ *Environmental Labelling*

Government environmental policy and the business community are focusing more and more on the environmental aspects of products and how to communicate product information to interested parties. The transfer of product information to third parties, including consumers, plays an important part of this process. In Subcommittee 3 to ISO/TC207, a number of initiatives in this field are being standardised under the broad heading of 'Environmental Labelling'. The subcommittee's work also includes the development of standards that provide a basis for the 'acceptability' of the environmental declarations and claims that usually accompany any form of 'eco-labelling'.

ISO/TC207/SC3 distinguishes between various types of labelling:

- Type I labels are based on a voluntary, multiple-criteria-based practitioner programme that awards labels claiming overall environmental preference of a product within a particular product category based on lifecycle considerations.
- Type II labels encompass informative environmental self-declaration claims.
- Type III labels include those containing quantified product information that has been subjected to independent verification using preset indices.

Currently, Types I and II receive the most attention and are usually indicated as environmental labelling programmes and manufacturers' self-declarations.

ISO 14024, *Environmental Labels and Declarations: Environmental Labelling Type 1: Guiding Principles and Procedures*, describes the principles on which environmental labelling programmes should be based and gives guidelines for drawing up testing criteria and the execution of the labelling procedure. In many countries, environmental programmes have been set up in recent years: with well-known quality marks such as the 'Blue Angel' from Germany and the 'White Swan' from Scandinavia. In 1993, the United States Environmental Protection Agency (US EPA) carried out a comparative survey into the environmental inspection programmes in 24 countries.[5] The conclusion of this survey is that the programmes differ considerably as regards set-up, selection of product groups, requirements that must be met for obtaining an environmental quality mark, and the way in which products are verified. This can result in uncertainty among consumers about the value of the various quality marks. In addition, it is inconvenient for producers who want to obtain a quality mark in several countries. It is hoped that ISO 14024 will be used to harmonise the criteria used in different national eco-labelling schemes and will provide some uniformity in the operating methods of environmental inspection programmes.

ISO 14021, *Environmental Labels and Declarations: Environmental Labelling: Self-Declaration Environmental Claims: Terms and Definitions*, is currently being prepared by ISO/TC207/SC3 and can best be described as an international environmental claims code. In this standard, guidelines are given for environmental claims by suppliers of products and services. In the case of environmental claims, one therefore has to think of information about the recyclability and/or re-usability of the product itself or the packing materials used. In the standard, basic conditions are formulated for the use of such claims. In addition to this standard, standards will also be developed for the use of environmental symbols (for example the well-known 'Möbius loop') and for methods of verifying environmental claims.

■ Lifecycle Assessment

In order to assess the environmental impact of a product properly and to be able to compare alternative products with one another, the lifecycle assessment (LCA) tool has been developed. An LCA is a systematic way of evaluating the environmental effects of a product using a 'cradle-to-grave' approach, in which all the life stages of a product from raw material usage to final disposal are taken into account. There is a great need for objective information to be able to make choices from an environmental perspective between products and materials and in product design. A well-known example is the choice between disposable plastic coffee cups or earthenware mugs in company canteens or between linen or cotton 'terry' and paper disposable nappies. Historically, LCA studies in this field have often produced as many different results as studies, because of the different LCA methodologies available. The need for harmonisation in this area has proved that it is a typical subject for standardisation.

Subcommittee 5 of ISO/TC207 is working on a series of standards on the subject of LCA. ISO 14040, *Life Cycle Assessment: Principles and Framework*, is a general document that provides the principles and general framework for LCA. ISO 14040 addresses issues such as:

- Terms and definitions in the field of LCA
- The various phases of an LCA (see Fig. 4)
- The methodological framework for the various phases
- Reporting and critical review of LCAs

ISO 14040 provides general guidance on the different phases of an LCA. Separate but detailed guidelines for the lifecycle inventory and the lifecycle impact assessment are included in ISO 14041, *Life Cycle Assessment: Life Cycle Inventory Analysis*, and ISO 14042, *Life Cycle Assessment: Impact Assessment*. Recent discussions have illustrated how difficult it is to reach consensus on the scientific methods to be used for inventory and impact analysis. It is

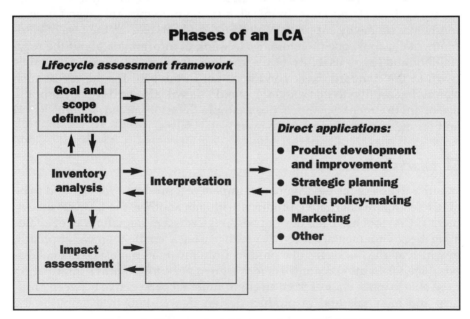

Figure 4: *Phases of a LCA according to ISO 14040*

therefore sometimes difficult to lay down a 'best-practice' or 'most acceptable' method. For the time being, the achievement of a clear understanding of the basic principles and assumptions of leading researchers and how these affect the LCA results, provides an important step towards the continuing harmonisation of LCA methodologies.

■ Terms and Definitions

When developing the standards, it is essential to ensure the clarity and consistency of terminology and definitions used throughout the ISO 14000 series standards. Subcommittee 6 of ISO/TC207 has summarised all the terms and definitions used in the various subcommittees and working groups of ISO/TC207 and has incorporated them in a terminology standard: ISO 14050, *Environmental Management: Terms and Definitions*. ISO/TC207/SC6 also performs the function of intermediary between the different subcommittees with the aim of achieving uniformity in terminology throughout the ISO 14000 series. The subcommittee also provides expert advice on how certain terms can best be defined in accordance with the ISO terminology standards.

■ Environmental Aspects of Product Standards

ISO is developing many standards for product specifications across a range of industrial sectors. This is carried out by many different technical committees in all sorts of industrial sectors. It is recognised that it is important when preparing such standards to review the environmental aspects. For example, it may still often happen that, for certain types of fire-resistant materials, the use of asbestos is specified, or that the use of virgin materials is required where perhaps recycled material would also suffice. ISO/TC207 Working Group 1 has developed a guide for standards writers to ensure that the environmental aspects of product specification standards are fully taken into account during the standards development process. This guide may be referenced in the ISO Directives on Methodology for the Development of International Standards. In doing so, ISO technical committees are urged to apply the guide, e.g. when reviewing their standards according to the usual five-year revision period.

■ The Significance of the ISO 14000 Series for the Business Community, Governments and Consumers

The international business community has been at the forefront of the development of the ISO 14000 series standards and has sought to develop standards for environmental management that are compatible with those previously developed in ISO for quality management (the ISO 9000 series). The final structure and content of management systems and auditing standards from the ISO 14000 series reflect the broad achievement of this objective.

ISO 14001 will have a significant influence on the field of environmental management that is comparable with the impact of ISO 9001 in the fields of quality assurance and quality management. There was a good reason why comparable numbering was chosen: ISO/TC207 realised that if a standard for environmental management systems was to be accepted by the business community, it must be compatible with ISO 9001. This does not mean that the two standards are identical, but they do show great compatibility as regards the management system principles on which they are based: 'plan', 'do', 'check' and 'act'. This so-called 'Deming Circle' basically means: ensure good control of critical business activities according to plan, check whether the predetermined requirements are met and, where necessary, make corrections. An important difference from ISO 9001 is that ISO 14001 obliges a company to improve continually the environmental management system to achieve increasingly better environmental performance. Hence, ISO 14001 describes a more dynamic system, with which it must be possible to meet the increasingly more stringent requirements that society places on the business community in the environmental field.

It is worth noting at this point that the business community is not solely responsible for development of the ISO 14000 series standards. The standards development process has also included participation from government agencies, consumer groups and non-government organisations (NGOs). Similarly, the standards are intended for use by a whole range of users, from commercial organisations to other social actors, such as governments and consumers who thus also have an interest in their content (see Fig. 5).

The ISO 14000 series actively promotes the adoption of sound environmental management practices by organisations and it is here that government can only benefit from the use of the standards. The standards themselves may, in turn, form a good basis for agreements between the business community, the authorities and the public; and aid the execution of environmental policy and legislation. Certainly, in some countries, this is already the case. For example, in the Netherlands, environmental management systems are important tools in the execution of so-called 'covenants' (voluntary agreements on environmental issues between government and industrial sectors) and the granting, implementation and enforcement of environmental licences. Such co-operative partnerships bring the goal of sustainable environmental and social development one step closer to realisation.

Consumers, too, derive advantages from the sound environmental behaviour of companies. For the neighbours of industrial enterprises, improved

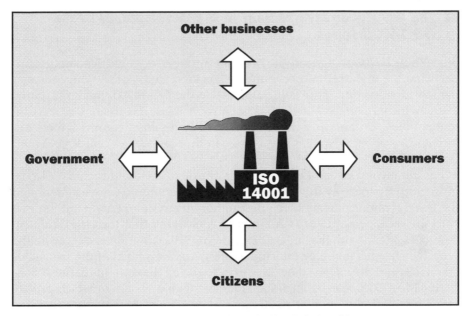

Figure 5: *ISO 14001 Stakeholder Relationships*

control of the release of environmentally-polluting emissions is important, as well as the minimisation of the risk of industrial accidents or incidents. The environmental management system of a company is an important tool for achieving these objectives.

In addition, consumers are the buyers of the organisations' products, and they benefit from products that have been produced with minimum impact on the environment and for which the environmental effects during and after use are as low as possible. Lifecycle assessments put companies in a position to trace environmentally-critical links in the lifecycle of a product, and thus to take environmental aspects into account in choices in design and production. Environmental labelling methods aim to give consumers reliable information about the environmental aspects of products. As a result, consumers are in a position to make informed choices in their purchasing behaviour.

It thus becomes clear that the ISO 14000 series fits in with new relationships between the business community, the authorities and the general public where environmental aspects are involved. The standards are important tools for social self-regulation in the environmental field where there is increasing talk of the government withdrawing; of companies taking initiatives themselves to achieve an improvement in their environmental performance; and of consumers making their choices partially on the basis of environmental aspects.

■ *The Worldwide Involvement in the Development of the ISO 14000 Series*

Since ISO/TC207 was set up in 1993, the number of countries and groups participating in the technical committee has increased rapidly. This indicates the worldwide importance being attached to the ISO 14000 standards. However, an increased membership has not always made discussions on the contents of the standards any easier, as inevitably different countries have different ideas about what they want to see in the standards, according to their cultures, politics, regulatory regimes, level of economic development, state of environmental awareness, etc. Their attitude to the standards may also be influenced by the way in which the business community, government and the general public interact with one another on environmental matters.

In the meetings of ISO/TC207 it is often striking that the non-European countries attend with the largest delegations. This situation is closely linked with the changes that they anticipate following the arrival of the ISO 14000 standards and the interest they have in influencing the contents of these standards. Most European countries are familiar with a system in which private standards and government regulations complement one another (e.g. the interaction of standards and regulations within the framework of the so-called 'new-approach' EC directives). In Europe, the ISO 14000 series fits in well with this approach and will not lead to significant institutional change.

Traditionally, the approach to environmental regulation and protection in the US has been very different from the European approach outlined above. However, more recently, the US EPA has signalled a possible shift from its previous dependence on a classical (and rather rigid) 'command-and-control' approach towards a more flexible business-oriented policy for environmental protection. This shift in thinking has arisen as a result of the considerable pressure being placed on the EPA by a Republican party with a majority in the US House of Representatives. The Republicans would like to see increased emphasis placed on de-regulation, with industry adopting self-assessment and voluntary responsibility schemes (such as 'Responsible Care') and the EPA is developing environmental improvement partnerships and programmes with industry. The use of the standards from the ISO 14000 series, particularly ISO 14001, to achieve these aims is already being considered by the EPA, and would fit in well in this new agency–industry relationship.

In other parts of the world, the newly industrialised and developing countries from South-East Asia, Africa and South America are concerned about possible trade restrictions arising from the use of the standards in contractual specifications. Some of these countries lack a well-defined and effective legislative and enforcement infrastructure (often as a result of insufficient government environmental policy), but wish to improve their record of environmental performance. It is in these countries, particularly where multinational companies operate, that the ISO 14000 series standards are already being used by organisations to improve environmental performance and prevent themselves from falling foul of potential trade restrictions. Perhaps this explains why it is these very countries who are coming to the ISO/TC207 meetings in great numbers.

Given the different approaches and the regulatory situations in all the countries involved, it is often no simple matter to reach consensus on the content of the standards!

A standard such as ISO 14001 must on the one hand be flexible enough to be used in many different situations by all types of organisations, but on the other hand strict enough to ensure that the system leads to an improvement in an organisation's environmental performance. Here, the European countries favour an ISO standard that can be used as a European standard within the Eco-Management and Audit Scheme (EMAS) Regulation issued by the European Union (see Fig. 6 for more information). In order to achieve this, the ISO standard had to meet a number of essential requirements mentioned in that Regulation, which various non-European countries saw as an unacceptable political influence of the ISO process. The different use of a standard such as ISO 14001 in a legal framework also results in different visions of its content. As previously stated, European standards are often prepared as a further technical development of legislation organised by private parties (for example, the European product standards for the benefit of the open European market). A sufficient degree of detail in the standards is then desirable: for example, to prevent interpretation problems in product certification and declarations of conformity.

In the US, standards play a part as state-of-the-art documents in legal liability procedures, and, as a result, everything that is included in a standard can be used against a company. In that context, great detail in a standard is in fact less desirable. In addition, in the US, certainly in the past, people are less interested in third-party certification. The US therefore favours a brief, non-prescriptive standard, which offers individual companies sufficient opportunities to fill in the system according to their own understanding.

In the case of Japan, there is concern that an ISO standard will not fit comfortably with the Japanese business culture and that problems may arise, particularly with the implementation of ISO 14001. Japan has had bad experiences in this respect with the ISO 9000 quality system, which fits in poorly with the Japanese approach to quality assurance. Under pressure from the European market, the Japanese business community was forced to implement the ISO 9000 system and obtain certification on a wide scale. Against

MANY OF THE standards that are developed by ISO/TC207 can be of benefit in the execution of European environmental legislation. With the Vienna agreement between CEN and ISO, there have arisen various possibilities for giving the ISO standards the status of European standards, and hence for ensuring that these standards are implemented in full, and uniformly, in the European countries that are members of CEN.

For the draft standards for environmental management systems and environmental auditing, the choice has been made to vote for them in parallel as European standards; this is the quickest way to implement ISO standards in Europe. The reason for this is that the European Commission has given CEN a mandate (assignment) to develop standards to support the execution of the EMAS Regulation (Eco-Management and Audit Scheme—a voluntary recognition system for good environmental management combined with external environmental reporting). CEN thereupon decided that it is not desirable to develop separate European standards for environmental management and auditing, but that it would be best for all parties to take over the ISO standards (i.e. ISO 14001, 14010, 14011 and 14012). In the preparation of the ISO standards, the European countries have ensured that, as far as the contents are concerned, these meet the requirements of the EMAS regulation.

To execute its mandate fully, CEN developed the following additional documents:

- A **Comparison Document** giving a clause-by-clause comparison between EMAS requirements on the one hand, and the corresponding ISO 14001 requirements on the other; in a third column, the related informative text of ISO 14001 Annex A and the ISO auditing guidelines is given.

- A **Bridging Document** identifying the areas where EMAS establishes detailed system and auditing requirements that are not specifically covered by the ISO standards, or where the agreements between EMAS and ISO may not be readily apparent.

- An **Explanatory Note** that includes some information on the possible use of the bridging document.

The most interesting and controversial document is the Bridging Document. Although some CEN members believe that the ISO 14001 requirements adequately cover the corresponding EMAS requirements, others do not. In particular, Germany, Denmark and Ireland are of the opinion that ISO 14001 has serious shortcomings compared to EMAS. The current Bridging Document can be seen as a compromise between the CEN members to bridge the perceived gap between ISO 14001 and EMAS. The Bridging Document contains guidance on a number of EMS-related elements that companies should take into account when implementing ISO 14001 to ensure that they also meet corresponding EMAS requirements. The main issues dealt with in the Bridging Document are: continual improvement and the application of EVABAT; the scope of environmental management system audits; audit frequency; environmental review; and good management practices.

Upon publication as European Standards, ISO 14001, 14010, 14011 and 14012 will be adopted by all the European Member States without modification as national standards, and any existing national standards for environmental management systems in Europe, including BS 7750, have to be withdrawn within six months of the publication of the European Standards. Therefore, after 31st March 1997, ISO 14001 will be the only standard for environmental management systems in Europe.

Figure 6: *The Relationship between the ISO 14000 Series and the European Standards*

this background, it is a credit to the ISO/TC207 membership that agreement has been reached on the contents of ISO 14001 in such a short period of time. (For an additional viewpoint on Japan, see Chapter 8.)

The achievement of worldwide agreement on the contents of the environmental auditing standards has proved equally problematical. The political and commercial sensitivity of environmental information plays an important part in this. In the case of a third-party audit, external personnel, often independent of the organisation, must observe extensively the operational activities and management practices of an organisation to verify that organisation's assertions that it is complying with the requirements of the standard and relevant legislation. Of course, such an auditor has a confidentiality obligation, but that does not detract from the fact that he/she prepares an audit report outside the organisation detailing the activities of the company and any non-compliances.

North American companies in particular are afraid that environmental auditors will have too much freedom, so that, in addition to reporting findings (established facts), they will go too far in drawing all sorts of conclusions about the environmental performance of the organisation. The existence of such sensitive information and judgements outside of the organisation may lead to a situation where the organisation may feel vulnerable to litigation.

■ Sector-Specific Guidance on the Use of the Standards

Following considerable speculation about the uniform implementation of the ISO 14000 standards, it has been suggested on several occasions that ISO/TC207 should address the needs of different industrial sectors by developing sector-specific guidance on the use of the standards, especially ISO 14001. To date, ISO/TC207 has indicated its reluctance to produce such guidance by passing several resolutions on the issue. It is generally accepted that, until practical experience has been gained in the implementation of the first edition of ISO 14001, no sector-specific guidance will be forthcoming from ISO/TC207 and its subcommittees. However, it is still possible that when the revision process for ISO 14001 commences in 1999, the needs of specific industrial sectors may be further addressed. (ISO stipulates that all of its standards should be subject to a five-year revision cycle to ensure their continuing relevance to industry.)

☐ Forest Management

However, at its meeting in Rio de Janeiro in June 1996, ISO/TC207 resolved to set up a new working group on forest management following intense lobbying from that sector. The establishment of this group follows the work of an Informal Study Group on Sustainable Forest Management which studied the application of ISO 14001 to the sustainable forestry management sector for about a year. Following fierce discussions after the group reported to ISO/TC207 during its 1996 plenary in Rio de Janeiro, it was granted permission to continue its work as a formally-constituted working group reporting to ISO/TC207. This new working group, ISO/TC207/WG2, was charged with preparing a report describing informative reference material to assist in the implementation of ISO 14001 by forest organisations.

■ Integration of Environmental Management with Quality and Occupational Health and Safety Management

Clearly, environmental management is not the only aspect of an organisation's activities that requires the day-to-day involvement of management personnel. Quality and occupational health and safety are also core management concerns which must be addressed and communicated to interested parties (see Fig. 7). The incorporation of quality management into the corporate culture has spread enormously since the 1980s with the introduction of the ISO 9000 series.

While environmental management has proved to be the focal point of the 1990s, occupational health and safety (OHS) management systems may soon receive similar attention. For example, the British standard BS 8800, *Guide to Occupational Health and Safety Management Systems*, was published in April 1996. This guidance standard is based on the management systems model proposed in ISO 14001. Similarly, in the Netherlands, a guideline for occupational health and safety management is also being prepared. Here, too, good harmonisation with ISO 14001 will be aimed for. However, when ISO held an international workshop to discuss the need for international standardisation in the OHS arena in September 1996, the general consensus appeared to be that ISO members believed that a discussion of the need for standardised OHS management systems was premature.

The challenge for standardisation in the coming decade is the effective harmonisation of these different management areas and their possible integration within a suite of integrated management systems standards. Switzerland has recently proposed to the ISO Technical Management Board that a Strategic Advisory Group on Management Systems be established to develop a strategic vision for ISO on the harmonisation and integration of standards for management systems and the related certification procedures. The ISO Technical Management Board (the management body for ISO) is due to come

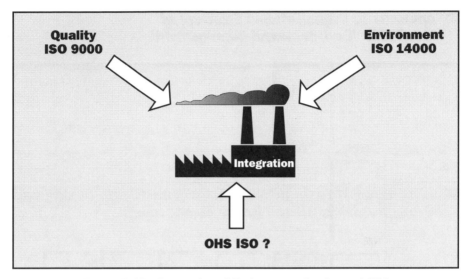

Figure 7: *The Integration of Environment, Quality and OHS*

to a decision on the need for integrated management systems standards in early 1997.

■ The Future of the ISO 14000 Series

It is clear that the ISO 14000 series standards have the potential to improve the environmental performance of organisations worldwide. ISO/TC207 certainly hopes that they will form the accepted model for management of corporate environmental activities in the future, and will be supported by a wide range of interested parties. ISO/TC207/SC1 and SC2 have resolved to commence the revision cycles for their standards in 1999—at this point it may be necessary for these core standards to address new issues, such as sustainable development, as a result of changes in the environmental awareness of governments, organisations and individuals.

◼ *Appendix A: Organisational Structure of ISO/TC207 'Environmental Management'*

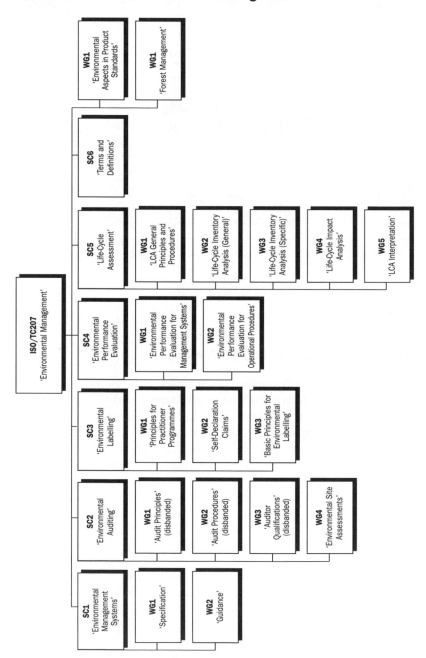

ISO/TC207 'Environmental Management'

SC1 'Environmental Management Systems'
- **WG1** 'Specification'
- **WG2** 'Guidance'

SC2 'Environmental Auditing'
- **WG1** 'Audit Principles' (disbanded)
- **WG2** 'Audit Procedures' (disbanded)
- **WG3** 'Auditor Qualifications' (disbanded)
- **WG4** 'Environmental Site Assessments'

SC3 'Environmental Labelling'
- **WG1** 'Principles for Practitioner Programmes'
- **WG2** 'Self-Declaration Claims'
- **WG3** 'Basic Principles for Environmental Labelling'

SC4 'Environmental Performance Evaluation'
- **WG1** 'Environmental Performance Evaluation for Management Systems'
- **WG2** 'Environmental Performance Evaluation for Operational Procedures'

SC5 'Life-Cycle Assessment'
- **WG1** 'LCA General Principles and Procedures'
- **WG2** 'Life-Cycle Inventory Analysis (General)'
- **WG3** 'Life-Cycle Inventory Analysis (Specific)'
- **WG4** 'Life-Cycle Impact Analysis'
- **WG5** 'LCA Interpretation'

SC6 'Terms and Definitions'

WG1 'Environmental Aspects in Product Standards'
WG1 'Forest Management'

■ *Appendix B: The International Organisation for Standardisation (ISO)*

☐ *Organisational Structure*

ISO is a non-governmental organisation established in 1947 for the purpose of developing worldwide standards to improve international communication and collaboration and to promote the smooth and equitable growth of international trade. ISO work results in international technical agreements which are published as international standards. Technical fields covered by ISO show a wide variety: from more 'classical' fields, such as screw threads, to new topics such as information technology and ergonomics. The members of ISO are national standards bodies, one for each country. Standards development work is carried out in technical committees (nowadays more than 200 technical committees are in existence within ISO), mostly subdivided into subcommittees and working groups, in which experts from many countries co-operate to reach agreement on the technical content of international standards. ISO member bodies can become 'P' (participating) or 'O' (observatory) members of a technical committee. The secretariats of committees and subcommittees are held by national standards bodies and each committee establishes liaisons ('L' [membership]) with international organisations that can contribute to its work.

☐ *Procedures within ISO*

A key principle in the ISO standards development process is reaching **consensus** with **all interested parties** on a **voluntary** basis.

Before a new work item is included in the working programme of a technical committee or subcommittee, a so-called proposal for a new work item is circulated for ballot among the members of the (sub)committee. Each proposal should be accompanied by a justification explaining why international standardisation of this item is considered necessary; this is to limit the working programmes to the most essential items. When an item is approved, it is allocated to an existing or newly-established working group. Members of working groups are experts nominated by the P-members and liaison organisations of the (sub)committee; to each working group a convenor or project leader is appointed by the (sub)committee to which it belongs. Those experts co-operate to reach an agreement about the technical content of proposed international standards. When agreement has been reached and the working draft of the group is accepted by the (sub)committee, it is circulated as a so-called 'Committee Draft' for ballot to the members of the (sub)committee.

When accepted and when comments made during this voting procedure are dealt with, the revised document may be circulated as a so-called 'Draft International Standard' to the members of the technical committee and all ISO member bodies for ballot. When accepted, the revised document can be published as a full International Standard. Each technical committee has the obligation to seek full backing of the liaison organisations for those international standards in which they have interest. The whole procedure takes at least several years. This may seem long, but ISO work is based on the principle of reaching consensus as far as possible and this takes time.

Every five years, an International Standard must be reviewed by the responsible technical committee to decide whether it should be confirmed, revised or withdrawn. This procedure should ensure that ISO standards are kept up to date.

☐ *ISO Activities in the Environment*

ISO took on work in the field of environment in the beginning of the 1970s. In 1971, two new technical committees were established: ISO/TC146 was charged with the task of preparing standards in the field of air quality and ISO/TC147 with a similar task in the field of water quality. Both committees mainly deal with standardisation of measuring methods, related statistical techniques and reporting formats. Currently, a large number of standards in these fields have been finalised. In 1985, ISO/TC190 'Soil Quality' was established completing the range of ISO committees for specific environmental compartments. With the establishment of ISO/TC207 'Environmental Management', ISO took a new approach in the field of environment-related standardisation. The ISO 14000 series of standards for environmental management will allow companies to manage their full range of activities that may have an effect on the environment.

■ *Appendix C: Working Programme of ISO/TC207 'Environmental Management'*

ISO number	Document title	Expected publication date*
SC1 'Environmental Management Systems'		
ISO 14001	*Environmental Management Systems: Specification with Guidance for Use*	1996
ISO 14004	*Environmental Management Systems: General Guidelines on Principles, Systems and Supporting Techniques*	1996
SC2 'Environmental Auditing and Related Environmental Investigations'		
ISO 14010	*Guidelines for Environmental Auditing: General Principles*	1996
ISO 14011	*Guidelines for Environmental Auditing: Audit Procedures: Auditing of Environmental Management Systems*	1996
ISO 14012	*Guidelines for Environmental Auditing: Qualification Criteria for Environmental Auditors*	1996
ISO 14015	*Guidelines for Environmental Site Assessments*	> 2000
SC3 'Environmental Labelling'		
ISO 14020	*Environmental Labels and Declarations: Basic Principles*	1998
ISO 14021	*Environmental Labels and Declarations: Self-Declaration Environmental Claims: Terms and Definitions*	1998
ISO 14022	*Environmental Labels and Declarations: Self-Declaration Environmental Claims: Symbols*	1999
ISO 14023	*Environmental Labels and Declarations: Self-Declaration Environmental Claims: Testing and Verification Methodologies*	1999
ISO 14024	*Environmental Labels and Declarations: Environmental Labelling Type I (Voluntary, Multiple Criteria-Based Practitioner [Labelling]) Programmes: Guiding Principles and Procedures*	1998
ISO 14025	*Environmental Labels and Declarations: Environmental Labelling Type III (Quantified Product Information Label based upon Independent Verification Using Preset Indices): Guiding Principles and Procedures*	1999
SC4 'Environmental Performance Evaluation'		
ISO 14031	*Guidelines for Environmental Performance Evaluation*	1998
SC5 'Life-Cycle Assessment'		
ISO 14040	*Life-Cycle Assessment: Principles and Framework*	1997
ISO 14041	*Life-Cycle Assessment: Life-Cycle Inventory*	1998
ISO 14042	*Life-Cycle Assessment: Life-Cycle Impact Assessment*	1999
ISO 14043	*Life-Cycle Assessment: Life-Cycle Assessment Interpretation*	1999
SC6 'Terms and Definitions'		
ISO 14050	*Environmental Management: Vocabulary*	1997
WG1 'Environmental Aspects of Product Standards'		
Draft ISO Guide 64	*Guide for the Inclusion of Environmental Aspects in Product Standards*	1997

* Members of ISO maintain registers of currently-valid International Standards.

■ Notes

1. In 1996, a second working group directly accountable to the technical committee was established, namely on forest management.
2. ISO 14001, 14010, 14011 and 14012 were simultaneously adopted as European Standards (ENs) by CEN, the European standards body, as they were subject to a parallel voting process under the auspices of the Vienna Agreement (see Fig. 6 for more detail).
3. Although many standards, such as ISO 14031 (EPE) and the ones on LCA, can be used as stand-alone documents.
4. It is also possible for organisations to make a self-declaration of compliance to the requirements of the standard.
5. US EPA, *Status Report on the Use of Environmental Labels Worldwide* (EPA 742-R-9-93-001; Washington, DC: Office of Pollution Prevention and Toxics, September 1993).

Neither International nor Standard

The Limits of ISO 14001 as an Instrument of Global Corporate Environmental Management

Harris Gleckman and Riva Krut

ISO 14001 is a new draft international standard for environmental management systems. It was conceived in response to important government, multilateral and industry initiatives, but it did not try to develop in an integrated fashion with these initiatives. Instead, ISO created a parallel industry initiative, and is now actively in the process of positioning this standard as a definitive 'green seal' of environmentally-sound business operations. At the same time, pre-existing governmental initiatives are being stereotyped as overly bureaucratic and burdensome, and presenting inconsistent 'standards' that hamper industry's capacity to be competitive.

ISO 14001 needs to be better understood by the public and industry, and especially by public authorities, so that it can be evaluated in the coming months as it is marketed to non-participating industry (particularly small and medium-sized businesses), to governments (including those in the developing world and Eastern Europe) and non-governmental organisations. This article presents four key questions from a public policy perspective, committed to environmental protection and to democratic international decision-making.

Like all complex rhetorical questions, each has a short and long answer. These are provided here. This article is not a complete nor a detailed presentation or analysis of the ISO 14000 series of standards, of which ISO 14001 is the only specification standard. Rather, it presents key public interest questions about ISO 14001 that relate to the ISO 14000 series as a whole.[1]

This new initiative needs to be understood in the light of the confluence of three histories.

☐ The History of ISO, and the Context of ISO 14001

ISO, the International Organisation for Standardisation, was established in Geneva in 1946 for the principal purpose of standardising industrial and consumer products moving across national borders: to ensure that pipes were of the same thickness, widgets made in standard sizes, telecommunications technology used the same band widths, etc. Its mission was to facilitate the exchange of goods and services, and to foster mutual co-operation in important areas of human endeavour, namely scientific, technological and economic pursuits.

Technical standardisation decisions are internal to industry, and ISO became the international standardisation-setting body, working consensually with national standardisation bodies, engineers from government departments and industry and consumer representatives—particularly with transnational corporations, for whom this issue is crucial.

In the 1980s, ISO departed from this technology and engineering mandate by working on a 'soft' management issue: quality management systems. The ISO 9000 series was the result: a standard that underpins a certification process that designates a company as having implemented a quality management system.[2] With the ISO 14000 series, ISO has once again moved into the area of organisational management, this time to establish the basis for an international certification process for environmental management systems, either at organisational or at site level. There is a difference, however, between the consequences of the ISO 14000 series and that of the ISO 9000 series. Introducing a policy of 'zero defects' and imposing this on suppliers may reduce the time and cost to produce a product, but ISO 9000 addresses business efficiency, not public policy. With the 14000 series, ISO leaves the industry–client realm and enters into a field of significant public interest: the environmental performance of companies.

With this initiative, a key section of industry has opted to avoid various multi-stakeholder initiatives and to work within ISO to develop a basically 'private-sector' definition of an international environmental management system. In so doing, they potentially undercut significant industry, government and NGO initiatives that were progressively raising the level of international corporate management.[3] For this reason, it is crucial that the mandate of the ISO 14000 series is carefully understood and its claims to technical competence are not casually extended.

In the full text of ISO 14001, the language is both precise and confusing. In reading the text, it is important to note that apparently non-technical terms were the subject of intense debate. Several major issues of linguistics and definition were crucial. The first is the emphasis on measuring environmental *conformance* (to an internal set of standards), not environmental *performance*. Related issues are the agreements to drop most references to corporate environmental 'impact' in favour of the term environmental 'aspects'; to select 'certifier' over 'verifier' as the designation of an ISO 14001 auditor, and to replace the commitment to 'pollution prevention' (which has legal consequences, at least in the US) with commitment to the 'prevention of pollution' (which has no legal consequences and includes end-of-pipe solutions).

This issue of language confusion is also seen around the translation of the acronym 'ISO' into English. The acronym of ISO is routinely translated into English incorrectly as the 'International Standards Organisation' instead of the 'International Organisation for Standardisation'. The mis-translation does not occur in official ISO materials or in technical journals, but it does occur in professional journals, expert NGO papers, and expert business materials.[4] The difference is crucial. In the minds of environmentalists, environmental standards imply that a floor is being set for environmental protection. Standardisation has to do with harmonising arrangements and procedures. The confusion over the proper name for ISO raises the expectation that the organisation is about performance standards rather than standardisation.

This confusion of 'standards' and 'standardisation', and the replacement of 'impacts' with 'aspects', and 'performance' with 'conformance', has significant consequences. It has effectively allowed the ISO 14001 system to be promoted as the most systematic and comprehensive means to global corporate sustainable development management and global environmental protection,[5] while actually reversing the trend of raising the level of national, international and corporate global environmental performance standards.

☐ The Context of the New International Trade Rules and the World Trade Organisation

The new World Trade Organisation (WTO) came into existence in January 1995. The WTO Agreement on Technical Barriers to Trade (TBTs) has changed the context in which international standardisation and performance standard-setting activities are conducted. Whereas international standards-setting was previously conducted by intergovernmental bodies in public arenas and subject to a significant degree of public accountability, it is now possible for international standards to be set largely by private industry bodies with no accountability to public processes in their decision-making.

Under the new Agreement on TBTs, the WTO is charged with harmonising the rules of trade and creating a predictable, more uniform environment in which to carry out global trade activity. International performance standard-setting bodies that meet certain criteria will be able to develop environmental, health and safety (EHS) 'technical regulations'. These 'regulations' will be used by the WTO and its members as the basis for ascertaining whether national and local environment and health and safety standards are technical barriers to trade. National standards-setting organisations are urged to utilise guidelines issued by international bodies, although this procedure reduces the scope of national political review and, thus, citizen input. This process has occurred in the creation of the ISO 14001 standard. Further, the agreement states that, if an international standard is 'imminent', it has to be followed, even if it has no formal status at that point. Thus, the draft international standard status of ISO 14001 gives it the status of a *de facto* international trade standard.[6]

The WTO will have the power to enforce new performance standard-setting criteria under its dispute settlement process. Member countries may challenge standards that they consider barriers to trade. The new General Agreement on Tariffs and Trade (GATT) places the onus on the defending country, with its more stringent environmental, health or safety measure, to defend the legitimacy of the measure on the basis of its trade impact, technical and scientific evidence and climatic or geographic need. Countries that cannot justify the use of higher standards on these specific grounds are faced with the choice of changing their national standards in order to come into line with the international standard, facing the threat of financial sanctions, or cross-retaliation on other export products. There is no dispute mechanism to challenge the use of lower standards, provided that these meet WTO standard-setting criteria.[7]

☐ The Context of Corporate and Industry Sustainable Development Management Initiatives

As all the stakeholders in this process know well, global corporate environmental performance standard-setting is immensely complex. Nevertheless, international industry has often shown a willingness to grapple with this issue and raise the level of environmental management at the same time. Several of the major national and international industry associations—including the Canadian Chemical Producers' Association, the International Chamber of Commerce, the Japanese Keizai Doyukai and the UNEP Advisory Committee on Banking and the Environment—have made commitments to use home country standards as the basis for their operations abroad (see Fig. 1).[8]

Companies that are grappling with environmental management must address the very difficult question of what material benefits can come from sound environmental management under existing market conditions. Businesses tend to make environmental investments where there is a short-term bottom line reward, or where there is some other economic benefit: for example

---

Canadian Chemical Producers Association: Responsible Care
The codes encompass member country operations both inside and outside Canada.

International Chamber of Commerce: Business Charter on Sustainable Development
To [. . .] apply the same criteria internationally.

Group of International Banks:
Statement by Banks on Environment and Sustainable Development
We will, in our domestic and international operations, endeavour to apply the same standards of environmental risk assessment.

Keidanren, Japan: Global Environmental Charter
It is a requirement that a corporation's environmental policies regarding atmosphere, water quality, waste products, etc. meet the minimum standards of the host country.

Keizai Doyukai, Japan: Recommendations on Measures to Apprehend Global Warming
Rather than abiding by the environmental standard of the local area, environmental management of the same high standards as in Japan should be pursued (including auditing by the headquarters).

Figure 1: *Examples of Industry Environmental Commitments to Use Home-Country Standards Abroad*

in marketing or public relations. At the same time, companies that are interested in this area are undertaking a range of exercises to address the larger question of what the relationship should be between industry and environment. For example, industry leaders have been experimenting with innovative ways to create global environmental performance standards without sacrificing local autonomy or corporate competitiveness.[9] The ISO 14000 series will reverse this trend, and will effectively discourage transnational corporate experimentation because it will grant an 'easy A' to companies with ISO 14001, even if they have low environmental performance standards.

■ Question 1: Does ISO 14001 help implement Agenda 21 or any international environmental convention?

☐ Short Answer

No. It reverses the direction of global environmental performance standard-setting, whether public or private.

☐ *Long Answer*

The international intergovernmental community has begun to build the initial elements of a system for global environmental management, integrating inputs from industry, governments and non-governmental organisations (NGOs). The degree of NGO involvement in the process of this international consensus-building exercise has been unprecedented and widely welcomed as indicative of a new spirit of democratic international decision-making. Although ISO 14001 claims Agenda 21 as its ideological parent[10] commitments in ISO 14001 are retrograde in comparison to Agenda 21.

The ISO 14000 series does not include any reference to the Montreal Protocol, the Basel Convention, the Convention on Climate Change, the Convention on Biological Diversity, the OECD Guidelines on Hazardous Technologies, or any other international environmental agreement. The only compliance aspects are to conform to applicable laws and legal regulations (see Fig. 2); and, although it cites Agenda 21 in its Appendix, the principles are not reproduced in the ISO 14000 series.

It should be noted that this standard does not establish absolute requirements for environmental performance beyond commitment, in the policy, to compliance with applicable legislation and regulations and to continual improvement.

Figure 2: *Draft ISO 14001: Introduction:*
Definition of Corporate Environmental Responsibilities

The implications of this for global environmental management are significant. A trend in international environmental performance standard-setting has been to generalise from existing national and industrial best practice. For example, Agenda 21 recommended that transnational corporations 'report annually on routine emissions of toxic chemicals even in the absence of host country requirements', drawing on the model of the US Toxic Release Inventory. In addition, Agenda 21 contained recommendations to transnational corporations to 'introduce policies and commitments to adopt equivalent or not less stringent standards of operation as in the country of origin'; and to 'be encouraged to establish worldwide corporate policies on sustainable development'. Moreover, Agenda 21 recommends that firms adopt standards for public reporting, improved environmental performance and full-cost accounting (see Fig. 3). None of these recommendations are cited in ISO 14001, even though many governments and leading international organisations are working towards full implementation of such programmes.

Reporting

- Report annually on their environmental record as well as on their use of energy and natural resources (Ch. 30.10[a])
- Report annually on routine emissions of toxic chemicals to the environment, even in the absence of host-country requirements (Ch. 19.51[c])

Sustainable Consumption Patterns

- Play a major role in reducing impacts on resource use and the environment through more efficient production processes, preventative strategies, cleaner production technologies and procedures throughout the product lifecycle (Chs. 30.2 and 30.4)

Full-Cost Accounting

- Work towards the development and implementation of concepts and methodologies for the internalisation of environmental costs into accounting and pricing mechanisms (Ch. 30.9)

Figure 3: *Sample Agenda 21 Recommendations for the Conduct of Sustainable International Business [chapter references in brackets]*

The past several years have seen increased commitments to sustainable development from international development banks such as the World Bank and the European Bank for Reconstruction and Development. After Agenda 21 was adopted by national governments, the World Bank, in its own words, underwent a 'greening' of its policies and operations. Not only has the Bank tried to integrate environmental concerns across its whole portfolio of activities, it has also expanded the definition of environmental management to include its social dimensions, cultural and social costs and benefits.[11] In the words of one US international environmental attorney, environmental initiatives among the multilateral banks are having the result of shaping the environmental policies of foreign countries as well as the environmental performance of companies contracting with these banks' overseas projects.[12]

ISO proponents have argued that registration should be a demonstration of environmental commitment and proof that they are delivering on the promises of the industry, environmental initiatives such as Agenda 21, the ICC Business Charter on Sustainable Development and the chemical industry's Responsible Care Programme.[13] They assert that ISO 14001 will provide demonstrable proof that the company is environmentally sound, and advocate that certification in itself should therefore reduce compliance and liability costs (see Fig. 4).

Demonstration of successful implementation of the standard can be used by an organisation to assure interested parties that an appropriate environmental management system is in place.

Figure 4: *Draft ISO 14001: Introduction*

The lower horizons of ISO 14001 might be acceptable if the ISO discussions remained within the historical bounds of ISO—to standardise technical and engineering procedures within industry. ISO proponents did not want to reference any international engineering standards or any intergovernmental standards. Yet they are moving to have ISO 14001 replace parts of many major international and national proposals for environmental management systems.

While the ISO 14000 series is seeking credibility as the international environmental standard, it fails to advance the goals and legal principles set over the last ten years through international agreements and conventions. ISO 14001 is trying to claim an intellectual and political legitimacy from the international community while it fails to include mandatory requirements for any external references to sound engineering or to sound public policy decisions from the international community, in the required environmental policy (as does the EU's Eco-Management and Audit Scheme [EMAS] requirement).

■ Question 2: Can ISO 14001 become an international trade standard without operative participation from governments or NGOs?

☐ Short Answer

Yes.

☐ Long Answer

ISO is a body that has typically produced technical standards for industry. Drafting and decision-making have therefore been largely undertaken by the trade and industry sectors. However, in leaving the area of technical standards-setting and entering the area of setting standards for global environmental management, ISO enters an area that is of substantial public interest. Moreover, under the new international trade rules, ISO is for the first time legally empowered to create international trade standards that will be used to judge the appropriateness of publicly-set national and local environment, health and safety standards. In other words, in principle, ISO 14001 could become an international standard without having actively integrated public comment. In practice, the process has belatedly invited delegates from governments and citizens' groups, but has used this invitation, and the limited participation that ensued, to claim an openness while ignoring their substantive input.

Decision-Making in ISO: Governments, NGOs, small and medium–sized enterprises, and developing countries play a negligible role in the ISO 14000 series drafting and decision-making. The public and governments may be invited to have observer status at ISO committee meetings, but decision-making in ISO is mostly by trade associations and industry representatives. Other participants, while they may be invited and are recorded as 'participants' in a 'consensual' decision-making process, do not have voting rights. This leads to the incorrect impression that delegates who attended specific meetings in order to lobby for certain elements in the standards are recorded as participants in a 'consensual' decision-making process. NGOs were not integrated into early ISO 14000 series discussions. Since then, a handful of environmental groups are liaising with TC207 or participating as observers in ISO meetings. In consequence, ISO drafting reached its final stages without active input from NGOs.

An industry magazine commented:

> Industry, which has dominated work on drafting the standard, now faces the task of selling ISO 14001 to regulators, environmentalists and other stake-holders in environmental management. Environmentalists and some national environment ministries have found drafts to date insufficiently ambitious.[14]

This highlights a crucial difference in decision-making processes in democratic as compared to closed private-sector systems. In a democratic system, if a citizens' group disagrees with a political decision, it can lose the technical debate but then take its policy position to the political process or the legal system. In an ISO debate, a decision, once taken, cannot be appealed. This effectively minimises pressure from the public or from public bodies. While this was not relevant within the historical engineering mandate of ISO, it has become crucial with the move of ISO into the policy arena. Moreover, because of the new international trade rules under the WTO, this special-interest decision-making process can result in the formation of a new international trade standard.

It is worth noting that the ISO process also tends to exclude the needs of small and medium-sized businesses because, in practice, they do not have the funds and specialist technical expertise to participate directly in the negotiation rounds.[15] The same point can be made for industry and governments from developing countries and countries in transition. Although ISO claims that the meetings have been broadly representative, in practice it is those who consistently attend meetings and take on the drafting work who actually decide the content of the ISO standard.

This means that the drafting committee is 'considerably less representative' than the technical committee membership, and made up principally of executives from large international corporations, national standards-setting associations and consulting firms.[16] TC207 itself solicited the participation of key developing countries only very late in the process, when the period of substantive review was largely over. Input from small and medium-sized businesses was also invited very late in the process. In the US these are participating on various technical advisory group working parties, and vote on issues before they are presented to ISO TC207.[17] Most of the key decisions regarding scope and content were already made, so their input will have only a minimal influence on the final product.

■ Question 3: How will an ISO-14001-certified company demonstrate that it has a good environmental, health and safety performance?

☐ **Short Answer**

It cannot.

☐ **Long Answer**

There are two interconnected elements to this answer. First, what is the implication for environmental, health and safety standards of ISO 14001 being a conformance, not a performance, standard? Secondly, what is the nature of environment, health and safety management in the ISO 14001 draft international standard?

ISO 14001 is a specification standard for conformance, not performance, and environmental aspects, not environmental impacts. A sound ISO 14001 EMS will give a firm the capacity to measure and monitor the environmental 'aspects' of its operations. In a speech at MIT, Joe Cascio (then Chairman of the US Technical Advisory Group on 14001) said that he did not care 'how much' waste an ISO-certified firm dumps into a river. What is important is that the company's EMS knows that it has happened.[18]

Environmental performance relates only to the measurable performance of the environmental management system (see Fig. 5). The environmental management system can be internally defined and system performance

Measurable results of the environmental management system, relating to an organsation's control of the environmental aspects of its activities, products, or services, based on its environmental policy, objectives and targets.

Figure 5: *Draft ISO 14001, 3.9: Environmental Performance*

results are confidential. The public and public authorities are being asked to depend on the corroboration of the certification bodies that ISO 14001 companies will:

1. Improve their environmental performance in accordance with their environmental policy
2. On discovery of an environmental problem, will correct and remedy the situation

However, one should remember Mr Cascio's anecdote about the waste in the river. Getting demonstrably and measurably better at the ISO 14001 standard does not necessarily improve the firm's environmental performance. Moreover, reference to environmental impact has been diluted and recast into commitments to examine environmental 'aspects'. Commitments to environmental performance are commitments simply to comply with 'applicable' regulations.

The nature of environmental, health and safety standards. There are no requirements for health and safety standards in the ISO 14000 series as it now stands. All work on health and safety was put aside in the ISO discussions. A meeting held in Spring 1996 decided that there was no need for an international guidance standard on environment, health and safety (see Fig. 6).

> This standard is not intended to address, and does not include requirements for, aspects of occupational health and safety management; however, it does not seek to discourage an organisation from developing integration of such management systems elements. Nevertheless, the certification/registration process will only be applicable to the EMS aspects.

Figure 6: *Draft ISO 14001: Introduction, Standard on Health and Safety*

The business argument for performance-oriented management systems. Conformance rather than performance, the orientation to input rather than output measurement of ISO 14001, runs against current business theory. There is a business truism that 'if you don't know where you're going, any road will get you there.' Management textbooks emphasise how important it is for any business to select a mission and targets and the measurable steps to achieve them. This view stresses that it is crucial to aim for business effectiveness rather than process efficiency, and that management systems and organisational design ought to work from the desired *results* to identifying performance measures that can be managed and improved. Furthermore, this argument insists that becoming more effective will lead an organisation to become more efficient, but that the reverse is not true: becoming more efficient does not necessarily make an organisation more effective. Although ISO 14001 requires companies to state an environmental policy giving its environmental intentions and principles, it offers no mandate to incorporate sustainable development aims—or, for that matter, any other environmental limit values—into the policy.

■ Question 4: How do governments, workers and the public get access to all the environmental information prepared by an ISO-14001-certified company

☐ Short Answer

They do not. It is company-confidential. Corporate disclosure is discretionary.

☐ Long Answer

Under ISO 14001, environmental information is gathered for purposes of helping track and manage the corporate environmental management system, and is viewed as company-confidential (see Fig. 7).

The organisation shall consider processes for external communication on its significant environmental aspects and record its decision.

Figure 7: *Draft ISO 14001, Section 4.3.3: Communications*

This moves against all international environmental agreements and leading voluntary corporate environmental management initiatives which have been moving progressively towards greater disclosure of environmental information in line with public requirements for corporate environmental reporting. Agenda 21 established a public 'right to know', and made several specific reporting recommendations (see Fig. 3). Since then, the OECD has sponsored several workshops for governments, industry and non-governmental organisations, to implement national pollutant release and transfer registers, based on the US Toxic Release Inventory. These will create a public information vehicle that can be used to benchmark corporate and governmental accountability for environmental performance, while stimulating and measuring the shift to cleaner production and products.[19]

Among international firms, levels of disclosure and reporting formats vary enormously, but leading industries and many international firms now routinely produce environmental reports.[20] At minimum, they indicate environmental investments, liabilities and expenditures. Increasingly, they include figures on performance and relative performance in reducing emissions and toxic emissions, decreasing waste, integrating lifecycle analyses into new projects, widening the circle of stakeholders to include local communities, and so on.

A core idea of the EU's EMAS, which does have a disclosure requirement, is that public pressure will motivate companies to improve environmental performance. However, in order for this to work, there needs to be disclosure of corporate environmental performance. Without external audit and public disclosure, self-monitoring is an oxymoron.[21]

■ Notes

1. A longer version of this paper was commissioned by the European Environment Bureau (EEB) for a seminar, 'Free Trade and Environment: Co-operation between NGOs of EU and Third World Countries: An International Workshop', Brussels, Belgium, 26–28 October 1995. The EEB used it as the basis for their presentation to the Meeting of European Environmental Ministers in Sofia in October 1995. Assistance in producing this paper was provided by Manuela Soler and Peter Thimme.
2. The precedent of using quality management system standards as the basis for environmental management standards was established by the British BS 7750, which was modelled after the BS 5750 standard on quality management systems. See Caroline Hemenway and James Gildersleeve, _What is ISO 14000? Questions and Answers_ (Fairfax, VI: CEEM Information Services, 1995), p. 5.
3. Even if some standard-setting bodies are quasi-governmental bodies, e.g. the French AFNOR or the German DIN, they have been represented at ISO meetings by firms and consultants. In addition, although NGOs have been brought into the process, this was done late in the day and to a very limited degree.
4. Examples from industry are: David Van Wie, Director of Environmental Services, Robert Gerber Inc., US: 'National and International Standards for Environmental Management Systems (EMS)' (presentation to the _International Caribbean Environmental Management Conference_, Port of Spain, Trinidad, 20–24 March 1995); ML Strategies, Inc., _Project 14000: ISO 14000: Key Questions and Answers_ (Boston, MA: ML Strategies, Inc., June 1995); and _International Environmental Reporter: Current Reports_, 10 January 1996, p. 1.
5. 'Sustainable Development' section in 'Standards: Draft ISO Document Standards Approved; To Be Circulated then Possibly Adopted', in _International Environmental Reporter_, 12 July 1995, p. 527.
6. See 'Agreement on Technical Barriers to Trade: Technical Regulations and Standards: Preparation, Adoption and Application of Technical Regulations by Central Government Bodies', in _Final Act Embodying the Results of The Uruguay Round of Multilateral Trade Negotiations_ (Version of 15 December 1993; MTN/Fa II AIA-6;

Washington, DC: Office of the US Trade Representative, Executive Office of the President).

7. Provided a new international standard can claim that it has an adequate justification, such as environmental objectives, 'it shall be rebuttably presumed not to create an unnecessary obstacle to international trade.' See 'Agreement on Technical Barriers to Trade', p. 3.

8. See United Nations Conference on Trade and Development, *Self-Regulation of Environmental Management: An Analysis of Guidelines Set by World Industry Associations for their Member Firms* (Environment Series No. 5; UNCTAD/29; New York, NY: United Nations, 1994).

9. See United Nations Conference on Trade and Development, *Environmental Management in Transnational Corporations: Report on the Benchmark Corporate Environmental Survey* (Environment Series, No. 4; ST/ CTC/149; New York, NY: United Nations, 1993).

10. In a presentation about ISO 14000 to the MIT Working Group on Business and the Environment, Boston, 25 June 1995, Joe Cascio, IBM Director of EH&S Standardisation, and Chairman of the US Technical Advisory Group (TAG) to ISO Technical Committee 207 on Environmental Management, and a leading US architect of ISO 14000, said that the UNCED was the catalyst for ISO 14000. This is reflected again in an internal ISO TC207 Committee Draft, *Guide to Environmental Management Principles, Systems and Supporting Techniques*, prepared by ISO/TC207/SC1/WG2, 2 September 1994, p. 1.

11. See The World Bank, *Mainstreaming the Environment: The World Bank Group and the Environment since the Earth Summit: Summary* (Washington, DC: The World Bank, 1995).

12. David Hackett, a Chicago-based international environmental lawyer, speaking at Baker and McKenzie's Sixth Annual International Environmental Conference, 17 October 1995, cited in *International Environmental Reporter*, 1 November 1995, p. 826.

13. ISO TC207 Committee Draft, *Guide to Environmental Management Principles, Systems and Supporting Techniques*, prepared by ISO/TC207/SC1/WG2, 2 September 1994, cites all 27 principles of Agenda 21, and the whole text of the ICC Business Charter on Sustainable Development, in its Appendices, pp. 51-59.

14. *Environment Watch: Western Europe*, 17 February 1995, p. 8.

15. This occurs particularly in the US, where small and medium-sized companies do not have a strong lobby. This point was made by Mary McKiel, Director of the EPA Standards Network, of US small and medium-sized businesses. She said that this inherently undemocratic process could not produce a standard that was viable across industry. Cited in 'EPA Says No to Standards in Rules', in *International Environmental Systems Update*, Vol. 2 No. 6, pp. 5-6, Naomi Roht-Arriaza points out that, while the intention behind convening meetings all over the world may be to make meetings accessible to people living there, the financial and time costs of going to meetings in Australia, South Africa and France soon add up. See Naomi Roht-Arriaza, *Shifting the Point of Regulation: The International Standards Organisation and Global Lawmaking on Trade and the Environment* (San Francisco, CA: University of California, Hastings College of Law, 1995), pp. 45-46.

16. Roht-Arriaza, *op. cit.*, pp. 45-46. 'In the ISO debates, multinational companies, particularly American multinationals like IBM, Du Pont, Allied Signal, Merck,

3M, and ARCO have had a substantial input.' (Whitman Bassow, 'The Appeal of ISO and World Standards', in *Environmental Protection*, 1994, p. 12).

17. See Hemenway and Gildersleeve, *op. cit.*, p. 20.

18. United Nations Conference on Trade and Development, *Self-Regulation of Environmental Management*; see also 'International Environmental Management Standards: ISO 9000's Less Tractable Siblings', in *ASTM Standardization News*, April 1994.

19. See Eric Howard, 'OECD: The Establishment of National Pollutant Release and Transfer Registers', in *RECIEL: Review of European Community and International Law*, Vol. 2 No. 2 (1995), pp. 195-96.

20. A World Wildlife Fund (WWF) report found 130 companies in Europe and North America that produce environmental reports. See Irwin *et al.*, *A Benchmark for Reporting on Chemicals at Industrial Facilities* (Baltimore, MD: WWF, 1995).

21. See Harris Gleckman, 'Transnational Corporations' Strategic Responses to Sustainable Development', in *Green Globe 1995* (Oxford, UK: Oxford University Press, 1995).

3 The ISO 14001 Environmental Management Systems Standard
One American's View

Christopher L. Bell

WITH THE final publication of ISO 14001, the cacophony of claims and counterclaims regarding the nature, use and value of the standard is growing louder. The commentary is wide-ranging, with some of the more noteworthy claims being that ISO 14001:

- Is a 'green passport' signifying environmental excellence
- Is a plot by industry to undercut more deserving international environmental initiatives
- Is a plot by consultants to make money out of industry
- Is a plot by governments and NGOs to push a 'green agenda'
- Should be a basis for eliminating environmental regulation or government oversight of industrial activity
- Is useless as a tool for improving environmental performance
- Does not represent the 'leading edge' of thinking in the EMS area
- Represents a completely new way of thinking about environmental issues
- Is too complicated and complex for implementation by small and medium-sized enterprises
- Will be used as a non-tariff trade barrier
- Will facilitate trade

The range of opinion on ISO 14001 provides public and private organisations with little useful advice on fundamental questions such as why, where, when and how to do ISO 14001, and what decisions about ISO 14001 might mean for purposes of environmental protection, public policy and trade.

This article is an attempt to sort out what ISO 14001 is and is not. The perspective taken is as neither proponent nor opponent of ISO 14001: ISO 14001 is not something to support or oppose, any more than one should support or oppose a hammer. Rather, just as one would do with any tool, one should clearly understand the standard and what it can and cannot do: where it is useful, use it, and where it is not, do not. Zealous proponents go too far when they imply that ISO 14001 is the answer to all the world's environmental problems, and harsh critics over-react when they claim that the standard has no use and fear that ISO 14001 will somehow undo decades of environmental progress.

This perspective on ISO 14001 comes from the United States, where stringent environmental requirements and tough enforcement have been a major influence on the development of environmental management systems over the past decade. However, this article does not purport to provide the definitive US view: the standard is too young and opinions too divided to come to the conclusion that there is a single US reaction to ISO 14001. However, some generalisations can be made based on the unique context of managing environmental obligations in the US.

The various perspectives on ISO 14001 in the US can be summed up with the phrase 'cautious interest'. Many companies are evaluating their existing EMS against the standard, but most industry is taking a 'wait-and-see' approach to third-party registration to the standard. While there is general acceptance of management systems approaches to environmental protection, there is concern that the standard may be too bureaucratic and a suspicion that registration adds little value other than income to consulting firms. Government is for the most part receptive to the standard, though it is not clear whether ISO 14001 will lead to any changes of governmental policy. Environmental groups are wary of the standard to the extent that they believe that it may lead to any relaxation of existing legal requirements.

The years of experience with EMS in the US underscores a fundamental point that seems to be lost in some of the more heated debate about ISO 14001: an EMS standard is only a framework for identifying and managing performance criteria that are set elsewhere. It is not the function of an EMS standard to set or replace such criteria and it is not the function of ISO, an international non-governmental organisation (NGO), to supplant the role of sovereign governments or the public in setting such criteria. An analogy can be drawn from the computer world: a personal computer must have two types of software to be of any practical use to the average user: operating systems software (e.g. MS-DOS or Windows®); and applications software (e.g. word-processing programmes). ISO 14001 is the equivalent of systems software that is useless without applications software, but makes the more effective use of the applications software possible. An EMS standard by itself will not improve environmental performance—it must be understood in the overall context in which it will be implemented.

This article begins with a short history of environmental management systems in the US, follows with an analysis of each of the major provisions of ISO 14001 and how they have been received in the US, and concludes with some general observations about the nature of ISO 14001 and how it might be implemented in the US.

■ An Overview of Environmental Management Systems in the US[1]

The US views on ISO 14001 are based on decades of very practical experience designing and implementing environmental management systems. This experience is based largely on the need to establish systems to assure compliance with applicable legal requirements and to respond to public concerns and pressures regarding environmental issues.

The US has issued tens of thousands of pages of statutes and regulations on environmental issues in the last thirty years, creating a detailed 'command-and-control' system. These laws are enforced not only by federal, state and local governments, but by individual citizens as well. Most of the environmental laws in the US operate on a strict liability basis: good-faith efforts to comply or a record of improved environmental performance typically are not a defence. Furthermore, it is very difficult, if not impossible, to obtain variances from the published regulations: complete and literal compliance is expected. This includes compliance with detailed documentation and record-keeping requirements.

The consequences of non-compliance are severe. For example, in 1993, the Federal Government alone collected over $140 million in fines and obtained 135 criminal convictions. These figures do not include the penalties obtained through state and citizen enforcement actions. In addition to enforcement consequences, companies that pollute the environment in the US must pay clean-up costs as well as compensation for damages to the environment, property and human health. This includes costs associated with past disposal activities, even if those activities were legal at the time of disposal. Consequently, facilities operating in the US must take the full range of environmental obligations very seriously, which has encouraged the development of EMSs to manage these obligations.

Public involvement has also had an important influence on how companies manage their environmental issues. Many companies in the US must by law regularly publish detailed reports on releases of chemicals to the environment, the generation of wastes and recycling activities; and the number of companies and chemicals subject to these reporting requirements are being expanded. These toxic release inventory (TRI) reports identify specific facilities, chemicals, and volumes of releases in accordance with specific requirements, and are not simply summaries or aggregations of data. The Government can and has taken enforcement action against companies that fail to submit reports or that submit inaccurate or incomplete reports. These reports are made available to the public in 'user-friendly' form, such as computer databases. The Government conducts training sessions for citizens' groups and the media on how to interpret the data.

In addition to the TRI requirements, there are numerous other avenues of public communication. For example, every permit application, in all of its details, whether for air, water or waste, is also available for public inspection and comment. Once permits are issued, the public has full access to all regular permit monitoring reports. Facilities must also submit regular reports on their waste generation and recycling activities. Companies must also make an annual report to the Government and investors of any 'material' environmental expenditures—which are generally defined as expenditures that equal 10% of the consolidated assets of the company, or if management believes that these are likely to affect earnings significantly, or affect the company's competitive position. In addition to the legally-required communications obligations, many US companies also voluntarily make environmental performance information available through periodic reports, using models such as those developed by the Public Environmental Reporting Initiative (PERI).

This information is a powerful tool for public involvement. Public availability of environmental information has become a powerful motivator for companies to engage in voluntary pollution prevention activities. Furthermore, private citizens regularly exercise the authority, independent of government authority, to enforce most legal requirements, including permits, by taking companies to court to seek penalties for violations of the law. Penalties obtained by citizen suits go to the Government: this is not a mechanism intended to enrich private citizens. Citizens also regularly and successfully take the US Government to court if they believe that regulations issued by the Government are not stringent enough.

The compliance and public scrutiny situation in the US has encouraged companies to manage their environmental obligations in a systematic manner. Failure to do so increases a company's exposure to significant penalties, public pressure, potential criminal action, or onerous clean-up costs. Therefore, many companies in the US have had environmental management systems in place for many years. Initially, these systems were not based on any formal standards or requirements; rather, companies applied normal management practices to environmental issues and developed EMSs that were integrated into and consistent with their existing management systems. The issue of external verification of these systems also was not an issue: the proof was in the measurable performance of the companies (as measured by legal compliance or clean-up expenditures), not whether the EMS itself conformed to a particular standard.

The US Government also recognises the relationship between EMS and environmental performance: at least performance as measured in terms of legal compliance. The result has been several documents outlining what the US Government believes the key elements of a legal compliance-assurance programme should be. One of the first formal government statements regarding management systems came in the area of occupational safety, with the Occupational Health and Safety Administration's (OSHA) Voluntary Protection Programme in 1982.[2] Under the OSHA VPP programme, a company may implement an occupational safety and health management system and, in exchange, receive a reduced enforcement and inspection response from OSHA. Since then, OSHA has also promulgated final regulations governing the process of safety management of 'highly hazardous chemicals', including requirements for process safety design, planning and employee involvement.[3]

The US Environmental Protection Agency (EPA)'s first formal foray into the EMS management arena was its 1986 Audit Policy, setting forth general guidelines for auditing programmes.[4] The EPA will give positive consideration to companies that implement such programmes in its enforcement decisions. It revised this policy in 1995, suggesting that companies that discover, promptly disclose and correct non-compliance may, under certain circumstances, qualify for significant penalty mitigation.[5] Among the criteria that EPA will consider is whether a company has exercised 'due diligence' in addressing its environmental obligations.

In 1991, the US Department of Justice (DOJ), which is one of the primary enforcers of environmental law in the US, issued a policy entitled 'Factors in Decisions on Criminal Policy Prosecutions for Environmental Violations in the Context of Significant Voluntary Compliance or Disclosure Efforts by the Violator' (1st July 1991). The DOJ policy views self-auditing, self-policing and voluntary disclosure of environmental violations as 'mitigating factors' when deciding between civil or criminal enforcement.[6]

That same year, the United States Sentencing Commission issued the Organisational Sentencing Guidelines, which establish mandatory criteria used by courts when imposing criminal penalties on companies. These criteria encourage companies to implement systematic compliance-assurance programmes, to report detected violations, and to co-operate affirmatively with the government investigators.[7] If an organisation implements a compliance-assurance programme and thereby demonstrates that it created a context that encouraged individuals to comply with the law, it can receive a substantial reduction in its scheduled criminal penalty under the Sentencing Guidelines. Though the Organisational Sentencing Guidelines do not formally apply to companies convicted of environmental crimes, they have nonetheless been commonly used by companies as a guide in the development of compliance-assurance programmes in the environmental area.

Taken as a group, these documents indicate that if a company has in good faith implemented an effective EMS, the US Government may take less severe enforcement action and the resulting penalties may be lower. This is a very attractive benefit, and many companies in the US have for several years been implementing EMSs that are modelled on the US Government documents.

The primary elements of the compliance-assurance programmes set forth in the US documents are familiar to those involved with EMS: a top management policy committed to environmental protection; implementing procedures; auditing programmes; and the ability to demonstrate to the authorities that one has such a programme in place. These policies also call for the disclosure of violations to the Government as well as full co-operation with the Government in investigating violations. It should be understood that these are policies, not full-blown management systems standards. Therefore, in comparison to documents such as ISO 14001, they are quite short and give companies significant flexibility in how to design and implement their programmes. In addition, since many in the US equate legal compliance with environmental protection, these policies focus narrowly on compliance.

These various US Government policies also do not require registration or verification by third parties, nor is conforming to the policies a legal requirement. However, companies that wish to use their compliance-assurance programmes to take advantage of the applicable governmental policies can expect that their systems will be subjected to close government scrutiny. This is particularly true in the case of actual or potential criminal situations, where the documentation and implementation of the compliance-assurance systems will be very closely reviewed if they are to influence government actions.

US regulations are also increasingly reflecting a systems approach. For example, many facilities in the US are subject to detailed process safety management and accidental release prevention regulations intended to ensure that companies managing larger quantities of certain chemicals have operating and maintenance systems in place to reduce the likelihood of chemical accidents and releases.

The US EPA has also been encouraging companies to adopt EMS outside of the traditional 'command-and-control' context, look beyond the traditional 'end-of-pipe' control technology approach to controlling pollution and move toward pollution prevention strategies that avoid creating the pollutants in the first instance. These strategies include raw materials substitution, energy efficiency and process and product redesign. For example, hundreds of US companies have committed to a voluntary programme to reduce emissions significantly, beyond what would be required by law, of a group of high-priority chemicals (33/50 Voluntary Toxics Reduction Programme), foster waste prevention and recycling ('Wastewi$e') and install energy efficient lighting and other equipment ('Green Lights'). Since these strategies typically require the involvement of product and process designers, their implementation usually demands a more systematic and integrated approach to environmental issues. The US EPA has also been encouraging EMSs through a variety of pilot programmes, in which it is selecting leading companies to participate in innovative EMS projects. Federal and state regulatory agencies are also exploring the potential applicability of ISO 14001 to regulatory policy.

The US Government has also begun to apply environmental management systems concepts to its own operations. The US EPA recently announced a 'Code of Environmental Management Principles' that will be applicable to the operations of all federal agencies. These principles are consistent with ISO 14001, and the policy statement indicates that it will be permissible for federal agencies to use ISO 14001 in lieu of the various principles.[8]

In addition to the government documents, recent years have seen the development of private industry initiatives in the US. These include the Chemical Manufacturers Association's 'Responsible Care' programme and guidance documents produced by organisations such as the Global Environmental Management Initiative (GEMI) and the Public Environmental Reporting Initiative (PERI). Many companies and industrial sectors have realised that EMS is beneficial not only for legal compliance, but also from the perspectives of overall environmental performance and community relations.

The result of this combination of regulatory pressure and voluntary programmes is that many facilities in the US have been practising EMS for decades. A survey of Standard & Poors 500 companies in the US conducted in 1996 by the Investor Responsibility Research Centre revealed the following with respect to the respondents in the manufacturing sectors:

- 94% had documented environmental policies.
- 65% indicated that environmental performance is a factor in senior executive compensation (e.g. bonuses), and 72% indicated it was a factor in operating manager compensation.
- 56% of the companies were applying total quality management approaches to environmental issues.
- 39% subscribe to voluntary codes of conduct developed by NGOs.

- Over 80% have a board of directors committee responsible for addressing environmental issues, and the senior executive responsible for environmental affairs was a vice president at 59% of the companies.
- 96% reported that they evaluated environmental factors when selecting suppliers.
- 56% publish an annual environmental report.
- Over 96% of the companies had environmental auditing programmes in place, with 75% of their US facilities and 59% of their non-US facilities audited in the last two years. The average age of the auditing programmes was eight years.

The conclusion that can be drawn is that companies operating in the US have had decades of very practical experience in designing and implementing EMSs in a very rigorous environment. Companies in the US operate in a situation where the requirements are extensive and detailed, the margin for error is very small, and the consequences of mistakes are very large. Therefore, facilities operating in the US take EMSs and the development of EMS standards extremely seriously. This experience has also produced a certain amount of caution in US industry regarding ISO 14001, as many leading companies are wondering if the cost and paperwork associated with the standard will actually improve their existing EMS and environmental performance.

■ ISO 14001: A Review of the Key Provisions

Many of the claims about ISO 14001 appear to be based on a misunderstanding of what the standard is and what it is intended to achieve. Therefore, it may be useful to assess the potential uses and abuses of ISO 14001 in the light of what the standard actually requires.

ISO 14001 is a comprehensive systems standard that calls for organisations to conduct their environmental affairs within a structured management system and be integrated with overall management activity. Organisations that intend to conform to ISO 14001 will have to implement an EMS that includes the following core elements.

☐ Top-Management-Level Environmental Policy

The top management of the organisation must define an environmental policy.[9] This policy must include commitments to legal compliance, prevention of pollution and continual improvement, be the framework for setting and reviewing objectives and targets, must be documented, implemented and communicated to all employees, and be made available to the public. There has been some criticism that the standard only demands a 'commitment' to compliance rather than simply demanding compliance. This language reflects reality: if conformance to ISO 14001 was based on 100% compliance 100% of the time, which is what 'full' compliance means, very few organisations could ever get certified. As a practical matter, most organisations, no matter where located, will state that they are committed to full compliance.

There are differences in the stringency of legal requirements in various regions, and perhaps even greater differences in how those laws are being enforced. Indeed, the latter is probably the primary concern for those worried about a 'level playing field.' As an NGO made up of national standards bodies, ISO does not have the authority to change the laws of individual countries. In fact, some critics find themselves taking both sides of this issue simultaneously: criticising ISO's inability to strengthen the laws in some countries, while at the same time expressing the fear that ISO 14001 might over-ride stringent laws in other countries.

While it cannot change laws, ISO 14001 may have a positive effect on the degree to which existing laws are implemented. Most countries have environmental laws on their books, but in many areas they are unevenly, if ever, enforced. It will be interesting to see what effect ISO 14001 will have on this situation. Though the standard does not demand 100% compliance and does not itself contain legal standards, it will require companies to demonstrate their commitment to compliance through the policy, which must be implemented, objectives and targets that must be consistent with that policy, and a procedure to measure the degree of legal compliance. There is no exception for countries that do not typically enforce the laws. Therefore, while ISO 14001 will not change legal requirements, it may have a positive effect on the degree of compliance with those laws.

Since legal compliance has been at the heart of EMS in the US for two decades, the relationship between compliance and ISO 14001 has been the topic of much discussion and debate. Industry does not want a voluntary standard such as ISO 14001 to be transformed into a comprehensive compliance-detection tool, which in the US brings with it significant enforcement concerns. On the other hand, government and NGOs want to be sure that ISO 14001 conformance brings with it an acceptable level of performance, as measured by compliance.

There is general agreement with the proposition that 100% compliance 100% of the time is not a prerequisite to conformance with ISO 14001, as this would effectively make conforming to the standard impossible in the US. There is also general agreement that the EMS must be able to achieve a certain level of compliance if the standard is to have any credibility. However, this leaves open the question of how much compliance is enough (or not enough), the degree to which a good auditing and corrective action programme will provide sufficient reassurance in the event non-compliance is detected, and how much should compliance be reviewed in determining conformance with the standard. Governmental officials and NGOs tend to want a closer review of compliance performance than does industry, with the latter emphasising the concept of systems reviews rather than detailed compliance reviews. The ultimate balance between compliance and ISO 14001 will be worked out over time.

The commitment to 'prevention of pollution' has also been the subject of commentary in the US. The definition of 'prevention of pollution' in ISO 14001 is broader than the statutory definition of 'pollution prevention' in the US, which excludes recycling and treatment from its definition of 'pollution prevention'. Needless to say, industry is for the most part comfortable with the ISO 14001 definition, particularly the inclusion of recycling, though some parts of the US Government are not.

This policy is intended to be top management's expression of its desires with respect to the environment. The balance of the standard essentially involves the deployment of this policy in ever-greater levels of detail as it moves through the organisation, including setting objectives and targets, establishing defined programmes to achieve those targets, and translating those programmes into specific and practical work instructions.

☐ *Significant Environmental Aspects*

Organisations must have a procedure to review those aspects of their operations, products and services that are related to the environment in order to identify those aspects that are related to significant environmental impacts.[10] Environmental aspects are elements of an organisation's activities, products or services that can interact with the environment. This process is basically the compilation of an inventory of inputs and outputs. For example, energy use, water consumption, raw materials inputs, air emissions and water discharges are environmental aspects. It is important to note that this procedure must cover services and products as well as facility-based aspects.

The use of the term 'aspects' rather than 'effects' or 'impacts' has been criticised as shifting the focus of the EMS away from actual environmental performance.[11] The term 'aspects' was used to be consistent with what most facilities can reasonably accomplish. Most facilities are not in a position, either financially or technically, to conduct scientific environmental impact assessments. What facilities can do is measure and understand their inputs (e.g. natural resource use, energy use) and outputs (e.g. air emissions, water discharges, product energy use). However, it should be clear that data on air emissions is not impact data: air emissions data simply tells a facility what it emits, not what impact that emission has had on the environment. Of course, ISO 14001 does not prohibit an organisation from conducting a full-scale risk assessment if it has the capability or need to do so.

The standard has been fairly criticised for not providing sufficient practical guidance on how to conduct the review of environmental aspects. The difficulty faced by the drafters was agreeing on auditable requirements that would be applicable to a wide variety of organisations around the world. Additional guidance may be found in the Annex to 14001 and in 14004.

The purpose of the aspects review procedure is to determine which of the aspects are significant and thus worthy of further attention in the management system. Significance is assessed in terms of actual or potential environmental impact. The standard does not establish baseline criteria for what constitutes significance: creating an internationally-acceptable definition of this term simply was not possible. That will be done by the organisations. This leaves the standard open to potential abuse by organisations that might define 'significant' in a self-serving manner that would exclude important issues from their EMS. The requirement that organisations must take legal and other requirements into their EMS may provide at least one driver to avoid this result: it probably would not be prudent to define aspects that have legal controls (e.g. air emissions controlled by a permit) as 'insignificant'.

In the US, this element of the standard has not been particularly controversial because many of the environmental aspects of most companies are already subject to strict regulation as well as data collection and reporting requirements. Accordingly, the environmental aspects identification process can take advantage of existing data and records. That is why the US opposed the inclusion of the concept of 'environmental registers' in the standard: there was a concern that this would require the creation of additional records regarding environmental aspects when existing records would be adequate. The exception to this observation is that there is a tendency in the US to not pay as much attention to inputs or outputs that are not regulated, even if they may be of environmental significance. Therefore, ISO 14001 will have the effect of broadening the perspective of many US companies to take non-regulated environmental aspects into account.

The defined significant aspects form the core of the EMS: they become the subject of objectives and targets, management programmes, operational controls, measurement and auditing.

☐ Legal and Other Requirements

Organisations must have a procedure in place to identify their legal obligations. Since this is an environmental standard, this obligation focuses on laws
and other requirements that are applicable to environmental matters, including environmental requirements related to services and products delivered
by the organisation. Legal requirements may range from local ordinances to
applicable international treaties (e.g. the Montreal Protocol or the Basel Convention). Thus there was no reason for the standard to attempt to highlight
or list potentially applicable international environmental law.

This provision is uncontroversial in the US, given that a clear understanding of legal requirements has always been at the heart of EMSs in the US.
However, it should be remembered that the US opposed the inclusion of a
requirement to maintain 'registers' of legal requirements in ISO 14001. This
opposition was based not upon an aversion to knowing what one's legal
obligations are, but rather because legal requirements in the US are so extensive, well developed and well catalogued that a register requirement was
viewed as a paperwork requirement that would not add value to the EMS.

The organisation's procedure must also address any other obligations to
which they voluntarily subscribe (e.g. ICC Charter on Sustainable Development, 'Responsible Care').[12] Companies that voluntarily subscribe to various
initiatives will, if they seek ISO 14001 registration, have to demonstrate that
they are taking those commitments seriously.

ISO 14001 does not list all potentially interesting non-binding documents
regarding the environment or demand that organisations subscribe to any
particular one of them. These are useful source documents that contain often
laudable goals. However, ISO 14001 should not transform a subset of such
documents into auditable requirements.[13] This would be an inappropriate
intrusion into the public policy realm, effectively picking 'winners and losers'
in the realm of non-binding international environmental documents.

ISO 14001 has been misunderstood as being inconsistent with the efforts of
multinationals that may wish to apply uniform performance standards worldwide.[14] Quite to the contrary, companies in the US with international operations and worldwide standards have tended to be among the most receptive
to ISO 14001 as a tool to further this approach. If a company wants to establish worldwide performance standards, ISO 14001 will support that strategy,
though it does not mandate it. A company that claims to have worldwide
standards will, under ISO 14001, have to demonstrate that it actually operates pursuant to those standards. That is the reason why the reference to
'other requirements' is included in the standard. 'Other requirements' include
requirements that are adopted by the organisation that may not be legally

required. The broad definition of the term 'organisation' also supports this result.

Unlike other EMS approaches, such as EMAS, ISO 14001 is not site or facility based. ISO 14001 is based on the broadly-defined concept of the 'organisation', which can be a site, but can also include an entire multinational with all of its sites.[15] Thus, some organisations have already publicly announced their intention to seek ISO 14001 certification on a company-wide basis, applying their corporate performance criteria on a worldwide basis. ISO 14001 also does not require organisations to design and maintain several systems to address different legal requirements around the world. This would be the outcome only for companies that wanted to do this.

It is argued that ISO 14001 is inconsistent with some 'other requirements' that may arguably be more stringent than ISO 14001 or that contain explicit performance requirements.[16] This concern is misplaced. A systems standard can be a framework for meeting performance criteria that an organisation has chosen to meet. Further, if an organisation chooses to adopt a particular approach to environmental matters, ISO 14001 demands that the organisation actually implements the commitment—ISO 14001 will not accept 'trophy' policies that only sit on the shelf. Thus ISO 14001 is completely consistent with programmes such as the Chemical Manufacturers Association's 'Responsible Care' or the ICC Charter on Sustainable Development. Indeed, if a 'Responsible Care' company seeks ISO 14001 certification, the certification will include a review of the company's 'Responsible Care' programme.

☐ *Objectives and Targets*

Organisations must establish documented and quantifiable environmental objectives and targets that are set taking into account the organisations' legal obligations, significant environmental aspects, the commitment to pollution prevention and the views of interested parties.[17] Setting objectives and targets is essentially the process of translating the generalities of the policy into defined goals.

The requirement to take the views of interested parties into account is vaguely stated. The general nature of this requirement has been, often quite fairly, criticised. On the other hand, it was not possible to reach agreement on a more precise formulation on issues such as the definition of interested parties and their role. For example, while the 'take into consideration' language is quite soft, it is unrealistic to replace with language requiring interested-party approval of objectives and targets: that would give interested parties veto power over company decisions. Providing more detailed guidance on the process of obtaining interested-party input proved difficult due to the vast differences between regions in how the public participates in the environmental affairs of individual organisations.

In one element, the standard does go beyond virtually all law and most other standards in that it mandates interested-party input into the decision-making process, not merely after-the-fact reporting. However, to make it work, one would at a minimum need to identify the interested parties whose views will be taken into account, determine what those views are, and establish some process to take them into account.

Along with uncertainties about the relationship between ISO 14001 and compliance-assurance, the issue of public participation is one of the most hotly-debated issues in the US regarding the implementation of the standard. The debate arises primarily in the context of whether the implementation of ISO 14001 will merit any changes in governmental policies or legal requirements. NGOs and the Government are relatively comfortable with the standard as written where it will be used for internal or commercial purposes. However, if it is to be applied to public policy related to the environment, NGOs and the Government are leaning toward a more explicit and comprehensive degree of public participation.

What they have in mind goes far beyond the concept of the EMAS annual environmental report, which only deals with providing summary information to the public. In the US, such information must already be provided to the public by most organisations in much greater detail than is required by EMAS. Any significant changes in how ISO 14001 companies are regulated may have to take additional public participation in their affairs into account. Decreases in the current level of public participation are unlikely to be accepted by the Government or NGOs. On the other hand, industry is unlikely to welcome additional public participation in their affairs unless there are significant changes in how they are regulated.

Another controversial point is that it is possible under ISO 14001 for companies to set objectives and targets that can be relatively easy to meet. It was not possible to set defined baseline objectives and targets in the standard. Including generalities such as 'sustainable development' would not have added any practical substance to the standard because it still would have left organisations with the discretion to set whatever specific targets they believed were consistent with this broadly-defined concept. The approach taken in EMAS, to drive environmental performance to the level achieved by the economically-viable application of best available technology ('EVABAT') was rejected because in many instances technology-based approaches are extremely expensive, resulting in 'treatment for treatment's sake' and are not always suitable for application in lesser developed countries or by SMEs. In ISO 14001, technology-based solutions are one of many options, not the only option.

There are, however, some constraints in the standard: the objectives and targets must take the significant environmental aspects and the views of interested parties into account, and must be consistent with the policy, which itself must include commitments to comply with the law, prevention of pollution and continual improvement. For example, in the US, companies will have to commit to 100% compliance—to put anything else in writing would be very risky. With such a policy, it would be problematic if the organisation had no compliance-related objectives and targets. These objectives would, at a minimum, have to be set at 100% compliance—the equivalent of 'zero-defects' in a quality system—unless the organisation wanted to take on significant legal risks.

In the final analysis, the primary drivers for setting aggressive objectives and targets will not come from ISO 14001 itself. This pressure will depend on the organisation's commitment to environmental issues, customer demands, and public and governmental inputs. Any effort by ISO, an NGO, to set specific performance standards would effectively have been an effort to override the sovereignty of national governments that set such requirements. Any performance requirements set by ISO would inevitably have been more stringent than those set in some countries, raising concerns about sovereignty, and would have been less stringent than those set in other countries, resulting in cries that ISO was lowering performance requirements. This underscores the point that ISO 14001 needs to be evaluated and implemented in the overall context in which organisations operate and that it is not particularly useful as a stand-alone document.

☐ *Environmental Management Programme*

Organisations must establish a programme for achieving their objectives and targets, including designating responsibility for achieving the objectives and targets and the time-frame within which they will be met.[18] This is the translation of the objectives and targets into concrete programmes for action. Objectives and targets that are not accompanied by defined programmes for achieving them are not acceptable under the standard. ISO 14001 explicitly encourages management to focus on the results, or performance, of the EMS.[19]

☐ *Structure and Responsibility*

Management must provide adequate resources to implement and control the EMS.[20] Furthermore, roles and responsibilities within the EMS must be defined and documented, including the role of reporting on the performance of the EMS to top management. This a practical requirement: who is doing what and where, and do they have sufficient technical and financial resources to carry out their responsibilities? A 'shell' programme that is not adequately supported will not pass muster under the standard.

This is a key issue in the US and the existing governmental policies regarding compliance-assurance systems. The US Government firmly believes that the key to environmental performance is top management commitment as reflected in having senior management directly involved in the design and implementation of the system.

☐ *Operational Control*

Organisations must establish documented procedures to control operations and activities associated with the identified significant environmental aspects.[21] These procedures, which include maintenance and must stipulate operating criteria, must cover situations where their absence could lead to deviations from the organisations' policy or objectives and targets.

This is the next step in deploying the top management policy: after objectives and targets have been set and implementation programmes developed, they must be transformed into specific operating controls and work instructions that direct employees as to what they must do so that the programme will be followed, objectives and targets achieved, policy implemented. These instructions and controls may apply to a wide range of employees, depending on the aspects to be controlled and the objective and targets that have been set. For example, if packaging waste is a significant aspect and waste reduction is an objective and target, instructions might range from directions for engineers that design packaging to the individuals responsible for managing waste.

☐ *Emergency Preparedness and Response*

Organisations must have procedures for identifying and responding to accidents and emergencies.[22] These procedures must be periodically reviewed, particularly after incidents, to determine if they need to be revised.

☐ *Training*

Organisations must establish training procedures to train their employees regarding the importance of conforming with the EMS, the significant environmental impacts (actual or potential) associated with their work activities,

their roles and responsibilities within the EMS, and the potential consequences of the failure to follow operating procedures.[23] The progression is
straightforward: employees must be trained in the work instructions that are
necessary to follow the management programme to meet the objectives and
targets in order to fulfil the policy.

☐ *Communication*

Procedures for internal communication must be established, as well as for
receiving, responding to and documenting relevant communications from
interested parties.[24] Relevant environmental procedures and requirements
should also be communicated to suppliers and contractors. This standard
does not require publication of an organisation's EMS, nor does it require the
publication of environmental statements, though organisations must consider whether to have programmes for external communication.

Much is made of the absence of a public communications requirement in
ISO 14001, particularly in comparison to the requirement of a verified public
report in the European Union's EMAS regulation. This was a difficult issue
during the negotiations due to the wide variety of national approaches to
public involvement in environmental issues based on political and cultural
differences. For the US, this issue is largely irrelevant due to the existing
requirements for public reporting and communication that exist under a
variety of public 'right-to-know' statutes. The US requirements go far beyond
any public reporting requirements that currently exist elsewhere in the world.
However, it would have been inappropriate to impose the US approach
through the ISO standards-writing process. This is an unsettled issue that
should be resolved in arenas more appropriate for setting public policy.

☐ *Monitoring, Measurement, Auditing and Corrective Action*[25]

Organisations must have procedures for regularly monitoring and measuring the key characteristics of their operations that can have a significant
impact on the environment. These procedures must include tracking the
implementation of operational controls, performance against objectives and
targets, and evaluating compliance with the law. Regular and comprehensive audits of the EMS itself must also be conducted. This is the 'what-gets-
measured-gets-done' element of the system. These various measurement
processes must cover more than simply compliance with legal requirements:
conformance with the organisation's EMS procedures and work instructions,
performance against objectives and targets, and environmental aspects are
also measured.

The standard has been criticised because it measures 'conformance' and inputs, not 'performance'.[26] The clear language of the standard refutes these assertions. 'Non-conformance' is often used as a general term denoting deviation from any planned arrangements, while 'non-compliance' is typically reserved for deviations from legal requirements. As noted above, ISO 14001 requires organisations to set objectives and targets that are quantitative wherever practicable. The standard then requires organisations to monitor and *measure* conformance with those objectives and targets, record information to *'track performance'*, and also evaluate compliance with the law.[27] Indeed, the term 'inputs' does not even appear in the section of the standards on measurement. Therefore, while implementing ISO 14001 is not a guarantee of a particular level of performance (and what is?), it has been explicitly designed as a tool to assist companies in setting performance targets and measuring their performance against them.

The primary debate in the US regarding this section centres around the requirement that organisations periodically evaluate their compliance status. Many NGOs and government officials criticise this language as too soft and would have preferred if there had been a direct reference to compliance auditing. As a practical matter, most facilities in the US conduct periodic compliance audits anyway that more than fulfil the requirements of ISO 14001.

Corrective action procedures, including defining responsibility and authority for handling and investigating non-conformance and initiating corrective and preventative action, must also be established. This points to the breadth of the concept of conformance: corrective action must be initiated when any non-conformance is detected, not only non-conformance that is also non-compliance with legal requirements. ISO 14001 demands that an organisation take non-conformance with its own procedures and expectation just as seriously as non-compliance. Any changes to procedures resulting from corrective action must be documented.

The robustness of the corrective action procedures will be a crucial component of determining the balance between ISO 14001 and compliance— non-compliance that has been detected and is being dealt with adequately by the corrective action programme should not create problems under ISO 14001. Such a result would parallel US Government policies that provide enforcement relief for violations that have been detected by an audit or due diligence, are promptly reported to the Government, and are being corrected.

The audit and corrective action requirements of ISO 14001 are distinct from third-party certification audits that must be conducted if an organisation desires to be certified as an ISO 14001 organisation. Such certification audits will be conducted by nationally-accredited and qualified independent auditors. The concern that these auditors will not be truly 'independent' is a legitimate one, but does not recognise the effort by most national accreditation bodies to prohibit certifiers from offering both consulting and auditing services to the same organisation.[28] However, the statement that certification is an 'internal audit' is on its face incorrect: certifiers must by definition be independent of the organisation being audited.

☐ Management Review

Top management must periodically review the EMS to ensure its continuing adequacy and effectiveness.[29] This review, based on all of the information collected by the EMS, must consider changes to the policy, objectives and elements of the EMS. This review must be conducted by top management of the organisation, not just environmental professionals. This is one of the 'continual improvement' elements of ISO 14001.

☐ Documentation and Record-Keeping

The EMS must be well documented.[30] Organisations must have defined procedures describing the function of their EMS. Typically, such procedures will have to define who is responsible, what they are to do, and how they are to do it. In addition, sufficient records must be created and maintained to demonstrate that the system is in place and operating.

A document control system must be maintained to ensure that documents are accessible, kept current and periodically reviewed. These procedures must also cover environmental records, such as training records and audit results. ISO 14001 is, in part, an information management system, encouraging the vertical and horizontal flow of environmentally-relevant information throughout the organisation.

Many organisations in the private sector in the US, particularly small and medium-sized businesses, are concerned about the potential 'red tape' that may be generated by ISO 14001. This is a legitimate concern that is based on the negative experiences that many companies have had with the ISO 9000 series of quality standards. The fundamental function of an EMS is systematically to manage environmental issues in order to achieve defined goals, not to create unnecessary paperwork. An effort was made in writing ISO 14001 to provide some flexibility on documentation issues. However, organisations will have to be vigilant not to turn ISO 14001 implementation into a paper storm. Particular attention will have to be paid to the certification process to ensure that excessive documentation requirements are not imposed that would serve only to make the job of the certifier easier, but would not actually help the company do its job.

■ Some General Observations about ISO 14001

This review of the standard reveals that ISO 14001 provides a framework within which companies can establish and identify their obligations, set their objectives and targets, and create a comprehensive system to achieve those objectives and targets. It encourages companies to include environmental considerations throughout the organisation, integrating environmental issues with normal business operations. This approach enhances a company's ability to operate in an environmentally-protective manner and comply with applicable laws.

However, it is important to understand that ISO 14001 provides *a* framework and is not the only possible approach to developing an EMS. Equally important, ISO 14001 is not self-contained: as a systems standard, the performance criteria that the EMS measures itself against are outside of the standard. This is not a weakness; it is a statement of fact. Some additional general observations can be made.

☐ ISO is an international non-governmental organisation, not an international legislative body.[31]

The primary participants in ISO standards-writing are national standards-writing bodies. ISO standards are not negotiated by governmental ministries, though there was significant government participation in the development of ISO 14001.

Many of the major participating standards-writing organisations, particularly from Europe, are quasi-governmental organisations. Many of the delegations included government representatives (e.g. Germany, the US). Further, as the vigorous EMAS–ISO debate demonstrated, there was significant input from the EU in the ISO 14001 development process.

Statements that ISO has 'member firms'[32] is inaccurate and misleading: individual private companies are not members of ISO. Similarly inaccurate is the description of ISO as an 'industry body'. US industry does not view ISO 14001 as 'its standard': it is evaluating the standard with equal portions of interest and suspicion. Given the cautious attitude of most industry to ISO 14001, the implication that industry somehow initiated the standards-writing process as a conscious strategy to undercut other international initiatives is quite remarkable.[33]

This is not to say that the ISO standards-writing process is a representative democracy. While participation is open to all ISO member national standards bodies from over 100 countries, as well as liaison bodies such as NGOs, it does take resources and time to participate. This is not because of any requirements imposed by ISO, but rather the simple fact that attending meetings costs money and takes time. Therefore, national standards bodies and organisations with limited resources were not able to play as active a role as they may have wished.

Those who are concerned about the difficulties of participation by lesser developed countries should be cautious in urging ISO to draft stringent standards that can only be met by countries or companies with significant resources. Increased participation at recent ISO meetings by less developed countries has revealed a significant concern that ISO 14001 is a standard written by the developed nations as a trade barrier. As those countries have increased their participation, there has been increased resistance to including more stringent or technically-based requirements in the standards. This was borne out by the very vocal complaints from developing countries at the 1995 TC 207 plenary sessions in Oslo regarding the lifecycle assessment standards.

☐ *ISO 14001 is an international voluntary standard, not a law.*

ISO 14001 is not a law unless some national governmental entity decides to make it such. Further, ISO does not require conformance with the standard. The decision on whether or not to conform to ISO 14001 involves considerations wholly outside of ISO: e.g. commercial, governmental or public pressure, a desire to improve environmental performance, etc. A key concern of industry in the US is to avoid governmental actions that might compromise the voluntary and market-driven nature of the standard.

ISO 14001 is not for everyone. Some companies may legitimately decide to take other approaches to implementing EMSs. Some highly-advanced companies in the US are already concluding that their existing programmes already meet or exceed the standard's requirements. Other companies in the US might continue to model their EMS after the existing compliance-oriented policies of the US Government. On the other hand, ISO 14001 does represent an internationally-agreed-upon baseline for designing and implementing EMSs around the world at a very practical, though not utopian level. Accordingly, US industry will discourage the US Government from developing EMS policies or documents that differ significantly from ISO 14001. Indeed, pursuant to a statute signed by President Clinton in 1996, governmental agencies are obligated to rely on international and national standards where possible rather than create their own criteria.

Organisations that decide to use ISO 14001 have considerable flexibility in that use. They may decide to use it for wholly internal purposes. Many companies in the US are highly sceptical about the value of third-party certification other than as a tool for bringing money into the coffers of consulting firms. Thus, many companies in the US have adopted a strategy of implementing an EMS that is consistent with ISO 14001, but declining to seek certification unless demanded by commercial, public or governmental pressure. There is a difference between an 'unsubstantiated claim' and 'self-declaration' for a product or an organisation: an 'unsubstantiated claim' is a claim for which there is no proof, while a 'self-declaration' is a claim that may or may not be unsubstantiated. Most claims are self-declarations. For example, a film manufacturer that places an 'ISO 400' label on the box to denote film speed is making a self-declaration and third-party verification or certification is not necessary. The key issue is whether the claimant has the facts to back up the claim, not whether a third party has verified the claim.

☐ ISO 14001 is not the 'state of the art' in EMS thinking: it is EMS for everyone.

Aggressive claims that imply that ISO 14001 represents the leading edge of EMS are misleading. Some companies already have EMS in place that are more sophisticated than ISO 14001. However, it would have been inappropriate for ISO to standardise such programmes. ISO is a standards body, not a research institution. The function of ISO is to harmonise national and regional standards, not serve as a 'think-tank' to create 'leading-edge' documents. While this may disappoint some, it is what standards organisations do. ISO standards must be applicable to all types of organisations, large and small, throughout the world. Further, standards should reflect what has been demonstrated can be done, not what some would hope could be or might be done. To do otherwise would transform standards-writing into an interesting exercise in social experimentation.[34]

If the standards are written to reflect the most recent 'best practices', the standards would be capable of implementation by only a relatively few very advanced organisations—precisely the companies that need ISO's help the least. SMEs would be completely left behind and many organisations in developing and less developed countries would also be severely disadvantaged. A leading-edge standard might bring some personal satisfaction to some parties and provide good fodder for press releases, but will not achieve the stated goal of improved environmental protection if most organisations in the world cannot implement it.

Ironically, the criticism that ISO 14001 does not sufficiently reflect the latest thinking on EMS is sometimes coupled with a complaint the ISO 14001 does not sufficiently take the concerns of SMEs and developing countries into account.[35] This does not take into account the fact that the primary concern of SMEs and developing countries about ISO 14001 is that the standard may be too advanced, expensive and complex to implement.

ISO 14001, despite claims to the contrary, does not discourage continuing work on advancing the 'state of the art' of EMSs. For example, there is absolutely no contradiction between ISO 14001 and documents such as Agenda 21.[36] Indeed, an organisation implementing ISO 14001 could adopt Agenda 21 as an 'other requirement' that it would then be held to implementing. However, transforming the general aspirational recommendations of documents such as Agenda 21 into immediately auditable requirements would limit ISO 14001 registration to a very small group of sophisticated companies that are probably already good environmental performers. Simply pointing to the fact that a document such as Agenda 21 exists and that it has garnered some support in public policy circles is not sufficient factual basis to conclude that it has been and can be implemented by a wide range of organisations, or that individual companies should be audited against it. As the state of the art is developed, implemented, and is demonstrated to be practical for international application for a broad range of organisations, ISO 14001 can be revised.

☐ *ISO 14001 is a systems standard that supports, but does not set or replace, performance criteria.*

ISO 14001 does not establish substantive performance obligations for organisations. ISO 14001 certification is not a performance certification, nor is ISO 14001 a performance guarantee. Actual technical environmental compliance or performance obligations are set by the countries within which particular facilities operate, international agreements, or other non-legal obligations to which an organisation might subscribe.

ISO's Technical Committee 207, which is writing the ISO 14000 standards, is barred by its mandate from setting such obligations. This makes the allegation that measurable performance standards were 'written out of' the ISO 14000 standards incorrect—the participants were not allowed to consider them in the first instance.[37]

It would also be an unacceptable intrusion on national sovereignty for ISO's TC207 to set specific performance criteria. It is precisely because ISO is not an international legislative body that is should not establish specific environmental performance requirements. Some of those who have complained about the unsuitability of the alleged undemocratic character of ISO proceedings for setting public-policy-sensitive environmental requirements have in the same breath complained that ISO 14001 did not set specific engineering standards.[38] The one time that TC207 came close to working on specific technical requirements, the initiative to write a standard on sustainable forestry provoked massive participation and resistance from the environmental NGOs.

The US Government and US NGOs would have been deeply concerned if ISO had attempted to establish technical performance standards since, given how stringent the US standards are, it is likely that the performance criteria set by ISO would have been less stringent. ISO is simply not the appropriate forum for setting public-policy-sensitive environmental performance criteria. On the other hand, countries with less stringent environmental requirements would have been very upset if ISO, an NGO, presumed to set more stringent requirements. Binding international environmental performance criteria should be set through the appropriate international bodies that involve official government participation, such as those that have created the Basel Convention or the Montreal Protocol.

Since ISO 14001 does not set specific performance criteria, it obviously cannot replace such criteria that are set by law, policy, associations, etc. Therefore, claims that conforming to ISO 14001 should eliminate the necessity for government oversight are silly. ISO 14001 is not, by itself, a guarantee of performance, nor should it be viewed as a certification of any particular level of environmental performance. Proponents who make such claims do the standard a disservice. The US Government and NGOs are coming to understand that ISO 14001 is only one part of the puzzle—it is a framework that can be used to meet performance criteria that are set outside of the standard.

This is not to say that ISO 14001 does not have any potential role in setting public policy on environmental issues. Several countries, including the US, Netherlands, Germany and Canada are exploring the relationship between environmental management systems and regulatory regimes. These initiatives range from the relevance of compliance-assurance systems to enforcement policies to exploring whether EMSs can be part of an overall effort to create regulatory systems that protect the environment while at the same time providing organisations more flexibility. The concept here is not that ISO 14001 would replace performance criteria that have been set by governments. Rather, the implementation of a third-party-verified EMS might serve as a basis for providing organisations more flexibility on how performance requirements are to be met, with the structure of the EMS possibly replacing some of the detailed procedural requirements that have been established by some countries.

Whatever role ISO 14001 may eventually play in setting public policy, it is incorrect to fear that ISO 14001 is inconsistent with or will somehow supplant environmental performance criteria that have been set at the national or international level, whether by legislation, regulation, formal treaty (e.g. Basel Convention, Montreal Protocol) or less formal agreements (e.g. Agenda 21, ICC Charter on Sustainable Development). Quite to the contrary, ISO 14001 may be used as a foundation for systematically identifying and managing such obligations.

One of the more remarkable criticisms of ISO 14001 is that national environmental laws could be 'over-ridden' by ISO 14001.[39] ISO 14001 demands precisely the opposite: a policy commitment to compliance that must be implemented by identifying legal requirements, setting objectives and targets, reviewing compliance status, etc. A facility cannot use ISO 14001 as the basis for setting lower standards for itself than those that apply in the nation where it operates. The theory underlying this criticism is particularly troublesome: that each element of the ISO 14001 standard should have been based on the most stringent equivalent legal requirement that could be found. This would have created a 'super standard' comprised of a collection of legal requirements that did not actually exist in any one country. Since one would have to conform to all of these requirements to obtain certification, this would effectively trump the sovereignty of nations by imposing international legal requirements from selected countries. Thus the claim that ISO 14001 'over-rides' national law is not only completely incorrect, but is also based on an approach that if implemented would itself violate sovereignty.

The concern that ISO 14001 may require changes to national environmental standards due to the GATT/WTO concerns is equally incorrect.[40] A national law imposing a specific requirement on facilities operating in that country could never be a non-tariff trade barrier. The issue of non-tariff trade barriers would arise for such requirement only where there was an attempt to apply them extra-territorially. For example, ISO 14001's relatively limited requirements on public disclosure or emergency planning cannot be the basis for an attack pursuant to GATT/WTO of the much more stringent requirements in these areas under US law. This is not to say that there are not any potential GATT/WTO issues related to ISO 14001. For example, if a country attempted to impose EMS requirements extra-territorially (e.g. demanding that products sold in that country come from facilities that operate in a particular manner, regardless of where the facilities are operating), that could raise GATT/WTO issues. In addition, there are a number of significant international trade issues related to the ISO 14000 standards on environmental marketing claims.[41]

Critics have to decide what they want. They cannot complain that ISO 14001 does not set any substantive performance criteria, then claim that ISO 14001 as written may over-ride the performance criteria that have been set by individual nations; they also attack the standard because it would allow organisations to choose to conform to laws in countries where the requirements may be less stringent than in those preferred by the critics, and on top of all this assert that ISO is not sufficiently democratic to make public policy decisions anyway.

☐ *ISO 14001 can be abused.*

ISO 14001 is not a 'fail-safe' system. It is a voluntary standard written with the intent that it will be used by organisations that sincerely wish to improve how they manage their environmental affairs. It was not written as many laws and regulations are drafted, with the intent to 'catch' wrongdoers and plug all potential loopholes contained in requirements. It does not contain its own performance criteria. Therefore it is possible that some companies may abuse the standard by attempting to design and implement EMSs that are relatively ineffective simply to get the certificate. This point has been a matter of some discussion with some US regulators who sometimes evaluate a document's value against how it might be able to control 'bad actors'.[42]

While this is certainly a limitation of the standard, it is a limitation that applies to any voluntary standard and is not a fatal flaw. This limitation should be explicitly recognised and be taken into account when evaluating some of the more ambitious claims that are being made on the standard's behalf. It will also be important for those involved in environmental issues in both the public and private sectors to guard against such abuses.

This limitation does not support the criticism that the standard cannot be used as a basis for improving and measuring performance.[43] As the discussion of the standard demonstrated, ISO 14001 is constructed on the premise that organisations should systematically identify and manage their environmental obligations and measure their performance against defined criteria. ISO 14001 does not provide those criteria—it simply establishes a framework in which to implement them. Any organisation that in good faith implements ISO 14001 will find a framework for setting performance criteria and measuring its performance against them.

■ *The Potential Public Policy Applications of ISO 14001 in the US*

There are several potential US domestic applications of ISO 14001. Many companies will turn to ISO 14001 as a guide to implementing an EMS for the purpose of improving their environmental performance. Others will seek third-party registration either because of customer demand or because of perceived competitive advantages. While these are important applications, this section will focus on the potential relationship between ISO 14001 and regulatory and compliance matters in the US.

There are two basic areas of potential application of ISO 14001 in the public policy context in the US. One is the opportunity to use the ISO 14001 standard as a framework or platform for identifying opportunities for alternatives to the 'command-and-control' approach to regulation. EPA is exploring these options in a number of programmes.

ISO 14001 projects have been included in the EPA's Environmental Leadership Programme, which is an initiative launched by the EPA in 1994 to encourage companies to adopt innovative approaches to improving environmental performance.[44] ISO 14001 projects have also been suggested in the context of the Common Sense Initiative, which is an ambitious effort announced by the EPA in 1993 to determine how it might streamline and consolidate the regulatory requirements applicable to whole industrial sectors.[45] A number of individual states in the US are also in the process of designing and implementing ISO-14001-based pilot programmes with industry.

The common thread in all of these initiatives is the hope that redundant or unnecessarily onerous regulatory requirements might be eased for companies that have instituted comprehensive EMSs. Another theme of these projects is one of experimentation. While there are many arguments raging about the value and efficacy of the standard, they are just that: arguments. What is missing is data. While most experts believe as a matter of common sense that systematically managing environmental obligations will improve environmental performance, this proposition has not been the subject of significant empirical testing.

The other potential regulatory application of ISO 14001 is in the area of enforcement discretion and penalty mitigation. The nature of a company's compliance-assurance programme is already taken into account by the Government in determining the type and intensity of enforcement response in the event that non-compliance is reported or discovered. Since the major elements of ISO 14001 are consistent with the major elements of compliance-assurance programmes set forth in the US Government documents, whatever enforcement discretion and penalty mitigation policies are adopted by the Government that entail the consideration of compliance-assurance systems, full credit should be given to companies that conform to ISO 14001.

However, since ISO 14001 is an overall EMS standard that contains many elements not directly related to compliance-assurance, such as pollution prevention, continual improvement, setting objectives and targets, etc., conforming to ISO 14001 should not be a condition of obtaining enforcement discretion from the government. Industry in the US is concerned that as the government considers the potential applicability of ISO 14001 to issues of public policy that the government may be tempted to revise or adjust the standard to reflect its compliance orientation. Though regulators should not adopt ISO 14001 as law or guidance for compliance-assurance programmes, guidance on compliance-assurance programmes should remain consistent with the ISO 14001 standard. As the internationally and domestically accepted voluntary EMS standard, ISO 14001 will probably become a leading benchmark against which organisations will evaluate their EMS. The elements of a compliance-assurance programme preferred by the government should be consistent with ISO 14001 because to suggest different compliance-assurance programme requirements for governmental purposes would force companies into the complexity and inefficiency of addressing multiple standards. It would also complicate the international competitiveness of companies, since to the extent that conformance with ISO 14001 becomes a condition of doing business abroad, companies would conform to ISO 14001 while, at the same time, prudence would dictate conformance to the policies of the US Government.

■ Conclusion

What many of the proponents and opponents of ISO 14001 have done is create a caricature of the standard, forming the basis for exaggerated and uninformed claims and criticisms. Some proponents claim that ISO 14001 is the answer to the world's environmental problems, creating a target that is too easy to resist. The opponents who cannot resist the easy shots are usually shooting not at the standard itself, but rather at the uninformed opinions of others.

Proponents of ISO 14001 need to be more modest in their claims. ISO 14001 is intended for broad application by a broad range of organisations and is not a 'state-of-the-art' standard. ISO 14001 will not, by itself, improve environmental performance. ISO 14001 is only one tool in the environmental protection toolbox, and must be implemented in the legal and public policy context that sets applicable performance criteria. ISO 14001 is a framework for supporting the implementation of performance requirements and criteria, but does not replace them. Proponents should also recognise that ISO 14001 is not for everyone: organisations may choose not to implement ISO 14001 for a number of legitimate reasons.

Opponents of ISO 14001 should be more precise in their analysis of the text of the standard and be more cautious in making claims about the motivation of those who drafted it as well as its relationship to other environmental initiatives. There are certainly legitimate criticisms that can be made of ISO 14001 (and of the claims that are being made by some proponents), but these do not include sweeping claims that ISO 14001 undercuts and pre-empts virtually all positive environmental initiatives of recent memory. This line of reasoning builds a false image of ISO 14001 and then, all too easily, knocks it down.

ISO 14001 is simply a voluntary environmental management systems standard that is intended to assist organisations in identifying and meeting requirements that are set elsewhere. While this may appear to be too modest a goal for some, the importance of systems approaches to managing an organisation's affairs should not be underestimated. Simply creating performance criteria or standards without providing mechanisms for achieving them may satisfy ideological needs but will not be of much practical assistance to the organisations whose environmental performance we all want to improve. Equally clear is that systems standards without meaningful criteria are empty shells.

ISO 14001 is based on the concept that better environmental performance can be achieved when environmental obligations are systematically identified and managed. Just as managers learned that product quality can be improved if quality objectives and systems are fully integrated into all aspects of a company's operations, there is a growing consensus that environmental protection and compliance can be best achieved if environmental factors are integrated into industrial operations in a systematic way: from design, raw materials selection, manufacture, sale, to the ultimate disposition or disposal of the product. The ISO 14001 standard provides an internationally-accepted framework for this approach. However, ISO 14001 is not 'the answer' to environmental protection and is not the only legitimate approach to EMS— it is only one more piece of the much bigger puzzle of attempting to improve environmental performance in a practical and cost-effective manner that is likely to actually achieve progress.

■ Notes

1. Parts of this section are drawn from Christopher L. Bell, 'Environmental Management Systems and ISO 14001: A US View', in *Environment Watch: Western Europe* (Cutter Information Corp.), 1 December 1995; and Christopher L. Bell, 'ISO 14001: Application of International Environmental Management Systems Standards in the United States', 25 ELR 10678 (December 1995).
2. 47 Fed. Reg. 29,025; 50 Fed. Reg. 43,804; 51 Fed. Reg. 33,669; 52 Fed. Reg. 7,337.
3. 57 Fed. Reg. 6356 (24 February 1991).
4. 51 Fed. Reg. 25,004.
5. US EPA, *The Exercise of Enforcement Discretion* (Washington, DC, 12 January 1994).
6. The New Jersey Prosecutor's Office has also issued Voluntary Environmental Audit/Compliance Guidelines (15 May 1992), which resemble the DOJ policy. The New Jersey Guidelines provide great detail about what a compliance programme must look like. Such programmes will be viewed as 'mitigating factors' in the exercise of the prosecutor's criminal environmental enforcement discretion.
7. 56 Fed. Reg. 22,762, Section 8A1.2, Comment K.
8. 61 Fed. Reg. 54062 (16 October 1996).
9. ISO 14001, Section 4.2. It is suggested that readers focus on the requirements section of the standard, Section 4, along with the definitions in Section 3, when reading

the standard. These are the auditable requirements. The introduction to the standard is not auditable and is not a significant component of the standard.

10. ISO 14001, Section 4.3.1.
11. Benchmark Environmental Consulting, *ISO 14001: An Uncommon Perspective: Five Public Policy Questions for Proponents of the ISO 14000 Series* (Brussels, Belgium: European Environment Bureau (EEB), November 1995), p. 4 (henceforth referred to as *'Uncommon Perspective'*). An abridged version of *Uncommon Perspective* was published in Issue 14 (April 1996) of *Greener Management International* by Harris Gleckman and Riva Krut under the title 'Neither International nor Standard: The Limits of ISO 14001 as an Instrument of Global Corporate Environmental Management' (see also Chapter 2 of this book). A revised version of this article has reappeared in the form of a draft UNCTAD report titled 'ISO 14001: International Environmental Management Standards: Five Key Questions for Developing Country Officials'.
12. ISO 14001, Section 4.3.2.
13. The ISO EMS guidance document, ISO 14004, does contain references to several of these documents. This is appropriate because ISO 14004 does not contain auditable requirements. For example, ISO should not lightly transform the 'recommendations' of Agenda 21 into 'requirements'.
14. *Uncommon Perspective*, p. 23.
15. ISO 14001, Section 3.12.
16. *Uncommon Perspective*, p. 9.
17. ISO 14001, Section 4.3.3.
18. ISO 14001, Section 4.3.4.
19. The criticism to contrary is inconsistent with the plain language of the standard. See *Uncommon Perspective*, p. 17.
20. ISO 14001, Section 4.4.1.
21. ISO 14001, Section 4.4.6.
22. ISO 14001, Section 4.4.7.
23. ISO 14001, Section 4.4.2.
24. ISO/DIS 14001.2, Section 4.4.3.
25. ISO 14001, Sections 4.5.1, 4.5.2 and 4.5.4.
26. *Uncommon Perspective*, pp. 15-17.
27. ISO 14001, Section 4.5.1.
28. *Uncommon Perspective*, p. 21.
29. ISO 14001, Section 4.6.
30. ISO/DIS 14001.2, Sections 4.4.4, 4.4.5 and 4.5.3.
31. To demonstrate the level at which the discussion of ISO 14001 takes place, one commentator has claimed that labelling ISO as the 'International Standards Organisation' rather than the correct 'International Organisation on Standardisation' is incorrect 'translation of the acronym "ISO" into English' and is a 'crucial obfuscation'! (*Uncommon Perspective*, p. 4). It is indeed uncommon to assert that the public will perceive significant differences between an organisation that works on 'standardisation' and one that works on 'standards'. The correct name for ISO has nothing to do with the proper 'translation' of the acronym, because ISO is not an acronym for the name of the organisation in any of its three official languages (French, Russian or English). For example, the official French name is Organisa-

tion Internationale de Normalisation, which certainly does not produce the acronym 'ISO'. I dwell on this issue not because it is a critical matter, but as an example of how a simple misunderstanding about ISO rapidly escalates into allegations of 'crucial obfuscations' which have absolutely no factual basis.

32. See *Uncommon Perspective*, p. 4.
33. See *Uncommon Perspective*, p. 11.
34. One critic even refers to many of the initiatives it claims ISO 14001 will discourage as 'experimentation' (*Uncommon Perspective*, p. 6). The entire point of experiments is to test propositions to determine if they are correct in a situation such that a determination that the propositions were incorrect is not catastrophic. What would discourage experimentation would be to conduct it through ISO standards that cannot easily be taken back and where the consequences of finding out that the proposition to be tested was incorrect are severe. Experimentation means that there a variety of potential solutions and approaches to a problem that remain to be tested—which, almost by definition, means that time is not yet ripe for a standard. ISO 14001 supports such experimentation by creating a systems framework within which organisations can implement leading-edge approaches.
35. *Uncommon Perspective*, pp. 7-13.
36. *Uncommon Perspective*, pp. 7-14.
37. *Uncommon Perspective*, p. 17.
38. See, for example, *Uncommon Perspective*, pp. 9 and 12.
39. *Uncommon Perspective*, pp. 20-21.
40. *Uncommon Perspective*, p. 21.
41. The most recent incarnation of the *Uncommon Perspective* article, the draft UNCTAD report, continues the misunderstandings about the relationship between the ISO 14000 standards and international trade. The draft UNCTAD report suggests that the ISO 14000 labelling standards would inappropriately prevent a government from establishing a labelling and market preference for facilities that improve their environmental performance (Draft UNCTAD Report, p. 57). This scenario would be true only if the government policy was inconsistent with the ISO standards. The current drafts of the ISO 14000 labelling standards are attempting to ensure that environmental labelling programmes are based on good science, consensus and transparency. Assuming that these principles prevail, a governmental programme that is consistent with these principles would be presumably acceptable under the Technical Barriers to Trade provisions of GATT/WTO. Therefore, the scenario in the UNCTAD draft report would be accurate only if the government in question established a policy that was not based on science, consensus or transparency: precisely the types of programmes that the participants in ISO 14000 want to discourage. The draft UNCTAD report also ignores the fact that it is standards such as those being developed by ISO that make legitimate national policies possible without running foul of GATT/WTO. Without the TBT provisions of GATT/WTO and internationally-accepted standards, national policies setting extra-territorial requirements that control process or production methods (PPM) have generally been held to violate GATT/WTO.
42. The issue of certifier liability is a difficult one. Financial auditors have been sued on the theory that investors 'relied to their detriment' on the assurances of the auditors that the company or financial institution was stable, solvent, etc. Such

reliance arguments will be more problematic in the environmental context. In addition, the certification is of the system, not the performance of a product or organisation at any given point in time, making it difficult to 'blame' the certifier for bad performance. Lastly, making certifiers liable for bad performance would rapidly dry up the supply of certifiers, since they cannot really control organisational behaviour.

43. *Uncommon Perspective*, p. 17.
44. See 59 Fed. Reg. 32062 (21 June 1994).
45. See 59 Fed. Reg. 55117 (3 November 1994).

Squaring the Circle
Fundamental Barriers to Effective Environmental Product Labelling

Mark T. Smith

THE CONCEPT of product environmental labelling is potentially very beneficial to consumers, manufacturers and government. It could eliminate confusing, unsubstantiated or extravagant environmental claims, clarify the position of a company in a market, and act as a lever to improve industrial environmental performance. A combination of both consumer and environmental protection led to the development of the first European Union (EU) Eco-label. However, too much is expected from this isolated and weak policy mechanism, which has become more ambitious and confused as the labelling scheme is now expected to deliver a wide range of goals.

Interest in labelling schemes has not been limited to the EU. Prior to the Rio 'Earth Summit' in 1992, the International Standards Organisation (ISO) was asked to contribute to sustainable development through international standardisation in environmental management tools, which included product labelling. ISO interpreted labelling as a tool for simultaneously preventing manufacturers from making false claims, providing consumers with independent information, and which was capable of moving markets in a sustainable direction. All this was to be achieved through the application of a voluntary standard, while avoiding the creation of unintended international trade barriers—an ambitious series of objectives. The apparent difficulties in meeting these objectives are written clearly in the current problems in reaching a final draft of the standards relating to environmental product labelling as part of the ISO 14001 series. This chapter examines the European experience and finds that the problems that have emerged in this context may simply be transferred to the international level without the benefit of resolution.

■ *The Origin of the EU Eco-label*

Over the last ten years, the use of unofficial (and often misleading) environmental claims or labels on a variety of consumer products resulted in growing consumer resistance to products promoted with 'green' endorsements, such as recycled toilet paper and washing powders.[1] In response, a number of nations introduced official, government-backed eco-labelling schemes in an attempt to reassure consumers. The German 'Blue Angel' scheme was one of the first such 'Eco-labels', and similar schemes now exist in most OECD countries. In preference to a national scheme, the UK supported the development of an EU-wide Eco-label. With its thorough product lifecycle assessment (LCA) of the environmental impacts of product design, the scheme was heralded as the ultimate in identifying the most environmentally-sound products on the market. Other schemes tended to focus on single, often emotive, issues.

The Eco-label was the first EU policy instrument to focus on the environmental implications of product design, in marked contrast to previous environmental initiatives, which concentrated on manufacturing processes, pollution abatement and clean-up 'end-of-pipe' technologies. The shift of regulation up the production chain should, it was reasoned, represent a more efficient approach to the prevention of adverse environmental impacts rather than attempting to clean up an already inherently polluting process.

A summary of the general features of the EU Eco-label can be seen in Figure 1.

■ *The EU Eco-Labelling Process*

The process of obtaining an EU Eco-label involves several key stages. First, a working group is established to determine a commodity group for consideration. Member state Eco-label Boards, appointed as the lead bodies for individual commodities, are responsible for setting criteria; they then recruit representatives from the relevant industry, consumer groups and environmental organisations. Boards commission an LCA, and each working group translates the results into criteria for awarding an Eco-label. The criteria are finally ratified by the European Commission (EC), and firms are able to apply to the national Eco-label Boards throughout the EU for acceptance of a product.

■ *Lifecycle Assessment (LCA)*

The development and acceptance of LCA has been crucial to the evolution of the EU Eco-label as a basis for the comparison of the quantitative environmental

---

1. It is applicable throughout the EU.
2. The scheme is voluntary.
3. The scheme is self-financing; applicants have to pay an annual fee.
4. It excludes food, drink and pharmaceutical products.
5. It involves an LCA.
6. It assesses individual products, not companies.
7. It is a simple pass/fail system.
8. It targets general consumer goods, for which alternatives exist.
9. Criteria are constantly revised to achieve higher environmental performance.

An unwritten general aim has been that 10%–20% of any particular product sector would be expected to meet the Eco-label criteria.

Figure 1: *General Features of the EU Eco-label*

Source: Environmental Business, *Eco-labelling Supplement* (London, UK: Information for Industry Ltd, 1992); Environmental Data Services, *ENDS Report*, No. 227 (December 1993), p. 26

impacts of different commodities. This technique identifies the environmental impacts of a product from cradle to grave, that is, from raw material extraction, production, distribution, to use and eventual disposal.[2] However, comprehensive and exhaustive LCAs are costly and time-consuming; information may be commercially sensitive, unreliable, or in an unusable form. There has been some consensus concerning the compilation of inventories, but considerable disagreement about techniques employed for comparing the impacts of different emissions. There has also been debate concerning the validity of using industry-standard or process-specific data, and any conclusions may be prejudiced by the sponsoring organisation.

■ *Squaring the Circle*

Despite six years of working group meetings, the commissioning of studies from consultants and preparing successive drafts of criteria for a number of product groups, only two EU-wide Eco-labels have been awarded. The first was in November 1993 for the Hoover 'New Wave' washing machine range, and the second was for recycled paper kitchen towels and toilet tissues.[3] The slow pace of approving product groups is indicative of the problematic nature of the scheme. The whole process of developing criteria for the Eco-labels for different products has become a tangle of debate between industry, consumer groups and individual national Eco-labelling boards.

The tactical difficulties of developing working criteria for the individual Eco-label groups are related to the strategic purpose of labelling in itself, as crucially the EU Eco-label was conceived as an instrument of consumer, not environmental, policy. In this context, 'tactical' issues are more concerned with the mechanics of the scheme, including its reliability, likely economic impacts, capacity for promoting 'state-of-the-art' technologies, consumer responses, marketing strategies and the need for internal reform. Even if these issues can be resolved, wider 'strategic' implications must be addressed. These include concerns that the introduction of any kind of scheme could be interpreted as an impediment to free trade across international boundaries. The precise nature and purpose of the scheme and alternative policy options should be explored, and its potential for co-ordination with other environmental policies considered.

■ Tactical Issues

☐ The Benefits and Cost of the Label

Washing machine manufacturers Hoover were awarded the first EU Eco-label for their 'New Wave' range, as the machines meet the required levels of electricity, water demand and detergent loss during use.[4] However, their competitors, including Hotpoint and Zanussi, have declined to apply, as they contend that the initial registration cost and licence fee of £500 is excessive, and are waiting for more convincing evidence of the success of the scheme. They are more likely to respond to consumer pressure, claiming that, at present, customers are unaware of the label, and Hotpoint have gone on record as believing that potential customers would find it irrelevant.[5]

The reason for Hoover's enthusiasm for the Eco-label reflects their recent marketing strategy. They have been particularly determined to move up-market and to challenge the dominant mainland European manufacturers. With the Eco-label, Hoover's share of the up-market washing machine market has more than doubled since 1993. Although not all the success can be attributed to the label alone, the German market is particularly receptive to environmental issues.[6] For Hoover, the Eco-label has fitted a particular marketing need. For other washing machine manufacturers, this is not the case and none have applied for the label, even though almost half the machines currently available would also qualify.

The Hoover example illustrates that the Eco-label can be commercially significant, but the situations in which a company would benefit from obtaining a label may be quite restricted. In contrast, Fort Sterling intend to apply for the second label for their 'Nouvelle' range of recycled paper toilet rolls and kitchen towels, although they acknowledge the marketing impact is uncertain.[7] These varying commercial tactics stem from the fact that, as the scheme is voluntary, there is no obligation to submit products for the Eco-label, and only if a product's environmental performance can command a price or advertising premium is it commercially worthwhile applying for an Eco-label at all.

☐ LCA Methodology

The core methodology supporting the LCA scheme is an evolving subject, and has been subject to criticism. For example, The Body Shop objected to the methodology employed because biodiversity and animal testing issues are ignored, and consequently withdrew from the scheme. Other LCA studies may have produced results biased in favour of the sponsoring industry.[8] ISO has accepted that LCA is an appropriate tool for identifying environmental impacts and is engaged in developing an internationally accepted methodology to resolve the contentious issues.[9] The organisation is monitoring other labelling schemes rather than generating and imposing its own methodology and targets.

☐ The Pass/Fail System and Product Standards

The pass/fail nature of the label raises several important and related issues. Such a crude device fails to permit further discrimination between competing goods bearing the label, particularly if it only applies to a restricted proportion of the market. It has been proposed that the Eco-label should be a graded index, similar in nature to the Energy label, although at the moment it seems that a simple pass/fail is generally favoured. Under the present conditions, only the best 10%–20% of products in a particular group would qualify for an EC Eco-label, which would presumably encourage the manufacturers of the other products to achieve the environmental standards of the leading producers. It could well be that even the best available fall well short of environmental standards that are possible or desirable. Thus, on this basis, clearly environmentally-damaging products may get an Eco-label simply because there are no really good products in that sector. Equally, a sector where a lot of work has been undertaken to reduce environmental impacts, such as washing machines and detergents, could end up with really good products failing to get an Eco-label simply because they fall outside the preferred group.

This approach could well end up *discouraging* environmental improvements. In a sector with a poor environmental record, an industry association may simply keep environmental standards low in order not to produce a more stringent target. Alternatively, an industry with a good environmental record will have no incentive to introduce any environmental improvements if a manufacturer is unable to get into the top 10%. If that sector is dominated by a different technology, then there is nothing to be gained through improved environmental practices.

Proposals to reform the label include suggestions for increasing the amount of product information on the label to provide consumers greater differentiation between similarly-labelled goods. There is, however, the risk that more information would in fact cause even greater confusion. Furthermore, the criteria for a product group could not be set lower than existing national standards without undermining the scheme. This raises the argument that industry sectors could abandon national schemes in favour of a weaker option.

Setting the technical criteria is also subject to industrial interests. For example, most emulsion paints already qualify for the paint label on the principle that the main criterion limits the permitted levels of volatile organic compounds (VOCs), but they have been deliberately excluded from the scheme, as most VOCs are concentrated in trim paints, which the manufacturers insist is the most relevant category of paint for labelling. However, such paint accounts for a very small overall market share. ICI claim they had to make small changes to the formulation of their range of paints in order to qualify for the label.[10] The company also already produces another paint that is eligible for the 'Blue Angel' scheme, and with minor modification would also qualify for the EU Eco-label; several other brands already bear similar European national labels.[11] One small UK-based paint manufacturer has pulled out of the scheme in protest against the domination by the large producers, claiming the standard has been set at a level that is currently achievable, rather than one optimised for environmental improvement.[12]

☐ Consumer Behaviour

A poor understanding of consumer behaviour and motivation has been cited as stalling progress towards a label. For example, the aerosol industry has been opposed to the label since an LCA study commissioned by the industry demonstrated that aerosols were comparable with non-aerosol technology in terms of the volume of the product delivered, but the Eco-labelling board applied a weighting factor against aerosols. The industry contended that it was impossible to determine how consumers used such products, and therefore what constituted a reasonable test.[13]

A further weakness is exposed by evidence that suggests consumers have a poor awareness of the relationship between initial purchase costs and operating costs. This is demonstrated by the poorly-constructed reward nature of the scheme which fails to recognise such a relationship, and accounts for the main barrier to the widespread use of Compact Fluorescent Lights (CFLs).

☐ Pace of Development

Several industries have complained about the apparent slow progress of the scheme. The European Commission has also been attacked by the EU Environment Council for refusing to implement the scheme, including the threat of legal action in the Court of Justice. It is alleged to have procrastinated over the adoption of criteria for soil improvers and tissue paper, and meetings of the Regulatory Committee, which oversees the scheme, have been subjected to cancellations. It has undergone several reviews designed to speed up the process of allocating labels.

Even the UK Eco-labelling Board's (UKEB) own publicity newsletter referred to the scheme's progress as 'slow and frustrating'.[14] For some time, the UKEB has tried to remain optimistic about the rate and timing of labelled products being available to consumers, but every prediction has so far proved inaccurate.[15] (Eco-labelled soil improvers were scheduled to be available from spring 1995; to date there is no sign.)

☐ Proposed Reforms and Industrial Domination

Consumer and pressure groups feel that industry trade bodies dominate setting criteria. Greenpeace complained that criteria are fixed to meet the needs of industry, rather than establishing environmental excellence.[16] Industries clearly resent the involvement of pressure groups, and are sceptical of the effectiveness of the scheme. The initiative to begin the process of developing Eco-label criteria for a product category lay largely with industry. Although it was the intention that the composition of a board developing Eco-label criteria 'is such to guarantee independence and neutrality',[17] there is no guidance in the EU regulations as to how this requirement may be achieved.

In practice, industrial considerations dominate; standards tend to be geared towards what is already technically achievable, rather than what is most likely to benefit the environment. This could influence new product innovation strategies, and particularly the commercial development of environmentally-benign technologies. Small firms generally tend to be more innovative than larger companies, but such firms generally have small market share. In contrast, large firms can rapidly influence a market through the diffusion of incremental improvements to existing designs.

Proposals for reform include measures geared to reduce the significance of industrial interests by changing the composition of the board, regulating the number of representatives, modifying voting rights and the general decision-making process. Such moves have been interpreted as the Commission attempting to distance itself from the scheme.[18] Even if the benefits, legitimacy of the methodology and independence of the scheme could be demonstrated, there remain some fundamental problems with the scheme.

■ Strategic Issues

☐ Trade Barriers

International trade laws are designed to regulate the import and export of products, but environmental legislation has evolved from a need to control process and production methods. The EU Eco-label, although voluntary, represents an expression of favoured processes, and discriminates against others.[19] Agreements relating to Technical Barriers to Trade (TBT) predate labelling schemes. This has generated friction between the Industrial Affairs (DG III) and Environment (DG XI) Directorates within the European Commission, because of claims that the introduction of the scheme may constitute an artificial trade barrier both within the Community and with other world trade organisations, therefore infringing one of the founding free market principles of the Union.[20] This has threatened the very existence of the label on several occasions.

This argument has been extended to the General Agreement on Tariffs and Trade (GATT). Several industries, including the US paper industry, have employed this objection, claiming the label imposes restrictions on non-European producers. They argue that non-participation implies that a product lies in an environmentally non-preferred category, and thus introduces substantial market disadvantages.[21] The industry has also sought the support of DGIII with complaints of unfair bias towards recycled paper, and alleges that the criteria lack transparency, a claim supported by other industrial interests. However, this rationale could be applied to any kind of product labelling; the same trade association does not oppose the US-based 'Green Seal' Eco-label using the same arguments! Conversely, the label could also be interpreted as a guarantee of quality, and provide a boost to trade.

The European Commission is investigating the compatibility of the EU Eco-label with the Technical Barriers to Trade section, which includes product standards, as part of the latest round of GATT. In response to criticisms that the label represents a standard, and therefore a barrier to trade, it has been defended on the grounds that it is voluntary, unlike energy labels. ISO are involved in this process of resolving the legitimacy of the scheme, and are also conscious of the label being interpreted as a trade barrier.

Two UN bodies—the Environment Programme (UNEP) and the Conference on Trade and Development (UNCTAD)—have both intervened with attempts to resolve the perceived conflict between the EU and GATT, by establishing multilaterally agreed, transparent criteria which do not disadvantage developing countries, or favour certain processes.[22] Brazilian paper pulp manufacturers first raised this issue, complaining that the criteria would disadvantage producers unable to meet those standards, even if the schemes imposed by northern countries are voluntary.[23] The joint venture aims to resolve the potential conflict between environmental initiatives and income derived from exports from developing nations.

☐ Policy Integration and Co-ordination

Apart from links with packaging ordinances, which generally constitute secondary criteria, the Eco-label is a very isolated policy mechanism. An almost accidental exception is its application to energy-efficient lighting, where it was anticipated that product labelling could contribute to the CO_2 stabilisation strategy, and simultaneously assist the EU 'SAVE' programme.[24] The lighting manufacturers have rejected attempts to introduce an Eco-label, and insist that because promoting energy efficiency is the main drive, this could be dealt with most effectively under an appropriate EU energy labelling directive.

☐ Selection of Product Groups

Because the label is voluntary, there has been a tendency to create working groups for products where there is a potential for commercial gain by identifying environmental performance and not where environmental impacts are a pressing concern. Thus working groups were set up to develop Eco-label criteria for cat litter and floor tiles, but no working group was set up for cars. This position is likely to be remedied, as the EU have proposed feasibility studies for at least two groups of products with major environmental impacts: passenger vehicles and personal computers.[25] The washing machine label is one of the few sectors where areas of popular green consumerism and real environmental impacts actually coincide.

☐ Confused Purpose

There is still much confusion and dispute concerning the purpose of the Eco-label, and thus its location within the European Commission. At the end of 1994, EU ministers rejected an attempt by the Commission to offload the scheme to the newly-created European Environment Agency. It appears the scheme does not fit happily within the environmental remit, and also may not be appropriate as an aspect of consumer affairs, handled by DGIII. The Environment Directorate (DGXI) have suggested setting up a new, independent European Eco-label organisation due to pressure from the EC to overhaul or abandon the system, as very few articles are likely to be labelled by the time the scheme is scheduled for a major review in 1997.

☐ The Greening of Industry and Alternatives to the Scheme

Even if all the strategic and tactical issues facing the EU Eco-label can be overcome, there are still broad movements and influences within industry that could render the entire scheme ineffective and inappropriate. It is possible that the label itself may become irrelevant in the general drive to 'green' industry. Other pressures acting on industry, including public concern, may derive greater environmental benefits irrespective of the introduction of an isolated product labelling scheme. Such changes may not be causally related to the scheme itself.

Alternatives to the label have been proposed. For example, the paper industry finally agreed to an Eco-label for copying paper, after considerable prevarication and objections stemming largely from the US and European paper manufacturers. The industry trade association has argued that the label addressed the production process, not the product, and were unable to agree to a satisfactory definition of 'sustainable forestry' as a natural resource. They were also concerned that the scheme will interfere with existing national and independent labelling schemes, and thereby reduce the impact of the EU label. Their preference is to adopt the Environmental and Management Auditing Scheme (EMAS), which they claim is a more reliable indication of the industry itself, and not limited to its products.[26] They do, however, acknowledge the potential for supply chain pressure which could influence future directions.[27]

An alternative approach has emerged in the UK. In 1993, the UK Group for Efficient Appliances (GEA) produced a study of national schemes designed to improve the energy efficiency of a range of electrical appliances. The report[28] concluded that minimum standards are the most cost-effective mechanism for achieving efficiency targets. They also concluded that information schemes acting in isolation, such as labelling, were severely limited. Any energy savings were highly uncertain, since schemes have to compete in a complex market where aesthetic and price factors dominate. However, the GEA recommended that a mix of policies, including mandatory energy labelling, could encourage widespread acceptance of the most efficient appliances.

Thus it was a short step, for the Eco-label became part of the UK Government's strategy for controlling CO_2 emissions. In the climate change strategy announced in January 1994,[29] the UK Government claimed that 5% of the saving of 10 million tonnes of CO_2 emissions needed to stabilise the UK's emissions at 1990 levels by 2000 would be achieved by labelling and energy efficiency standards. This is the key argument for promoting sales of CFLs.[30] CFLs are up to five times more efficient at converting electricity into useful light, and last up to ten times longer than conventional tungsten lights. However, the domestic lighting market remains dominated by traditional tungsten filament lamps, although CFL sales have grown rapidly, reaching 114 million/year in 1991.[31] The benefits gained from using CFLs to reduce energy costs has been widely appreciated by industry and commerce. Some estimates calculate that the substitution of incandescent lights for CFLs could reduce household electricity consumption by 23%.[32] The lighting Eco-label working group was set up in 1992 and eventually proposed a minimum energy-efficiency performance for electric lighting which effectively excluded all tungsten lights, but included all CFLs.

The core rationale of the Eco-label is that a small proportion of the total market (20% or less) comprise 'green' consumers that are sufficiently committed to pay a premium for more environmentally-benign products. The chances that this policy mechanism can work is remote when there is a major price premium and there is a need to affect the purchasing behaviour of all consumers. This has proved to be the case for CFLs, which retail in the range £8–£15, compared to tungsten lights costing £0.50 or less. Consumers have a very poor understanding of the relationship between the first cost and running cost of a product, and are generally unwilling to pay a price premium for energy efficiency. The price difference involved with CFLs is generally unacceptable to domestic consumers, even though over the life of the bulb it is cheaper than continuing to buy tungsten lamps. In such circumstances, the Eco-label is an entirely inappropriate policy instrument.

While this EU Eco-label was being debated, an alternative scheme intended to promote the virtues of energy efficiency in the UK was being established. This was the Energy Savings Trust (EST), funded by the energy industry. The inability of the Eco-label to stimulate demand for CFLs has been tacitly recognised in that a separate initiative to promote CFLs was launched in 1993–94 by the EST.[33] This was effectively a subsidy programme which had the effect of temporarily reducing the price of CFLs. By addressing the key barrier of price, the EST scheme succeeded in increasing sales of lamps at the low price, and producing a modest post-scheme boost in CFL sales.

Meanwhile, the lighting Eco-label has failed to materialise, mainly due to opposition from the lighting industry. It appears that this industry prefers a comprehensive, all-product energy label rather than the CFL-only Eco-label. The energy label, they argue, could be a standardised part of product information.[34] Such an energy label has been in operation in the EU since 1995, and applies to all refrigerators and freezers.[35] In contrast to the Eco-label, this EU government programme is a compulsory feature for all products and there is no registration fee. The information is presented as a seven-point scale, with appliance efficiency inversely proportional to the length of an 'arrow'. The complexity of the label may limit its effect, but this may improve with retail staff training.

The insistence of the lighting industry in favour of an energy label has put the future of the lighting Eco-label into doubt. The industry has shifted its arguments to embrace a quite contradictory stance. They favour energy labelling, but object to the Eco-label on the grounds that there is insufficient space to fit the logo onto the packaging. They would prefer to see a separate label for standard tungsten lights, which the industry argue are not the 'functional equivalent' of CFLs, and will never match the efficiency of fluorescent lighting. However, CFLs were originally devised and marketed as efficient replacements for such lighting, and two identical labels for apparently similar products with enormous initial price differences would produce considerable confusion for consumers. The board responsible for devising the lighting criteria for the Eco-label is also dominated by the industry, and all the members are producers of all types of lighting, including CFLs. As an environmental measure, Eco-labelling appears to be redundant in this sector. Consumers are unlikely to respond as long as CFLs remain prohibitively expensive, and the EST subsidy scheme has not resulted in a price reduction sufficient to produce a significant shift in purchase behaviour.

■ Conclusions

Since the EU Eco-labelling process was set in motion in 1990, the need for effective environmental policies has become increasingly apparent. Even though the EC does not know where to place the Eco-label, the UK Government now views it as an instrument of environmental rather than consumer policy.[36] This move may appear to be a perfectly understandable policy adaptation, but had the Eco-label been designed as an instrument of environmental policy in the first place, the key features of its voluntary nature, commercial-financing and industry control would have been seriously questioned.

Furthermore, related consumer policy options, such as amending the Trades Description Act, have not been vigorously pursued. At an EU level, the links between the Eco-label and other environmental labelling schemes, particularly energy labelling, appear tenuous and confused. Other product-level policy instruments have also been developed at the same time as the Eco-label, in particular minimum energy efficiency standards, with no clear connection between the initiatives. Environmental policy instruments, such as energy labelling and the development of minimum efficiency standards, have been characterised by their compulsory nature applied to the whole product range and independent control. As an instrument of environmental policy, the Eco-label's industry-dominated and voluntary nature appears to imply critical weaknesses.

Product labelling can play a role as an instrument of environmental policy but, as in all policy areas, the existence of several unco-ordinated initiatives, each with their own bureaucratic and industrial vested interests, in no way makes for effective policy-making. An integrated environmental policy would involve a combination of reinforcing instruments involving information, fiscal incentives and standards that address the 'levers' with most environmental impact. A labelling scheme might be more appropriately used to reward new innovation, where radical environmental improvements gained from a specific technology could be promoted and simply administered, and this would remove the need for a complex and unwieldy LCA. This would also simultaneously reduce the need for intervention by ISO. A complementary scheme could be introduced to support the diffusion of the best technology and products, but focusing on a mandatory, graded system such as the energy label, which also considers cost savings. Such measures could then be integrated with other policy instruments geared towards cost savings for the consumer, such as tax rebates. This may be a greater motivation than concern for the environment. ISO has similarly failed to recognise the limitations of environmental product labelling, quite apart from its doubtful capacity to stimulate moves towards sustainable development.

In its current form, the EU Eco-label is unlikely to make any significant contribution towards achieving higher environmental standards of product performance. There is very little evidence for its effectiveness, outside the rare and rather exceptional circumstances as exemplified by Hoover. However, an important role for an Eco-Label is to identify how close a product is to the environmental 'state of the art' and if it is innovative to the extent that it is advancing the 'state of the art'. It can only become relevant if its operation is reformed and it takes a place in a truly co-ordinated set of environmental policies, including minimum standards, fiscal incentives and penalties plus strengthened laws on advertising. As well as making the transition to becoming a useful part of environmental policy, the Eco-label would be enabled to fulfil its original function, that of providing reliable environmental credentials for consumers to make informed choices. If these issues cannot be resolved at European level, international initiatives such as the work by ISO will be doomed to repeat such recent history.

■ *Notes*

1. See National Consumers Council, *Green Claims: A Consumer Investigation into Marketing Claims about the Environment* (London, UK: NCC, 1996).
2. See Society of Environmental Toxicology and Chemistry, *A Technical Framework for Life-Cycle Assessment*, Report from SETAC Workshop, held at Smugglers Notch, VT, 18–23 August, 1991.
3. See Environmental Data Services, *ENDS Report*, No. 250 (November 1995), p. 24.
4. See Environmental Data Services, *ENDS Report*, No. 226 (November 1993), p. 25.
5. See Environmental Data Services, *ENDS Report*, No. 234 (July 1994), p. 27.
6. See The Open University, *Video 2: Green Product Development* (Milton Keynes, UK: Open University Course T302 Innovation; Design, Environment and Strategy, 1995).
7. See Environmental Data Services, *ENDS Report*, No. 250 (November 1995), p. 24.
8. See Environmental Data Services, *ENDS Report*, No. 238 (November 1994), pp. 9-10.
9. See UK Eco-labelling Board, *UKEB Newsletter*, No. 10 (1996).
10. See Environmental Data Services, *ENDS Report*, No. 258 (July 1996), p. 26.
11. See Environmental Data Services, *ENDS Report*, No. 252 (January 1996), p. 24.
12. See M.T. Smith, R. Roy and S. Potter, *The Commercial Impacts of Green Product Development* (Design Innovation Group, DIG-05; Milton Keynes, UK: The Open University, July 1996).
13. See Environmental Data Services, *ENDS Report*, No. 242 (March 1995), p. 27.
14. See UK Eco-labelling Board, 'Ecolabel Criteria', in *UKEB Newsletter*, No. 6 (March 1994), p. 1.
15. See Environmental Data Services, *ENDS Report*, No. 238 (November 1994), p. 23.
16. See K. West, 'Ecolabels: The Industrialisation of Environmental Standards', in *The Ecologist*, Vol. 25 No. 1 (January/February 1995).
17. *Ibid.*
18. *Ibid.*
19. See Environmental Data Services, *ENDS Report*, No. 252 (January 1996), p. 26.
20. See Environmental Data Services, *ENDS Report*, No. 236 (September 1994), p. 27.
21. See Environmental Data Services, *ENDS Report*, No. 237 (October 1994), p. 25.
22. *Ibid.*
23. See Environmental Data Services, *ENDS Report*, No. 252 (January 1996), p. 26.
24. See Environmental Data Services, *ENDS Report*, No. 239 (December 1994), p. 25.
25. See UK Eco-labelling Board, *UKEB Newsletter*, No. 10 (1996).

26. See Environmental Data Services, *ENDS Report*, No. 243 (April 1995), p. 33.

27. See Environmental Data Services, *ENDS Report*, No. 257 (June 1996), p. 26.

28. See Environmental Data Services, *ENDS Report*, No. 235 (August 1994), p. 24.

29. Department of the Environment, *Climate Change: The UK Programme* (London, UK: HMSO, January 1994).

30. Department of the Environment, *This Common Inheritance: The Second Year Report* (London, UK: HMSO, October 1992).

31. See E. Mills, 'Efficient Lighting Programs in Europe: Cost Effectiveness, Consumer Response, and Market Dynamics', in *Energy*, Vol. 18 No. 2 (1993), pp. 131-44.

32. See J. Kelsey and V. Richardson, 'Residential Appliance Saturation Survey: A Profile of the Residential Lighting Load in Northern California', in *Conference Proceedings of American Council for an Energy Efficient Economy* (Asilomar, CA, September 1990).

33. 'Little Bill: Large Savings', *EST Press Release*, 10 October 1993.

34. European Lighting Companies Federation, 'The Position of the Lamp Industry Regarding the EC Eco-label for Light Sources', from speech given by Ludo van Kollenburg at *The Future of Eco-auditing and Eco-labelling in Europe Conference*, 4–5 November 1993.

35. Commission of the European Communities, *Proposal for a Council Directive on Energy Efficiency Requirements for New Household Electric Refrigerators, Freezers and their Combinations* (Brussels, Belgium: Commission of the European Communities, 1995).

36. See *Official Journal of the European Communities*, L99, Vol. 35 (11 April 1992).

5 Trade, Competitiveness and the Environment

Donal O'Laoire

*T*HE PURPOSE of this chapter is to find a useful way of examining the relationship between trade, competitiveness and the environment in such a manner as to be applicable to the industrial and commercial world. My starting point is to suggest that the word 'environment', with its baggage of universal meanings, perceptions and prejudices, is an obstacle to our project and has the potential to add confusion and obfuscation. For the purposes of this chapter, I would like to remove it from our lexicon and to replace it with a simple equation.

$$E = CH = C + M + L$$

- Where **E** is **Environment**, but used as a code or a shorthand for rapid and unprecedented global change
- Where **CH** is that unprecedented global **Change** that is sweeping the commercial and industrial world and affecting trade and competitiveness
- Where **C** is economic **Cost**, the parameters and elements of which are manipulated and respond to global change
- Where **M** is consumption patterns and **Market** forces, which are responding to threats and opportunities created in the changing climate
- Where **L** is the **Legislative** framework, which circumscribes the commercial world, including not only legislation but also codes of practice, sectoral guidelines and product specifications

The hypothesis is that, to the business, commercial, financial and industrial communities, environment is shorthand for the sum of all the above and more. This hypothesis will hopefully allow us to restate and to explore the environmental issue in the language of business, commerce, finance, industry, trade and competitiveness.

■ *The Context of Change*

The underlying premise of this paper is that the Western industrial economy, which has had such a profound effect on human society, is undergoing a process of change and restructuring. In this period of transition to a new socio-economic order, it is prudent to examine the underlying structures and social values of the industrial society and to pose the question of their validity and appropriateness to present circumstances. This question is posed to the business community who, through their own experience, will realise more than most that periods of transition and change offer both constraints and opportunities.

Deep-seated and accelerated change is now taking place in the world economy, requiring nations and industries alike to adopt a global approach to their economic policies and business strategies in order to remain competitive. The continuing evolution of multinational and multi-domesticated corporations has transformed the competitive base of national economies and international trade. This has led to a dramatic change in the volume of world traded goods and services, information flows and the growing mobility of capital and technology. The interplay of global forces means that seemingly invulnerable competitive positions of major domestic firms, and even national economies, will change. At the same time, manufacturing firms are coping with the challenge of integrating competitively into a changing world economy while also facing serious and unabated pressure to improve their environmental performance, both from consumers and legislation. The pressure to improve environmental performance should not be taken in isolation: it should be seen as part of the changing landscape. It may be useful for the purposes of our discussion to consider some of the forces of change that are combining to forge the current and future commercial reality.

■ *Population*

In 1850, the world population was 1.5 billion. In 1995, the world population is 5.4 billion, and is projected to grow to 8.5 billion within 35 years. The process of fulfilling the needs and desires of this population to a level commensurate with the values of Western industrial societies is stripping the earth of its biotic capacity to produce and reproduce life. As one species among millions on this planet, humans currently consume 40% of the calculated primary production (i.e. that which directly involves solar energy and services as a base of life) on an annual basis. North America alone, representing a mere 5% of the world population, consumes 25% of the world's calculated energy and produces 25% of the world's CO_2 emissions. Compounding the pressure of population growth is the uneven pattern of the consumption of resources. In the North, human longevity—and therefore the capacity of each individual to consume during the course of a lifetime—far outweighs the phenomenon of falling birth rates in that part of the world. In the South, the pursuit of Western development models have put enormous pressure on environmental resources and have thereby posed major questions for the medium- and longer-term economic sustainability of some countries.

■ *The Role of Science*

Historically, the purpose of science was to impose order on nature and to manage risk by imposing predictability, but now science itself is perceived as a source of risk and threat in its own right. Science and the scientific community no longer enjoy the esteem they had ten or twenty years ago. While the pre-industrial era was characterised by hazards and risks that originated outside the economic and scientific communities, modern societies face hazards and risks no longer external to, but rather generated by, society and the scientific community itself.[1] Examples of these risks would include:

- The threat of nuclear contamination
- The risks of global warming
- The loss of lakes, rivers, land and seas to pollution
- The destruction of the ozone layer
- The compromising of the human immune system
- Biodiversity depletion

Far from contributing to the control of human anxiety, the scientific communities are increasingly perceived to be the generators of anxiety, and, as such, the pressure is on to look for a more benign use of science. The litany of scientific stigmatisms, resulting in faulty claims, environmental damage and human health disorders, has removed science from its pedestal of credibility. The idea of assuaging human anxiety, by purporting scientific truth or scientifically-proven safe limits of exposure, becomes less credible to an increasingly sceptical public.

■ *Insurance, Finance and Capital*

Risk in the industrial society was dealt with through compensation, precautionary after-care, limitation and accountability. The term 'litigious society' is a description of the current climate of spiralling litigation demands by the citizen.

Lloyd's underwriters in London have suffered severely in the last decade because of compensation payments resulting from environmental disasters. Ensuring against environmental mis-management or disaster is becoming untenable. Insurance companies are becoming more and more loath to provide insurance cover against environmental mis-management or disaster.

Lending institutions and banks in Europe are becoming conscious of the risk of lending money to companies that may not be capable of managing environmental risk. Land and property—the traditional security against bank loans—can now actually become liabilities, e.g. in the case of contaminated land.

■ *Infrastructure and Technological Developments*

One of the main benefits of the industrial society is that it has facilitated global developments in infrastructure and telecommunications. Development in the industrial economy was dependent on available energy resources, on-hand infrastructure, on-hand labour and other attributes. Large concentrated workplaces were the consequence of this requirement, e.g. the industrial heartland of the North of England.

However, the 'buzzwords' of the nineties are 'downsizing' and 'right-sizing'—the company shedding everything but its core activity. While, on the face of it, this may seem an unmitigated disaster, it is more a symptom of the development of a decentralised production chain. Various economic and production functions can be decentralised and subcontracted to smaller dedicated production units, now facilitated by global transportation and communication systems. The growth of intra-industry trade at the global level is perhaps the most significant development in the world economy during this decade. This is a phenomenon that industry and policy-makers alike must take into account. However, entry into intra-industry trade demands that equal standards of production are maintained at every point of the chain, whether in London, the Philippines or Tasmania. In this context, international technical standards and management system standards are becoming market-entry criteria.

■ Ecological Enlightenment

The ecological movement really began in Western society in the 1970s, specifically in Germany. This fringe political movement has, in a very short period of time, generated a momentum which has led to a position where much of its doctrine and philosophy are now integrated in the established political parties all over the world.

The ecological enlightenment is a shorthand for the growing consensus that economic and production systems that have perceived the environment to be outside the economic equation are simply not sustainable. In this respect, it challenges classic economic theory on which industrial society is based, and, in doing so, also challenges the institutional framework that Western societies have put in place to facilitate economic development. This institutional framework has focused on protection of the environment solely through command and control. In recent years, there has been a rapid growth in the nature and number of environmental regulations and standards with which companies must comply. This ever-stringent legislative framework is politically legitimised by the growing awareness and concern of the consumer and citizen. Apart from impacting on the legal domain, ecological enlightenment equally impacts on the market, where market choice has been governed more and more by the informed and concerned consumer.

■ Individualisation

Individualisation is a term that has been coined to describe the transition of Western society from one based on family and community units to one that is becoming more and more a community of individuals. With the transition of social institutions, such as family, community and extended family, individuals are becoming dependent on the market as the ultimate institution to provide not only the physical needs of sustenance, but also the culture and lifestyle values formerly provided by the family and the broader community. The implications of individualisation are that individuals become, ironically, not more independent but more dependent on the market to which they look to relieve anxiety and fears and provide comfort and security. In terms of market choice, price remains a key determinant, but is now only one of an increasing array of criteria that come into play.[2] Choice is balanced by a complex web of assurances that are demanded by this increasingly isolated individual. The pursuit of security has resulted in an ever-increasing array of demands from ethical values to health, welfare, environmental issues, pollution, animal welfare, etc.

■ *Bi-Polar to Multi-Polar Global Balance*

The political framework that supported the Western economy throughout much of the 20th century—particularly since 1945—collapsed dramatically with the fall of the Berlin Wall. I believe it is a mistake to perceive this collapse to be the same as the collapse of the Soviet Union. Immediately after World War II, the United States had a very strong relative position within the international system. The dominant power centre within the West during that period moved from the UK to the US. The hegemonic or near-hegemonic position of the US after the war was based on the relative dominance of its economy.

The general trend in the international position of the US in the decades after the war was a relative decline in its hegemonic powers. Economically, the US share of the world GNP in 1950 was estimated at 34%; its share of global industrial production was about 60%; its share in total foreign currency reserves at about 50%. In 1980, the corresponding figures were:

- Share of world GNP: 23%
- Share of global industrial production: 30%
- Share of total foreign currency reserve: 6%

The US's share of the global GNP by 2000 is forecast to diminish further to 17% or 18%. The reason for the relative decline is not the collapse of the US economy, but rather the relative economic growth of Western Europe, Japan, South-East Asia and the newly industrialised countries.

■ *A Redefinition of the Wealth of Nations*

Industrial society was based on the relative value of primary resources. Indeed, the European Union was founded upon its resource-base of steel and coal. In the society that is now emerging, the creation, organisation and use of knowledge is becoming the primary activity of the economy.[3] We are moving from a society based on nations and states organised around industry and production to a global society organised around knowledge, learning and communication. The dangerous implication of this new society is that a two-tier order will emerge, one level of which possessing the foresight and skills necessary to capitalise on the new market, and one level possessing inadequate skills and not having the capacity to change. Put simply, the wealth of nations can no longer be determined in terms of natural resources and military infrastructure, but rather on the creative capacity of its population and careful utilisation of resources. This redefinition raises more questions than it answers, particularly with reference to global trade, manufacture, and the control of resources on which society depends.[4]

■ The Implications of Change

The result of this changing context is that the parameters of competition and corresponding market access are changing fundamentally. At one level, we can see them as constraints, but on the other as opportunities.

We can look at these changes with respect to five parameters, for illustrative purposes:

- Production
- Regulation
- Organisation and human resources
- Management
- Market

□ Production

- With an increased awareness of society's relationship to a closed ecological system, there is a realisation that production practices must change in favour of resource and production efficiency.
- The future of production systems will be characterised by a precautionary or preventative approach.
- Against the emerging reality, science and technology are under scrutiny and have come under pressure to be used in a benign and assuring way, working with ecology rather than dominating or controlling it.
- The views that there are scientifically-demonstrable safe limits for anything is increasingly losing credibility.
- With the increasing freedom of environmental information in many parts of the world, advanced societies' production systems must be accountable and open to consumer enquiries.
- In order to retain its relevance in the market, production must be based on principles of continuous improvement. It must, therefore, go through a cycle of continual renewal. Production efficiencies are achieved by the combination of procedural and manufacturing technologies, including clean technologies.

☐ *Regulation*

- There is a growing realisation that regulation alone will not solve environmental issues. With the change in consumer demands, as discussed above, the market can be conceived as the ultimate regulator.
- Industrial activity is being asked to take ownership of its environmental effects and to internalise them as part of its production system.
- The regulatory domain now includes the interests of the stakeholders, i.e. all of those interests that will either facilitate or constrain the corporation.

☐ *Organisation and Human Resources*

- Within the new emerging socio-economic reality, the measurement of nations with reference to natural resources and military infrastructure is becoming less valid. In its capacity to create and innovate, the human resource is potentially the measurement of the primary resource of the new order. While this is not an established position, it is nevertheless reflected in the policies of the European Union.[5]
- At an organisational level, there is a movement from hierarchical structures towards flat organisations where peer relationships, participation and teamwork are the keys.

☐ *Management*

- Within the emerging socio-economic order, management is moving from a directional approach to one focused on systems management that does not rely on the firefighting of individuals. It is characterised by continual improvement and the continual reassessment of the situation through the setting of new objectives and targets.
- Management is ideally market-driven and clearly focused on the changing needs of the market it is serving.
- In these circumstances, management will be acutely aware of the requirements to provide an increasing array of assurances in the market that it is serving, including quality, environment, health and safety, hygiene, etc. Food safety and the beef crisis in Europe has been a spectacular example of the collapse of consumer confidence in the assurances provided.

☐ *Market*

- The market, in an advanced society, is becoming the ultimate institution, correspondingly demanding higher and higher standards from a society that is becoming more and more aware of its interdependence with ecological systems.
- The market for many activities will not necessarily be the final consumer. In many cases, it will be the intermediary in a production or service chain.
- The entry ticket into these markets is the ability to provide assurances to the consumer, whether that be the final consumer or within the production chain.
- In a developed welfare society, such as the OECD countries, price is no longer the primary determinant of choice. The market is going beyond price to seek assurances in other areas, such as quality, environmental protection, animal welfare, hygiene, ethics, values, etc.

■ *ISO 14000: A Coping Strategy?*

In the context that we have defined, the ISO 14000 series on Environmental Management Systems provides a framework that can be used to facilitate change. As we have discussed, use of ISO 14001 as a facilitator of change will require a fundamental revision of the notion of 'environment'. For this we refer back to the equation presented at the beginning of this paper. By using ISO 14001 to manage the newly-defined parameters of cost, market and legislation, we can go some way to ensuring a viable business. By adopting this approach, we would also have an outcome that contributes to sustainable ecological development. Figure 1 demonstrates the use of ISO 14001. To achieve efficiency and effectiveness, objectives must be set relating to the functions of:

- Production
- Legislative compliance
- Organisation and management
- Market access

This structural framework will allow for the incorporation of other systems, such as ISO 9000, Health and Safety, HACCAP, etc. The over-riding objective is to use the ISO 14001 framework against the background and in the context of changes we have described to drive down production costs, to

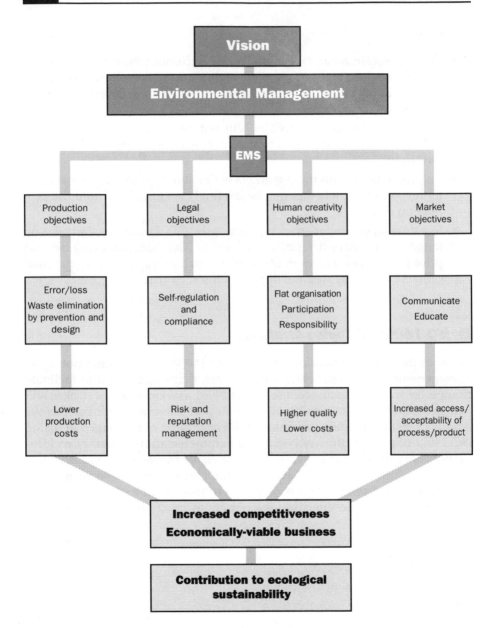

Figure 1: *Demonstrating the Use of ISO 14001 as a Framework*

eliminate error and resource loss, to comply with statutory laws and regulations, industry norms and market specifications, to use the human resource as the primary resource by tapping into creativity and productivity; and the ultimate objective is to increase the access to the markets and to increase the acceptability of product and process to a continually more discerning and more demanding marketplace.

■ Global Change, Competitiveness and Trade Patterns

The relationship between environmental goals and industrial competitiveness has normally been thought of as involving a trade-off between social benefits and private costs. The issue was seen to balance society's desire for environmental protection with the economic burden on industry. The notion of an inevitable struggle between ecology and the economy grows out of a static view of environmental regulation, in which technology, product, processes and customer needs are all fixed. In this static world, where firms have already made their cost-minimising choices, environmental regulation invariably raises costs and will tend to reduce the market share of domestic companies on global markets.

However, in the last 20–30 years, the paradigm defining competitiveness has been shifting away from this static model. The new paradigm of international competitiveness is a dynamic one based on innovation. Competitiveness at the industry level arises from superior productivity, either in terms of costs that are lower than those of rivals or the ability to offer products of superior value that justify a premium price. Detailed case studies of hundreds of industries reveal that internationally-competitive companies are not those with the cheapest inputs on a larger scale, but those with the capacity to improve and innovate continually. (Here we use the term 'innovation' broadly to include a product or services design, the segment it serves, how it is produced, how it is marketed and how it is supported.) Competitive advantage, then, rests not on static efficiency, nor on optimising within fixed constraints, but on the capacity for innovation and improvement that shifts the constraints. Figure 1, which incorporates the ISO 14001 framework, addresses this issue. It suggests that not only can this approach lower the net cost of meeting environmental regulations, but it can even lead to absolute advantages over firms in other countries not subject to similar regulations. Innovation offsets will be common, because reducing pollution is often coincident with improving the productivity with which resources are used; in short, firms can actually benefit from properly-crafted environmental regulations that are more stringent than those faced by their competitors in other countries. By stimulating innovation, strict environmental regulations can actually enhance competitiveness. There is growing support and evidence for this proposition (see Fig. 2).[6]

■ The Effects of Trade Liberalisation

It is indisputable that outward-looking trade policies have had significant environmental effects. Trade expansions have led to rapid growth and export-oriented industry. The composition of exports has varied across countries, depending on the resource endowment and the stage of industrialisation. At the early stages of export expansion, internationally-competitive industries have been mostly labour-intensive, processing and assembly operations or downstream processing of local raw materials. Extractive and processing industries generate large quantities of waste. At later stages of industrialisation, exports have included a larger proportion of machinery, industrial materials and products with higher technology content.[7]

In Thailand, for example, rapid industrial growth has raised hazardous waste generation to 1.9 million tonnes per year in 1990, and industry share has doubled to 58% in a decade. A fourfold increase in the volume of hazardous

---

- An Irish ceramic manufacturing company is in the process of implementing environmental management systems. As part of its approach, it specifically tackled energy. It has focused on its production systems and energy consumption and has reduced its energy bill by 18%. The company was faced with expanding the size of the factory to facilitate the scale of ceramic curing required. By capturing and recycling heat, it managed to increase the temperature and reduce the curing time, thereby delaying the necessity for capital investment.

- In its approach to environmental management, the Fruit of the Loom Company, Inc., in the north-west of Ireland, is researching the feasibility of recycling bleach and dye liqueurs. Research was carried out in late 1994 with the installation of the pilot project costing £50,000. The objectives were to minimise waste and to reduce water consumption. Results from the pilot facility exceeded the company's expectations and yielded a more technically- and economically-viable process as a result of reduced energy cost, reduced emissions and processed waste. The figures show that, if the process was extrapolated into a full-scale operation, the system would yield a yearly reduction of energy of 1.773×10^{11} kjs, electricity of 2.239×10^6 kW and a corresponding reduction in solid liqueur and gaseous discharges. This basic innovation is now incorporated into the environmental management system at the plant.

- A major Irish milk producer, faced with demonstrating the integrity of the product to a major cream liqueur company (which is its major customer), went down the road of environmental management systems to provide demonstrable customer assurances. While customer assurance was the primary objective, the company noted significant savings in its production cost. The company has achieved market security in the process.

Figure 2: *Examples of Environmental Management System Usage in Ireland*

waste is expected by 2001. Conventional, biodegradable industrial wastes are also accumulating rapidly, severely polluting rivers and estuaries. Until recently, the Government of Thailand did not insist that new investments include adequate emission controls. Energy consumption in Thailand is growing at 8% per year faster than GDP, and the country is shifting towards domestic lignite and very dirty fuel for electricity generation, with unfortunate implications for air quality and human health.

Rapid industrialisation in China, much of it associated with increased openness to international trade, has generated similar problems. Industrial waste water discharges have more than doubled in the later half of the 1980s, far outstripping treatment capacities and heavily polluting surface and ground water. Consequently, most of the urban Chinese population depends on unsafe drinking water, with severe health implications. Rapidly-increasing energy generation from coal, three-quarters of which is for industrial or electric power use in China, has led to some of the world's highest concentrations of fine particulates and sulphur oxides, some of the most acidic rainfall in the world, and chronic obstructive pulmonary disease five times more prevalent in urban populations in China than in the United States.

These growing environmental problems in newly-industrialised countries by no means imply that trade liberalisation and its associated outward-looking development strategy have been a mistake or are inconsistent with sustainable development. Outward-looking strategies, especially in the Asian region, have dramatically reduced poverty and raised living standards for a large fraction of the world's population. However, the challenge is to ensure that newly-created resources and capabilities do not result in increased environmental pressure.

There is a growing realisation that economic efficiency and environmental performance go hand in hand and that, far from impairing competitiveness, environmental controls can enhance it. This is recognised by many of the multinationals investing in these regions who are insisting on the highest standards of environmental performance for their branch plants. Similarly, there is increasing pressure on Western industries to be more concerned about their procurement policies and to ensure that their environmental credentials are not impaired by environmental inconsistency in their supply chain.

■ *The Effects of Trade Restrictions*

Continued inward-looking, trade-restricting development policies have produced equally serious environmental problems, along with significantly lower living standards.

Certainly, China, in the years prior to economic reform, experienced severe environmental degradation. Inefficient state-owned heavy industry has generated enormous pollution. Misguided centrally-planned management of agriculture, forest and other sectors has led to severe resource degradation. Similarly, India, which has only begun to dismantle its inward-looking development regime, has experienced low growth, low income and substantial environmental degradation. In the industrial sphere, obsolete technologies, over-emphasis on highly-polluting heavy industries, financial constraints and lack-of-perspective environmental controls have combined to produce pollution problems.

Trade restrictions in the OECD countries also have adverse environmental and economic consequences for their own society, as well as for their third-world trading partners. Tariff escalation by the processing stage inhibits the development of finishing industries that add value to raw materials produced in the South. The 'Multi-Fiber Agreement' and other trade barriers impose serious quantitative restrictions on exports of labour-intensive manufacturers from developing countries. Such barriers effect not only textiles and clothing, but also footwear and other relatively labour-intensive products.

These restrictions substantially lower incomes in developing countries and raise consumer prices in industrial countries. For example, in the 1980s, US consumers paid about $16 billion per year in excess costs just for clothing and textiles. At the same time, trade barriers exacerbate environmental pressures in developing countries by forcing them to intensify exports of natural resource-based commodities. Most newly-industrialised countries have a comparative advantage in the production and export of labour-intensive or resource-intensive commodities, but cannot compete in high-technology or capital-intensive industry. In the late 1980s, about half of all developing country exports still comprised fuels, minerals and other primary commodities. By impeding exports of labour-intensive and downstream processing industries, these trade barriers virtually force developing countries to raise exports of natural-resource-based commodities, especially when pressures on developing countries to meet high-debt servicing requirements are intense. Eliminating these trade barriers would have significant economic and environmental benefits.[8]

■ Trade Standards:
The 'Polluter Pays Principle' and the Terms of Trade

Firms in OECD countries favour competitors in developing transitional economies where environmental standards are less stringent or less strictly enforced and derive an advantage in the marketplace from lower compliance costs. However, it is safe to predict that international differences in process standards will have little competitive impact in world trade, because, even in the US, where regulatory standards are strict but not particularly cost-effective, pollution control costs average only about 1.5% of the value of the total sales of manufacturing industries. Only in a very few sub-sectors do they rise to about 3% of the value of the sales. Even if environmental controls brought no benefit whatever to the firm itself, the resulting cost of disadvantage to US firms would be less than 2% of sales prices for the large majority of industry. Compared to other competitive factors in international trade, such as differences in labour, transportation or material costs, or productivity or product quality, differential environmental control costs stemming from varying environmental process standards are unlikely to be noticed—let alone be decisive.

Governments of OECD member countries agreed to the 'polluter pays principle' twenty years ago. This was to avoid trade displacements and distortions that might result if some governments subsidised industries towards costs of compliance with environmental standards while others made the polluters pay. This principle has been useful, even though applied inconsistently within the OECD. Non-OECD countries have not universally adopted the principle, let alone the practice. However, non-OECD countries should look seriously at this. Governments of developing countries do not have the fiscal capabilities to subsidise pollution control expenditures to any great extent, and there are far more worthy potential beneficiaries for limited governmental funds. The polluter pays principle will complement market liberalisation programmes under way in many developing countries by ensuring that prices include the full incremental costs of production, including environmental costs. If developing countries collectively adopt reasonable environmental process standards in commodity-producing industries, and adopt the polluter pays principle, the damage to their own natural resources would be curtailed and the cost of environmental compliance would be internalised in the prices of their exports. Consequently, their terms of trade would improve because northern consumers, whose demand is relatively insensitive to price, would be paying a larger share of the environmental costs associated with their consumption pattern.

To illustrate the point: if environmental control costs averaged roughly 1.5% of production costs, as they do in the US, the $500 billion in annual exports from developing countries would include payments of up to $7.5 billion by importers, mostly in the North to help defray the costs of environmental control. This sum is far greater than the annual flows of development assistance to the South for an environmental programme. One would think that it should be a high priority to promote agreements among third-world commodity producers that they adopt environmentally-sound and sustainable production standards and apply the polluter pays principle. The ISO 14001 framework would facilitate this approach.

■ Conclusion

☐ Competitiveness Revisited

Against the background of change that we have discussed, the whole concept of competitiveness has been transformed. The key issue is the positive relationship between economic efficiency, market acceptability and environmental performance. Indeed, there is growing acceptability of the notion that environmental regulations (including self-regulation by industry) can dramatically increase competitiveness by driving down production costs and increasing the market acceptability of both the product and its production process.

☐ Trade Revisited

We have argued that both trade liberalisation and trade restriction can have equally damaging environmental impacts. We have also argued that developing countries should look to environmental regulation (whether legislative or self-regulated) as a means of increasing their competitiveness and as a basis for increasing their terms of trade with the North.

☐ Green Protectionism

At the conclusion of the Uruguay round of GATT, the Director of GATT suggested that the environment would be the trade issue of the next century. So-called 'green protectionism' is undoubtedly emerging under all kinds of guises, most of which will not remotely include the word 'environment'. There are, however, broad and powerful economic forces at work to discourage the manipulation of product standards for protectionism purposes. They are summed up in the phrase 'globalisation of the world economy'. A remarkably large and growing fraction of world trade consists of shipments between one branch of a company and another or between the company and its foreign affiliate. The Ford motor company, for example, has no incentive to keep the components made in its Mexican plant out of the United States since it built the Mexican facility precisely to supply those components to its factories in the US and probably in other parts of the world as well. As long ago as the mid-1980s, 52% of US imports and 57% of Japan's were intra-company transactions of this kind. Intra-company trade is buffered against the protectionist manipulation of product standards. The option to use green protectionism in such circumstances is severely diminished.

☐ *Think Globally—Act Locally—Think ISO 14001*

In summing up, we can look at the ISO 14001 framework as a structure to address, manage and control the key factors that affect our competitiveness. The design of that framework and that management system must take account of the global situation if actions at local company level are to be efficient, effective and appropriate.

■ *Notes*

1. See Ulrick Beck, *The Risk Society* (London, UK: Sage Publications, 1992).
2. See Ulrick Beck, *Ecological Enlightenment* (Atlantic Highlands, NJ: Humanities Press, 1994).
3. See P. Drucker, *The New Realities* (Oxford, UK: Butterworth-Heinemann, 1994).
4. See Commission of the European Union, *White Paper on Growth, Competitiveness and Employment* (Brussels, Belgium: Commission of the European Union, 1993).
5. See Commission of the European Union, *op. cit.*
6. See Michael Porter and Claas van der Linde, 'Toward a New Conception of the Environment–Competitiveness Relationship', in *Journal of Economic Perspectives*, Vol. 9 No. 4 (1995).
7. See Robert Repetto, *Trade and Sustainable Development* (New York, NY: UNEP/World Resources Institute, 1995).
8. See John Stonehouse and John Mumford, *Science, Risk Analysis and Environmental Policy Decisions* (New York, NY: UNEP, 1994).

Environmental Management Standards and Certification

Do They Add Value?

Tim J. Sunderland

NVIRONMENTAL management standards are encouraging companies to adopt effective systems to identify and control environmental
impacts that they may be creating through their activities, products and
services. Certification systems have been established to allow companies to
seek third-party certification of their systems, but how have companies
responded to these standards and certification? Some have welcomed them
and sought certification; others are keeping a close eye on developments.
However, industry is asking questions about the value of these standards
and of certification. Companies will need to think carefully before deciding
whether these standards and certification meet their needs.

Traditionally, environmental management has focused on achieving compliance with applicable environmental regulations. These regulations are
required when industry is unwilling to take voluntary action to protect the
environment. However, responsible and progressive companies have seen
an opportunity to break free of the burden of prescriptive regulations by
demonstrating to regulators and other stakeholders that continuous improvement can be achieved more effectively through voluntary programmes. In
responding to industry's desire for self-regulation, environmental management systems (EMS) standards provide a voluntary mechanism through
which companies can demonstrate their commitment to environmental protection without the need for increasingly stringent regulations.

Standards for EMSs have proliferated over the last three years. National
environmental management standards now exist in many countries including the UK, Spain, France, Ireland, Israel and Australia. The first standard of
this kind, British Standard BS 7750, has achieved far-reaching recognition

with companies seeking certification worldwide. The Eco-Management and Audit Scheme (EMAS), although a European Regulation as opposed to a standard, is gaining particular popularity in Germany, where the number of registrations is increasing rapidly.

In view of the number of these standards and the level of interest shown, the International Organisation for Standardisation has issued ISO 14001 in an attempt to achieve harmonisation, so as to avoid potential conflicts and trade barriers. While still in draft status, a number of companies certified to the draft ISO 14001 standard, including Philips Components in Austria, SGS-Thomson Microelectronics in California, Emballages Laurent, SA Canon Bretagne and Lexmark International SNC in France.

Although these standards are relatively new, some companies have been operating effective environmental management systems for many years—Arthur D. Little have been helping national and multinational companies develop and implement such systems since the late 1970s. So why is there so much interest in these new standards? Does ISO 14001, EMAS or any of the other national environmental standards really add value to business? In 1996, Arthur D. Little conducted a survey of industry views on the EMS standards in the US and UK, in which 115 blue-chip companies in the US and 84 in the UK, from a wide range of industry sectors, took part. Highlights from this survey are outlined below.

This chapter provides a personal view of the values that companies are placing on these standards and certification. These values differ between companies, and it is clear that EMS standards and third-party certification may not meet the needs of all businesses and their stakeholders.

■ The Case for EMS Standards

☐ *The EMS standards provide a useful environmental benchmark for leading companies.*

Companies that have been operating environmental management systems for many years, and which regard themselves as leaders, are finding that these standards provide a useful benchmark. By comparing their existing systems with the requirements of these standards, companies are able to identify potential weaknesses in their own procedures.

☐ *For companies that are not 'leading edge', the EMS standards provide a useful management model.*

Adopting effective management systems will help a company minimise its environmental liabilities. The management model provided by the EMS standards requires the development of systematic processes for the identification, evaluation and control of environmental impacts and risks relevant to a company's activities, products and services. Where the risk is high, the response is to adopt rigorous systems of control; where the risk is more acceptable, there is less need for formal control systems. Consequently, the standards provide a useful model to help companies without their own formal EMSs develop systems commensurate with the risks that they are facing.

☐ *The EMS standards have focused industry attention on environment as a business issue.*

Companies are realising that industrial pollution of the environment is an inefficient use of resources as well as causing environmental damage and economic losses. Despite this, companies are still finding it difficult to integrate environment into their business processes. Consequently, these companies are failing to see how environment can be used to improve their competitive position and contribute to growth and profitability. EMS standards are helping to reinforce the message that environment is a business issue and needs to be managed effectively. While the systems required by these standards may not necessarily tell a company anything new, they should ensure that environmental issues reach a sufficiently important agenda to receive the level of attention they merit.

☐ *Companies are reporting financial benefits resulting from the adoption of formal environmental management systems.*

Companies are beginning to see financial returns resulting from the adoption of EMSs in the form of cost savings, risk aversion or reduced business interruption. With respect to cost savings, Akzo Nobel Chemicals in Gillingham, UK, reported an 18% annual reduction in energy consumption, and National Power in the UK claims to have turned its £200,000 annual bill for general waste disposal into a £26,000 profit.[1] Although many of the benefits may be difficult to measure in financial terms, leading companies are recognising that effective environmental management can impart strategic business advantage. In addition, integrating environmental considerations into product development through 'design-for-the-environment' tools has helped companies reduce redesign and rework, shorten time to the market and lessen the chance of product recall.

■ The Case against EMS Standards

□ *Meeting the requirements of EMS standards is not essential to having a strong environmental management system.*

Environmental management systems are not new, and a number of major companies had already built strong, effective systems for managing their environmental activities prior to the introduction of EMS standards. Some of these companies have shown outstanding environmental results, and their individual management approaches have gained a wider industry acceptance. Consequently, EMS standards are of less value to 'leading-edge' companies who consider the cost associated with the level of formality, and the amount of documentation and bureaucracy required by the standards, unjustifiable. Many of these companies consider that they are managing their environmental impacts effectively without meeting all of the requirements set out by these standards.

□ *Some standards may not guarantee continual improvement in environmental performance.*

Continual improvement is a requirement of BS 7750, EMAS and ISO 14001 and is also a requirement of specific stakeholder groups. However, there seems to be some difference in interpretation. Whereas BS 7750 requires continual improvement in performance, the interpretation within ISO 14001 focuses on 'the process of enhancing the EMS to achieve improvements in overall environmental performance'. In this instance, the question arises as to whether a company actually has to achieve improvements in performance to maintain certification under ISO 14001, and what level of improvement the certifiers would expect. Standards that are not performance-based are likely to receive less support from key stakeholder groups (particularly regulators), as they provide no assurance of improved performance. The concern is that EMS standards may not be sufficiently demanding to reassure regulators that continual improvements will be achieved more effectively than through command-and-control regulations.

□ *Some EMS standards (including the EMAS Regulation) vary in their requirements.*

There are notable differences between BS 7750, EMAS and ISO 14001. Although ISO 14001 is an attempt to harmonise existing standards, it falls short of the requirements of EMAS. In Europe, ISO 14001 is seen as being less stringent in that there is no reference in the specification to a technology requirement (such as economically-viable application of best available technology) and no requirement for the publication of an environmental statement. The proposed bridging document between EMAS and ISO 14001 highlights these differences and states clearly that self-certification (an option under ISO 14001) is not acceptable. The variation in these requirements produces an uneven playing field and only serves to confuse industry further as to the value of the different standards.

☐ *EMS standards may discourage the integration of environment into existing business processes.*

Although the EMS standards address important environmental management processes, they may discourage business process integration. Some companies find that the easiest route to meet the requirements of the standards is through developing stand-alone environmental management systems that are separate from existing business processes. Some certifiers have responded to this by offering integrated ongoing surveillance for companies with both ISO 9000 and ISO 14001 certificates, although the initial assessments would remain focused on either quality or environmental management system certification.

■ *The Case for Certification*

☐ *Certification provides a mechanism through which good corporate citizenship can be demonstrated.*

The certification process (or verification under EMAS) provides a mechanism through which companies can demonstrate their commitment to environmental protection. The actual need to demonstrate this commitment will vary from one company to another and will depend upon their position in the supply chain, stakeholder requirements and the vulnerability of the company to environmental exposure.

☐ *National governments may require certification.*

Governments are examining the need to require certification as a condition of contract and may wish to set an example to industry. In the US, both the Department of Energy and Department of Defense may extend preferential buying to qualified suppliers who are certified to ISO 14001.[2] A number of governments in Latin America and Asia Pacific are considering making certification to ISO 14001 a requirement for foreign oil exploration and production operators. The Korean Government requires leading Korean companies to achieve ISO 14001 certification by the end of 1997. Early this year, the Provincial Court of Alberta fined Prospec Cdn for exceeding sulphur emissions and also ordered the company to become certified to the draft ISO 14001 standard.[3] The certification approach seems to offer governments a low-cost alternative to building expensive command-and-control structures for environmental regulation.

☐ *Certification could improve relations between industry and the regulators.*

EMS standards may be encouraging a new regulatory regime between industry and government based upon trust and self-regulation. A reduction in prescriptive regulations and government interference is an attractive proposition to industry. In the US, there is a view that the Environmental Protection Agency and Department of Justice could be lenient to a company with an environmental management system in place, should an environmental incident occur. As a result, regulators could play a crucial role in determining the value of certification to business; should regulators view certification as a credible mechanism for ensuring due-diligence and regulatory compliance, certification could become regarded as a regulation by default.

☐ *The insurance industry is beginning to recognise the value of environmental management systems in controlling liability.*

Commitments to continuous improvement, regulatory compliance and pollution prevention may help to lower insurance premiums, but it is likely that insurers will need to see real evidence that these commitments are being achieved. Certification to an EMS standard may help in demonstrating this, as long as the insurers are confident in the credibility of the certification process.

☐ *Companies with major export markets may benefit from certification.*

Certification may prove to be important in meeting environmental requirements introduced by customers into the supply chain. This concern is motivating industry in Latin America and Asia, who fear that international markets may begin to differentiate on the basis of EMS certification.

☐ *Certification can provide a valuable incentive for operational staff.*

Companies struggling with the problem of gaining environmental commitment and support from line management and operational staff can use certification as an incentive to help encourage the successful implementation of the EMS throughout the organisation.

■ *The Case against Certification*

☐ *Differences in accreditation criteria and enforcement will not ensure a level playing field.*

Individual accreditation bodies have established or are currently establishing their own criteria for EMS certification, in particular for ISO 14001. There is a view within industry that central co-ordination of these criteria is essential to provide a level playing field for certification. For example, with EMAS, the criteria and approach for accrediting verifiers differs between Germany (where individuals are accredited in accordance with Annex III of EMAS) and the UK (where organisations are accredited with more emphasis placed on meeting the requirements of EN 45012). In contrast, France accredits both individuals and organisations. Without mutual recognition of accreditation criteria, multinationals wishing to use the same certifier in different countries will face additional bureaucracy and costs in ensuring that the certifier meets the different accreditation criteria.

The International Organisation for Standardisation's Committee on Conformity Assessment (CASCO) intends to provide guidelines for accreditation bodies to encourage consistency in the criteria. Further efforts towards harmonisation are being attempted by the European Accreditation of Certification and the International Accreditation Forum.

☐ *The value of certification depends upon the credibility of the certifier and the certification process.*

If stakeholders—in particular, shareholders, insurers, lenders and customers—are to see the real value of certification, and if the regulators are to view certified companies with more leniency, then the certification process needs to be credible. Although the accreditation criteria lay down the rules for the certifiers, these are open to abuse. Key areas of concern include:

- The certifiers' interpretation of the requirements of the EMS standards, particularly where the accreditation requirements provide no guidance (for example, the Drax Power Station in the UK was awarded EMAS despite failing to meet all of the requirements of BS 7750)
- The competency of the certifiers, especially their understanding of environmental impacts and regulatory requirements particular to a specific industry or country
- The impartiality and integrity of the certifier, particularly where the same organisation provides both consultancy and certification

It is evident that, without effective control by the accreditation bodies, certification could be easier to achieve by choosing one certifier over another. This poses a dilemma for companies: should they seek the easy route for certification or use a certifier that claims to have the most rigorous assessment approach? Although the latter will ultimately add more value, companies will need to select their certifier carefully.

☐ Certification does not mean full compliance with environmental regulations.

EMAS requires a 'provision for full compliance' with regulatory requirements, while ISO 14001 requires a 'commitment to comply with relevant environmental regulations'. The question arises whether a company that is not in compliance, and has only the intention to comply, should be certified. Views on this issue differ and some companies such as Ciba Clayton have been awarded EMAS despite reporting continued regulatory 'excursions' in their environmental statement.[4] The draft US Registrar Accreditation Board criteria state clearly that 'the EMS registration process does not evaluate an organisation's degree of compliance with legislation and regulations'.[5] However, under EMAS, a company's registration can be suspended should a breach of the environmental regulations occur (Article 8), although it is unclear as to what scale of breach will be considered sufficient. Certification bodies can ultimately withdraw certification under an EMS standard should a breach of regulations become apparent, although the application of such a policy may vary between certifiers. Unless certifiers (and Competent Bodies under EMAS) operate a clear and rigorous policy in this area, regulators may find little to reassure themselves that a certified company is more likely to be in compliance that an uncertified company.

☐ *Certification will not differentiate companies in terms of their environmental performance.*

As ISO 14001 is not viewed as a performance standard, companies with very different performance objectives and levels of performance can be certified. Consequently, certification may not be a reliable mechanism through which companies can demonstrate performance improvements to their stakeholders or differentiate themselves from their competitors. Additional mechanisms to demonstrate performance improvements, such as environmental reporting, will be necessary.

☐ *Certification will add additional operating costs.*

Some companies view the requirements of the standards to be bureaucratic, with a focus on paperwork rather than management effectiveness. To achieve certification, companies may be exposed to additional administrative and financial burdens through adopting formal systems of management, without necessarily contributing directly to real environmental benefits; they may find it difficult to justify the extra expense.

Initial certification assessments and ongoing surveillance activities will pose additional financial costs, with assessors charging between US$600 and $1,200 per day, depending on their level of expertise (verifiers for EMAS generally need to have greater technical expertise than for the EMS standards). In addition, certification assessments can take between six and twenty person-days depending on the size and complexity of the site and the experience and competency of the certifiers. Companies with ISO 9000 certification will find themselves paying twice for surveillance visits unless their certifiers are encouraged to provide integrated quality and environmental surveillance assessments. Small and medium-sized companies may find the costs of maintaining certification prohibitive. However, schemes such as the UK Department of the Environment SCEEMAS (Small Company Environmental and Energy Management Assistance Scheme) and the Industrial Research and Technology Unit Support Scheme in Northern Ireland have been established to provide financial assistance to smaller companies seeking to register for EMAS.

■ *EMS Survey Highlights*

Arthur D. Little's recent survey showed that companies were following closely the development of EMS standards: 86% of the respondents in the US and 85% of those in the UK thought that consistency between their own EMS and ISO 14001 was important; 90% of the respondents in the UK had conducted an assessment of their existing EMS against the requirements of the standards. Sixty-two per cent of respondents in the US and 60% in the UK saw certification as an important issue to their business. In the UK, however, 50% of the respondents said that they were not actually seeking formal certification at this time.

The main reason for companies seeking certification in both countries was to show due-diligence to their stakeholders. In neither the US nor the UK was there a strong view that certification would reduce costs of environmental management or improve environmental quality. The view that environmental issues could be integrated into existing management systems—in particular, existing quality, health and safety systems—was much stronger in the UK. This may be as a result of the longer history in the UK of companies operating formal quality management systems in accordance with ISO 9000. In addition, 64% of UK respondents indicated that they intend to integrate their health, safety and environmental management systems. The principal concerns relating to certification included increase in documentation and bureaucracy and additional costs.

■ Conclusions

It is becoming accepted that environmental issues are real business issues that no industry can afford to ignore. The need to adopt appropriate systems of management to ensure that environmental issues are under control is vital for business survival. EMS standards and certification offer an opportunity for businesses worldwide to demonstrate their environmental commitment. However, the value of these standards and of certification will be particular to each business and the challenges they face. Whether the EMS standards provide the most effective model or whether certification is justifiable will be a strategic decision that can only be made once the implications are understood fully at all levels. The standards may have little to offer to companies that have already achieved total integration of environmental management into their business thinking. For those companies beginning to manage their environmental responsibilities proactively, these EMS standards provide a valuable tool. The value added through certification, however, will depend on the perceived views of the credibility of the certification process.

The issues raised here may help companies evaluate the benefits of the standards and certification, although the associated costs need to be considered carefully. For any company to attempt a cost–benefit analysis of these standards and certification, understanding and balancing the requirements of their stakeholders will be key. New environmental issues are constantly emerging and, consequently, stakeholder expectations will be changing. Today, certification may not be considered important, but tomorrow it may be essential simply to do business. Today, ISO 14001 certification may offer business advantage, tomorrow it may not. Even if the standards or certification do not meet a company's current business needs, one thing is guaranteed: no one can afford to ignore the changes taking place.

■ Notes

1. See Environmental Data Services, *ENDS Report*, No. 247 (August 1995), p. 18.
2. See J.S. Willson and R. McLean, 'ISO 14001: Is It for You?', in *PRISM*, January–March 1996, p. 27.
3. See *Business and the Environment*, Vol. 7 No. 3 (March 1996), p. 2.
4. See *ENDS Report*, No. 247, p. 18.
5. See Registrar Accreditation Board, 'Criteria for Accreditation of Registrars for Environmental Management Systems' (Draft 1; 15 February 1996), published in *International Environmental Systems Update*, Vol. 3 No. 3 (March 1996; Supplement).

● This article represents the views of the author and not necessarily those of Arthur D. Little.

Part 2 Coping Strategies
Important Trends

7 Environmental Management Systems
Challenges for Russian Manufacturing Industry

Jim Hutchison, Anatoly Pichugin and Ann Smith

A PARTNERSHIP involving the University of Hertfordshire, the British Standards Institution and Norsk Hydro (UK) Ltd, and funded by the UK Environmental Know-How Fund, delivered the first environmental management training course for the Urals Economic Region of Russia during November 1995. The course, largely based on the British Standard BS 7750, was attended by senior and middle managers representing industry, local authority, regulators and academic organisations.

Using examples from the course, this chapter examines how industrial plants in the city of Kamensk-Uralsky, prompted by local community demands for a cleaner environment, are facing these new environmental challenges. Local authority targets for improved air and water quality and more acceptable solid waste disposal have become major environmental issues driving the introduction of measures to modernise existing industrial operations. Although the public health and environmental problems caused by pollution are well documented for the city, there was little evidence of an understanding of the benefits to be derived by industry from managing these issues in a proactive and systematic manner.

■ Environmental Context

Market forces and private ownership have brought sharp changes to the economic climate for Russian industry. Large industrial plants are no longer protected by low-cost energy and raw materials and, because of outdated technology, the plants cannot meet existing environmental standards, let alone the requirements of sustainable development. Although Russian environmental laws are comprehensive by Western standards, the associated environmental standards are largely unachievable and the penalties so lenient that payment of fines remains economically more favourable than compliance and the introduction of new cleaner technology.

Management in the Former Soviet Union (FSU) large manufacturing industry is fundamentally different to the Western European experience. Social, economic and environmental dimensions create obstacles to environmental management that are both complex and specific to particular situations. Assessing the benefits of environmental management along strictly economic lines, as is the practice in Russia, inevitably leads to inappropriate outcomes. The word 'management' in Russian is usually linked with technological innovation, and this 'technology' is associated with the introduction of new tools or machinery, with physical and social contexts not taken into consideration. There is a lack of understanding that the choice of new technology involves decisions based on more than just economic factors. Interest groups, such as shareholders, local planning and regulatory authorities, seek to use their power to influence the outcome of redevelopment and privatisation programmes for a variety of economic and social reasons.

The changes accompanying the introduction of new technologies are often poorly managed. There are many examples in the FSU of the failure of new technologies—despite market opportunities and despite their advantages over conventional technologies—due to lack of Government support, ignorance and a prejudice in favour of large-scale production.[1] Despite privatisation, most property is still controlled by a bureaucracy, many members of which are former Communist Party leaders. They control all transactions and keep huge inefficient plants operating as part of their own political power structure.

Environmental matters do not feature high on the agenda of reform programmes in Russia and this is a direct result of Government policy. The Russian strategy for natural resource management has a long history of exploitation in preference to sustainable use. Russia's response to The Rio Summit, The First National Plan Towards Sustainability, published in 1994, is enshrined by the key statement that there should be a reasonable compromise between environmental legislation and economic development.[2] By the end of 1995, the Russian cabinet of ministers viewed the requirements of sustainable development as a threat imposed on Russian economic reforms by the West. Despite favouring economic reform at the expense of sustainable development, there is still no real acceptance of the management practices required by the world economic market.

Figure 1: *Location of the Urals Economic Region*

■ Industrial Impacts in Kamensk-Uralsky

The Urals Economic Region (see Fig. 1) is probably the most important industrial region of the Russian Federation, dominated by metallurgical, chemical, heavy manufacturing and military industries. Many towns and cities were formerly restricted, even secret areas, because military and nuclear plants were located there. It is no surprise that the region is considered to be the most polluted in Russia, and environmental conditions in a number of cities are critical, with serious risks to public health due to deteriorating quality of air, drinking water and locally-produced food.

Kamensk-Uralsky, 100km south-east of Ekaterinburg, is the third largest city in Sverdlovsk Oblast, with a population of a quarter of a million. There are eight large industrial plants, mainly metal works and power plants, which were built before or during the Second World War and are typical of former Communist large-scale industrial projects. The plants use wasteful technologies based on old and obsolete equipment. Discharges to air and water are largely untreated, and solid waste is mainly disposed of by dumping in uncontained sites. Groundwaters and rivers are contaminated, air quality is poor and many public health disorders are above average for

Russia. Thirty-nine substances are found at concentrations greater than the maximum permissible concentration (mpc) for air quality; fruit and vegetables grown in private plots may have up to eight times the mpc for cadmium, lead, nickel and chromium; and over 43 million tonnes of solid waste have been dumped over an area of 1,200 hectares.

■ Sinarsky Pipe Plant

Large industrial plants built in the FSU during the 1940s and 1950s not only contributed to full employment for the region but also provided housing, schools, hospitals and infrastructures such as roads and water and heating supplies. Some plants also maintained farms and greenhouses which supplied food to factory canteens. Water and power was heavily subsidised, and most of the social benefits provided by the plant were unaccounted in economic terms. The Sinarsky Pipe Plant in Kamensk-Uralsky, reviewed during a recent environmental management training course, is typical in its management structure and in its environmental impacts.

The Sinarsky Plant began operating in 1934, producing iron and steel pipes mainly for the oil industry. It has seventeen workshops, each with its own canteen, and at the height of production employed 12,000 people. The plant still provides housing, kindergartens, schools, a hospital and a sanatorium as well as leisure facilities for its workforce. Maintenance of these facilities has become increasingly difficult in the current economic climate and the plant is seeking assistance from the local administration to keep the hospital running. For the first time, an estimate has been made for the energy used to heat the residential housing provided by the plant, and this probably amounts to 54% of the plant's energy consumption. A number of workshops were not operating due to lack of orders and workers were laid off on a rotational basis.

The plant was identified as one of the main contributors to pollution of the city, and air and water discharges generally exceed the environmental standards. Solid waste was dumped within the grounds of the plant at a dumping site used since the plant began operation—the local administration closed the dumping site on environmental grounds, but the plant pays a recurring fine to continue dumping there. Much of the residential housing falls within the so-called 'sanitary protection zone' adjacent to the plant.

A preparatory environmental review took place in what was considered to be one of the worst workshops in the Sinarsky Plant, which had been shut down for up to six months due to lack of orders. Despite the Deputy Chief Engineer's own perception that a very strict safety policy had been implemented, first impressions of the workshop identified health and safety as an important issue requiring urgent attention. The workshop was dark and dirty, with poor ventilation and lighting, and uneven floors littered with pieces of discarded machinery. Rubbish was burned in open fires on the workshop floor and there was no guard on the large open furnace. Much of the machinery was also unguarded, and welding repairs were taking place while the machinery under repair was operating. There was food and a bottle of vodka on a table close to the pipe-casting process which involved pouring molten pig-iron into casts. Some workers (especially the women) were without hard hats, and only one worker was seen to be wearing a protective mask. Many workers were operating machinery with one hand while smoking. There were no signs in the appropriate places to warn of dangerous activities, no labels on equipment and barrels, and there were no audible warning signals, not even for skips of scrap iron and molten pig-iron transported by overhead cable to and from the furnace. There had been one fatality in the workshop during 1995 and this worker was said to have failed to follow a workshop procedure.

Each workshop had a member of staff responsible for overseeing the process equipment and this person was answerable to the environmental protection department. The Sinarsky Plant's own environmental standard specified the procedures to enforce environmental protection measures and this was mandatory for all departments in the plant. The standard included a register of regulations, all relevant documents for the control procedures and verification, and the allocation of responsibilities and liabilities. Staff were given training to deal with hazards and emergency situations. There were comprehensive environmental monitoring programmes for air emissions and waste water discharges, and the plant had an 'environmental passport' according to statutory requirements. There was a system of penalties for non-compliance, and heads of workshops 'were punished materially and morally'. Although the environmental protection procedures were in place, it was acknowledged that implementation was poor and legislation poorly enforced—the solution to this was believed to be pressure exerted by the plant administration. The Sinarsky Plant had set its own targets for pollution reduction, although these were not quantified.

■ *Kamensk-Uralsky Environmental Strategy*

The City Administration, with the Institute of Industrial Ecology (Urals Division of the Russian Academy of Sciences), has surveyed the city and its immediate area and submitted the results to the Federal Government for designation as an environmental emergency zone. Kamensk-Uralsky is one of the first cities in Russia to prepare an environmental strategy that includes targets for improved environmental performance for all the industrial plants. The proposed programme is based on an integrated approach, including pollution reduction and remediation and public health improvements, with particular emphasis on improving the standard of living and quality of life for the local community.

Environmental improvement targets set by the city environmental strategy will require the Sinarsky Pipe Plant to construct new facilities for the treatment of gaseous emissions and acid waste water and for solid waste processing for recycling, as well as improvements to the hot water distribution system. Industry in Kamensk-Uralsky is expected to fund 45% of the cost of the environmental improvement programme. The programme does not include funds for education, public awareness or training.

■ The Environmental Management Training Course

Course delivery was via a combination of lectures, interactive task group sessions, group presentations and site visits. Delegates were divided into three groups, each comprising a mixture of civil servants, industrialists and academics and were required to plan and carry out a preparatory review of a workshop in an industrial plant to include identification of the personnel to be interviewed, questions to be asked and documentation to be examined. The site visits included aluminium production, pipe manufacturing and non-ferrous metals processing plants. Managers from the industrial plants co-operated in the exercise and delegates were given access to key personnel, relevant documentation and a tour of the selected workshop within the plant.

The three groups of delegates presented the results of the preparatory reviews in a plenary session and each group received individual feedback from the UK team on the conduct of the review and on the interpretation of their results. General comments arising from the three reviews were given during an overview of the course. The interactive sessions and the preparatory reviews were well received and delegates participated enthusiastically. The 29 delegates came from local government, regulatory bodies (health and nature protection), industry, consultancies and academic organisations. More than half of delegates had been trained as chemical engineers; other delegates had trained in physics, mechanical and construction engineering, environmental health (sanitary specialists), energy production, radioelectronics, forestry and economics. Most of the delegates held current positions as head or deputy head of a department and had environmental responsibilities within their organisation. Three of the delegates were postgraduate research students from the Institute of Industrial Ecology, part of the Urals Division of the Russian Academy of Sciences.

Most delegates came from organisations with a top-down management structure and little lateral liaison between departments, which may be why many had difficulty adjusting to the proactive approach of environmental management systems. This was also reflected in the preparatory review presentations where the groups concentrated on gathering environmental data rather than focusing on whether the correct data was being collected and subsequently used to inform the management of environmental impacts.

Delegates were very knowledgeable about environmental issues as well as regional and local pollution problems requiring urgent attention. Many of the delegates had worked on the submission to the Federal Government for the region to be designated as an environmental emergency zone. There was a genuine desire to address the urgent environmental problems, but insufficient understanding of how to deal with them and little finance available. The implications of shifting towards a market economy and private ownership were not fully understood.

Delegates indicated that they would like to have been given more material on management systems in general and case-study examples of environmental management systems in practice. There was particular interest in quality systems, such as ISO 9000. The Deputy Chief Engineer from the Sinarsky Pipe Plant was hoping to implement ISO 9000 with the assistance of a German customer.

■ Environmental Management Hurdles

It is clear that many of the environmental problems facing Russian manufacturing industry could be overcome through the introduction of suitable quality and environmental management systems. The training course provided the opportunity to introduce environmental management systems through the framework of British Standard BS 7750, which, as an acknowledged example of best practice in this field, was considered an appropriate approach to take. Implicit in this approach was the provision for delegates of the detail needed to understand the requirements of ISO 14001, which at the time was still forthcoming, and which it was assumed would take precedence over national standards.

The difficulties in preparing and delivering management training in Russia go far beyond the need to present well-translated materials in a professional manner. The Soviet system is fundamentally different to that practised in the West, and what has worked well in the UK will not necessarily work in

Russia. The challenge is to ensure that management tools and techniques, such as ISO 14001, are adapted to fit the existing system and not vice versa. Despite the current programmes of reform, based on market-oriented theories, much of Russia still operates within a collectivist culture. It was within the context of the existing top-down management approach of Russian organisations that the introduction to environmental management systems took place.

It became obvious as the course progressed that pressures on industry to address environmental concerns were very different to those experienced in the West. While environmental problems have long been recognised in the UK, it has taken subsequent legislation and market forces to galvanise industry into action. In Kamensk-Uralsky, the dominant concern in addressing environmental issues currently focuses on regional health problems caused by industrial pollution. This is not surprising considering the close inter-relationship between industrial plants and the local community.

The fact that, despite comprehensive environmental laws, industrial organisations in Kamensk-Uralsky still seem unable to address pollution problems is symptomatic of the top-down regulatory system, which is openly acknowledged to be ineffective. The introduction of a voluntary self-regulatory system, in the form of an environmental management system standard, introduces a bottom-up approach, which is alien to the existing management culture. In the West, where legislative controls are tighter, but supported by many government initiatives to help industry, the benefits of voluntary measures are more tangible. In Russia, with poor legislative control and the remnants of a culture that frowned on individual initiative, new approaches will be more difficult to introduce. So where are the incentives for the development of environmental management systems?

The implications for organisations in Kamensk-Uralsky are that regional health problems are likely to be key factors in the short term influencing the design of any environmental strategy. It remains to be seen how influential market forces will be in formulating an approach to improving environmental performance. An indication of market influences has emerged with the stated intention of the Sinarsky Pipe Plant to develop a quality management system in accordance with ISO 9001, as stated, an initiative instigated by a German customer. The change in trading relationships will expose Russian companies to direct pressures from customers interested not only in their products, but also their processes. However, this can also mean benefits for organisations no longer obliged to operate through state agencies. They are now in a position to take advantage of expertise and guidance from customers looking to build mutually-beneficial working partnerships. The Sinarsky Plant's German customer will assist them in their efforts towards implementing

ISO 9001, and the Deputy Chief Engineer of the plant was aware that this project could play an additional role in environmental management. His reason for attending the course was clear: 'To learn a more systematic approach to environmental protection'.

The Deputy Chief Engineer at the Sinarsky Pipe Plant recognises that the expertise required to implement a quality management system does not currently exist within the organisation. In fact, there was little knowledge about quality management among delegates on the training course. This provided an indication of some of the difficulties that may be encountered when implementing an environmental management system based on ISO 9000 models in a Russian organisation.

In the West, proactive management of an organisation's processes has evolved through a systematic and integrated managerial approach, while feedback from delegates placed the emphasis, in Russia, on external control with little concern for internal systems. Results from course questionnaires and preparatory environmental reviews of the plants show that there are difficulties in looking at the situation from a systems perspective. Delegates tended to be very data-focused and did not succeed in taking the extra step of identifying what policies, procedures and processes were in place. There was a reluctance to follow their work through and ask questions about what they could do to improve the situation.

The Russian companies represented on the course came from traditional function-based organisations experiencing the full range of problems typical of such structures—exactly those problems that management system standards, such as ISO 14001, are intended to address. Different departments within the plant rarely communicated with each other, and treatment of environmental issues from the same process were regarded as discrete activities. Typically, environmental responsibilities were held by many departments and data-gathering was inconsistent and unreliable. A large number of documents was generated, but the system of monitoring and reporting was not regarded as a prestigious activity and was often carried out by poorly-qualified individuals. An 'environmental passport' was required by each industrial plant, which is a document containing information covering the history of the site, discharges, land use, monitoring and limits. While required by law, it was not regarded as a working document and delegates (inspectors and industrialists alike) were sceptical of the accuracy of the information contained within it.

The environmental framework for management is obviously compliance- or issue-based rather than process-oriented. Typical responses of delegates from industry on the environmental management approaches used by their company included:

Working with regulations and legislation. Methodology and guidelines for designing environmental documentation.

Pollution monitoring and control. Establishing priorities for pollution reduction. Planning or environmental activities. Implementation of the plan and taking measures.

Various techniques are used to treat waste waters (sedimentation, filtration) and air emissions (filtration, scrubbing, dust settling, solid waste treatment).

However, the course did enthuse some delegates, most notably the Deputy Chief Engineer from the Sinarsky Plant, who left with the intention of 'introducing management into process technologies'. Recognition that environmental issues are also management issues was an important element of the course. While responsibility for environmental management systems in the UK is often seen as an extension of quality or health and safety management, it is important to note that delegates on this particular course were largely trained as chemists. This provided an impressive understanding of the technology of environmental protection, while lacking the broader outlook of their operations and the economic and political climate in which they were operating.

The concept that business management skills can be as important as scientific knowledge made an impression on some delegates. The Energy Manager of the Kamensk-Uralsky Non-Ferrous Metals Plant stated that

> . . . [the] environmental policy of our plant is focused mostly on technical matters. I think commercial, financial and economic matters should also be closely involved. I shall try to make environmental programmes more open and publicly accountable. Environmental objectives of our plant need to be closely reviewed and interlinked with economic objectives.

The question that must be asked is: how can practices be changed that have been accepted as the norm for many decades? Implementing an environmental management system represents a process of change that reaches into all corners of an organisation. It involves questioning not only existing practices, but also values. This is difficult enough for companies currently practising a systematic approach, but consider the difficulties for those only just beginning to view their organisation as a business. In the past, production goals and targets were set externally. In environmental performance terms, these were imposed through largely unachievable legislation with little guidance or support to attain the targets. Much of this now has to be achieved internally and, in order for this to happen, Russian organisations need access to information and management tools denied them in the past.

The benchmark for progress will be the commitment of organisations to apply such tools meaningfully. The difficulty for senior management in a

Russian company is their new-found responsibility for managing their plant on a commercial basis. The Russian industrial manager is only now being required to face the challenge of how to allocate resources in a way that enhances profitability of the company. This involves setting new priorities, and how much environmental issues come into consideration will depend on the vision and commitment of senior management. The introduction of an organisation-wide effort to improve environmental performance through changes in structure, practices, systems, and, above all, attitudes will not happen overnight, and the costs of adjustment will be significant. Russia is steadily being introduced to the problem, also faced by the West, of the environment as a business issue and determining the level of commitment that can be allocated to addressing environmental problems.

Even for Russian organisations committed to improving their environmental performance, there are still substantial hurdles to clear should they wish to compete in global markets. Pressures to comply with recognised international standards on environmental management will surely follow the introduction of ISO 14001. ISO 14001 does not make any demands on environmental performance levels beyond that of complying with relevant legislation, and, while this may not appear a particularly onerous requirement, it has significant implications for international trade. National differences in environmental laws and expectations are great, and while Russian laws are comprehensive by Western standards, the support systems, technology, information and management awareness—available to their international competitors—are lacking.

Russian companies are disadvantaged due to the lack of a suitable infrastructure to support the introduction of environmental management systems. The legislative and regulatory framework places little pressure on companies to try to attain even the minimum requirements of ISO 14001; there are few economic and financial incentives. The fines imposed for non-compliance are small and, in the current economic climate, the threat of plant closure as a result of failure to comply with environmental legislation is remote. Against this background, there is the opportunity for Russian companies to achieve certification by meeting the minimum requirements of ISO 14001 and thereby attain a position comparable to foreign competitors who have achieved far greater environmental performance levels. This may represent a form of competitive advantage, albeit of a dubious nature.

Russia also lacks the infrastructure to support a certification scheme to ISO 14001. While companies may rely on external organisations to provide third-party certification, this will be costly and certainly does little to promote environmental management within Russia. As mentioned earlier, ISO 14001 represents a shift towards the use of voluntary self-regulation. The use of the word 'voluntary' may be viewed with scepticism from some quarters if, as with ISO 9000, implementation of ISO 14001 becomes a contractual necessity for suppliers. However, the credibility of the standard largely depends on the introduction of meaningful certification schemes, without which ISO 14001 would become a simple aid to self-declaration. With Russia developing closer trade links with the European Union (EU), industry will be looking for support to overcome such potential barriers.

Within the EU, the quality management system standard ISO 9000 has been widely adopted across a broad range of business sectors. It has become increasingly difficult for a company to do business as a supplier unless it has implemented a quality management system standard certified by a third party. Clearly, those organisations in the region already familiar with the procedural and documentation requirements of ISO 9000 will be in a very strong position to incorporate the system elements of ISO 14001 requirements into their overall company management structure. This point reinforces previous concerns about Russian companies having the ability to apply systematic approaches to environmental management. To what extent will an organisation's existing systems support an effective environmental management system? Is it feasible for Russian companies to adopt this approach to environmental management?

From the evidence gained during the training course, it is likely that Russian organisations will have to adopt an incremental approach to the introduction of environmental management systems. There are far more fundamental problems to be addressed within Russia before management tools such as ISO 14001 are widely accepted. Barriers to this change operate on two levels. At a national level, both the legislative and economic infrastructures are not able to support the voluntary self-regulatory approach, while, at the organisational level, existing cultural values and management structures are incompatible with a systematic approach.

The environmental management training course was delivered with the expectation that it would raise awareness rather than galvanise delegates into immediate action. An additional objective was to integrate the course into existing training programmes, and Russian nationals have been selected to continue running it with support from the UK team. It is anticipated that the Russian team will be able to train managers throughout the Urals Economic Region, with their base in Ekaterinburg becoming a 'transfer node' for environmental management training. The course can then be introduced through similar 'nodes' in other important industrial regions. Delivery of the first training course in environmental management in the Urals Economic region represents an extremely small but important contribution to improved environmental performance in Russia.

■ *Notes*

1. See Barbara Peitsch, 'Foreign Investment in Russia', in *The OECD Observer*, No. 193 (April/May 1995), pp. 32-34.
2. See 'Joint Ventures between Russian and Foreign Businesses Should Benefit Environment', in *Environment Business Briefing*, April 1995, pp. 6-9.

8 Attitudes and Experiences of the Japanese Business Community vis-à-vis EMS Standards

Tomoko Kurasaka

EMS (environmental management systems) standards emerged as a completely new framework for the Japanese business community. When considering the Japanese response to EMS standardisation, three factors should be considered.

☐ The Geographical Factor: The Importance of Exports

The first is the importance of exports: important because of Japan's geographical location, being an island country with limited natural resources. For example, most of Japan's raw materials are imported (99.7% of oil, 94.7% of coal, 100% of iron ore, and 99.4% of copper in 1994). To earn foreign currency in order to import precious natural resources, Japanese businesses are compelled to work hard on the export side of their businesses. From the Japanese business community's standpoint, certification in accordance with EMS standards received attention because it was regarded as a passport to exporting. When the ISO 9000 series was introduced, many Japanese companies were forced to spend highly, including translation fees, in order to be certified by a European assessor in accordance with ISO 9000 so they could sell their products on the European market. When discussions on the ISO 14000 series arose, many Japanese companies were in the middle of struggling for ISO 9000. The Japanese business community was very concerned and wanted to avoid an assessment system as inconvenient as the one they faced under ISO 9000.

☐ The Cultural Factor: Bottom-Up; Less Documentation

The second factor is the unfamiliarity with the idea of standardising management systems, which relates to Japan's cultural background. The traditional Japanese management style is characterised as a 'bottom-up' approach. In addition, a relatively low turnover has enabled Japanese management to neglect documentation. Individual employees' areas of responsibility are often ill-defined, due to a culture of joint responsibility and co-operation. Though some Japanese companies implemented environmental management practices as early as the 1970s, these were not structured as ISO 14000-type EMSs.

☐ The Political Factor: Government Supervision and 'Do-as-the-Neighbours-Do'

The third factor is government supervision. Respect for the Emperor or the authorities is a feature of traditional Japan, and the Japanese are, by nature, inclined to follow the crowd. Japanese economic growth after the Second World War is said to be due to the Government's supervision with strong leadership from the Ministry of International Trade and Industry (MITI). Keidanren (the Japanese Federation of Economic Organisations), the most famous Japanese organisation representing the business community, has been holding discussions with MITI to reach a realistic consensus regarding industrial policy. Issues relating to standards are also within the realm of MITI. Unlike other countries, the JISC (Japanese Industrial Standards Committee), Japan's ISO member body, is organised by the AIST (Agency of Industrial Science and Technology) of the MITI. However, members of the Japanese business community, especially large companies, are beginning to feel that the Government's involvement may not be necessary. The response to international EMS standardisation is the product of a mixture of government supervision and business-community initiative.

■ Experiences and Attitudes

☐ Historical Overview

The Japanese response to EMS standards so far can be categorised into the following three phases:

Initial reaction. The first official response to international standardisation started with the establishment of a committee. In May 1992, the Environmental

Management Standardisation Committee was organised at the Japan Standard Association (JSA) under the supervision of the Standards Planning Office of the AIST at MITI. This committee discussed how to respond to the proposal issued by ISO SAGE (Strategic Advisory Group on Environment).

At that time, neither the Government nor the business community was well informed about EMSs, but the significance of ISO 9000's impact had been noted. Business leaders warned that they should try to avoid repeating the mistakes made with ISO 9000.

In December 1992, Keidanren formed the Environmental Auditing Working Group within its Global Environment Committee in order to conduct research and to collect opinions about EMS and audit standards from the business community's standpoint. Active members of the Working Group come mainly from large companies exporting to Europe and other countries in the world.

Keidanren also decided to send delegates to the ISO meetings, and in June 1993 sent eighteen delegates to the first TC207 meeting. This was the largest group from a total of 23 delegates from Japan. That same month, the Environmental Management Standard Consultation Committee was formed under the JISC in response to TC207, which has been set up to decide Japan's official attitude towards ISO discussions. There have been times when the Japanese business community has advocated that business people should take the lead and try to formulate Japan's own draft international standard, but this has been seen as unrealistic. Though Japan continued sending delegates and trying to be involved in the ISO discussion process, it appears that the Japanese business community's focus shifted on to how to deal with the EMS standards that were likely to become the international standard.

Certification to BS 7750 and ISO (DIS) 14001. Without waiting for publication of ISO 14001, Japanese companies started seeking certification in accordance with existing EMS standards. This started with BS 7750. On 6th March 1995, the first BS 7750 certification in Japan was reported at two sites belonging to Canon, renowned as one of the world's largest copier manufacturers. Canon's certification body was SGS, the worldwide certification body based in Switzerland. In December 1994, SGS announced that it would start EMS certification in Japan, in addition to its ISO 9000 certification business. SGS was accredited by NACCB (now UKAS) as a certification body for BS 7750 on 3rd March 1995.

In addition to the overseas certification bodies, Japan produced her own. In November 1994, JACO (Japanese Audit and Certification Organisation for Environment) was founded by the electronics industry. JACO's shareholders include ten large companies, such as Matsushita (Panasonic), Hitachi, NEC,

Sharp, Sanyo, Sony, Toshiba, Mitsubishi, Fujitsu, and Fuji Electric, as well as two industry associations: JEMA (the Japanese Electrical Manufacturers' Association) and EIAJ (Electronic Industries Association of Japan). The Japanese electronics industry is known to have been the most seriously affected by ISO 9000. JACO advocated that Japan should establish its own certification system for ISO 14001 in order to avoid heavy payment to foreign certification bodies, and its first certifications were given to Hitachi and Toshiba in July 1995, which formed a part of the trial for its own accreditation. JACO was accredited by UKAS in November 1995. As of May 1996, there were more than 30 sites in Japan certified and registered in accordance with BS 7750, 23 of which by JACO and 11 by SGS.

Japanese companies have also started achieving certification in accordance with ISO (DIS) 14001. On 9th November 1995, Sony Chemical, a magnetic tape manufacturer, issued a press release and announced that they had obtained ISO (DIS) 14001 certification as the world's first. This certification was awarded by DNV, a certification body based in Holland. It is not confirmed whether this was truly the world's first, but it seems at least it was the first certification as far as the Dutch accreditation body are aware.

Japan now has several certifying bodies for ISO (DIS) 14000. In addition to SGS and JACO, JQA (Japan Quality Assurance)—the largest certification organisation for ISO 9000 in Japan—has also started awarding certification to ISO (DIS) 14001.

Towards a national programme for ISO 14001. As discussions at ISO proceeded, Japan started preparing for its national programme. On 7th October 1994, the *Nihon Keizai Shimbun (Japanese Economic Journal)* reported that the Japanese Government was preparing JIS (Japanese Industrial Standard) to be equivalent to ISO standards. Under the supervision of the AIST/MITI, the JSA started translation work. As soon as the ISO 14001 standard became official, an equivalent JIS was published in October 1996.

An accreditation system for ISO 14001 is also being established. The Japanese Accreditation Board (JAB), Japan's accreditation body for ISO 9000, announced in February 1995 that it would start a trial for EMS accreditation, and three certification bodies and two training bodies were chosen for the trial which started in August 1995. The JAB's draft accreditation standards were examined during this trial period, and in May 1996, the JAB's accreditation standards were authorised by the JISC as pilot standards; in June, the JAB started accepting applications for accreditation, and, as of July, it had seven certification bodies and four training bodies under its pilot programme.

In June 1996, under the supervision of the AIST/MITI, the Japanese Environmental Management Association for Industry (JEMAI) established a new organisation to register assessors. It was to follow Guide 64, which requires the accreditation body and registration body to be separate. For ISO 9000, the JAB had been working as a registration body as well as an accreditation body, and this arrangement is to be reorganised in the near future. As the ISO 14001 becomes an official standard and an equivalent JIS becomes effective, the JAB's pilot will become the official accreditation programme. Unlike ISO 9000, this schedule causes no delay to ISO, a fact that is appreciated by the Japanese business community.

■ The Experience of Industry

The response to EMS standards varies between industries. In general, industries with a strong focus on exports show greater interest in EMS standards. There follows a summary of the experiences of the major industries.

□ Electronics

As previously stated, this industry is very active with JACO, Hitachi and Toshiba being the first to be certified by JACO in July 1995; 200 sites are expected to be certified by the end of 1996.

Some companies have their sites certified by certification bodies other than JACO. For example, in April 1996, NEC, one of the largest electronic manufacturers in Japan, achieved certification to ISO (DIS) 14001 from JQA for its subsidiary NEC Kansai's Ohtsu plant, which made it the first semiconductor manufacturing site in Japan to be certified. NEC plans to certify all of its sites in Japan.

□ Machinery and Office Equipment

Office equipment manufacturers exporting to Europe are known to be very active. Konica and Canon, two of the world's largest copier manufacturers, registered their sites in Germany in accordance with the EU's EMAS in Autumn 1995. In March 1995, Canon achieved the first certification in Japan as described above and has certified all of its sites in Japan. The Japanese Machinery Federation is planning to prepare an industry manual.

□ Chemical

The Japanese Chemical Industry Association (JCIA) has been promoting Responsible Care (RC), but also paying attention to ISO 14001. In May 1996, the JCIA revised its RC standards and guidelines in the light of ISO 14001. The Japanese Responsible Care Council (JRCC), established in April 1996, proposes that chemical companies gradually implement EMSs based on RC standards and guidelines, provide information to the public through the JRCC, and certify to ISO 14001 when ready.

Some companies have already certified their sites. For example, Hitachi Chemical certified its Shimodate Plant in accordance with BS 7750 in April 1996, and will be switching to ISO 14001 certification as soon as possible. The company is planning to certify all of its six plants in Japan by the end of 1996. Another example is Idemitsu Petrochemical, which obtained certification in accordance with ISO (DIS) 14001 from UK-based BVQI for its two plants. This was the first certification of ethylene centres in Japan. Japan Chemical Quality Assurance Ltd, a corporation established as an ISO 9000 certification body for the chemical industry, is expanding its business into the field of EMSs.

☐ Steel

The steel industry changed its attitude and decided to go for certification. According to a report from the Medium and Small Business Research Institute,[1] steel companies were at first reluctant to certify in accordance with EMS standards because they were confident about their environmental programmes and did not feel they needed third-party certification. However, since industries with self-controlled EMSs, such as the chemical industry, were moving towards third-party certification, steel companies could not ignore this trend, and held discussions at the Japan Iron & Steel Federation (JISF), resulting in a decision to certify in accordance with ISO 14001. In March 1996, the Nippon Steel Corporation certified its Nagoya iron mill in accordance with ISO (DIS) 14001. JIC Quality Assurance Ltd, a corporation established as an ISO 9000 certification body for the steel industry three years ago, is now expanding its business into the field of EMSs.

☐ Automobile Industry

No agreement has been formed at the industry association level, but all companies are reported to be implementing EMSs based on their individual guidelines. In March 1996, Toyota achieved ISO (DIS) 14001 certification for its Takaoka plant as the first in the automobile industry in Japan, and other major automobile companies are also reported to be planning to achieve certification. According to *Asahi Shimbun* (*Asahi Newspaper*) on 18th August

1996, a new organisation will be established in 1996 as an ISO certification body for the automobile industry.

☐ *Pharmaceutical*

The Japanese Pharmaceutical Manufacturers' Association has been active in publishing guidance documents (sector application guides) for implementing EMSs, such as *Internal Environmental Auditing System and Standard for Pharmaceutical Companies* (March 1993) and *Guidance for Environmental Protection Plans for Pharmaceutical Companies* (June 1994), although it does not require certification in accordance with any EMS standard.

☐ *Oil Refining*

In April 1996, Nippon Oil Co. Ltd announced that it was receiving, from SGS, certification in accordance with ISO (DIS) 14001 for its Negishi refinery. It was reported to be possibly the first certification in the world for an oil refinery.

☐ *Construction*

No commitment towards certification is observed, but EMS-supporting tools have been provided by the industry association and a voluntary research group. In September 1994, the Japan Federation of Construction Contractors (JFCC) formed a working group on environmental management/audit systems and another working group for research on the Environmental Protection Action Plan as a follow-up for a guide that the Association issued to encourage member companies to prepare the Plan in November 1992. In November 1995, the Association formed a research group on ISO issues.

The CSD Research Group, a voluntary research group focusing on the construction industry and sustainable development, including members from eleven major general construction companies, published a guidebook for environmental management in the construction industry in February 1996. It contains a self-assessment programme modelled for the construction industry based on the ESAP (Environmental Self Assessment Programme) of the GEMI (Global Environmental Management Initiative) in the US.

☐ *Retail*

The retail industry is active in terms of environmental programmes, but less involved in EMS standard issues than the manufacturing industry. Co-ops are more aggressive than general retailers in implementing EMSs, although they are not seeking certification in accordance with any EMS standard so far. The Japanese Consumers' Co-operative Union invented a model framework for co-ops based on input from BS 7750, EMAS and ISO (DIS) 14001. The Co-op Tokyo and five other co-ops have implemented this programme so far. In the financial year 1996/97, thirty member co-ops are going to implement this, and another fifty will join the following year.

■ *The Experience of SMEs*

It is not only large companies that are interested in EMS standards, but smaller companies as well. The Tokyo Chamber of Commerce and Industry formed a working group on environmental auditing in April 1994 and issued a guide in October 1994 mainly for small and medium-sized enterprises (SMEs) on implementing EMSs. The first SME certification was given by SGS to LALF (Laser Application Laboratory and Factory) of Shinozaki, an engineering company with twenty employees, on 20th August 1996. According to the Medium and Small Business Research Institute's report,[2] Shinozaki at first thought about certifying in accordance with ISO 9000, but decided to choose ISO 14001 with the intention of improving its management system in general rather than only in terms of environmental aspects.

■ *The Attitude of National and Local Governments*

☐ *National Government*

The Japanese Government has not announced any intention to make certification to any EMS standard mandatory for Japanese companies. However, MITI has been reported as having been asked to prepare new environmental management plans that conform with requirements contained in ISO 14001. As described above, AIST of MITI has been working on ISO issues. It is also working to introduce EMS standards in accordance with ISO 14000 in JIS, the Japanese industry standard system. When the JIS equivalent to ISO 14001 is published, there is a possibility that certification in accordance with the standard might be required for government contractors. The Japanese Ministry of Construction is reported to be considering such a requirement for ISO 9000. Some say it is expected to expand this to ISO 14001.

The Japanese Environment Agency published *Kankyo ni Yasashii Kigyo Kodo Shishin* (a guide on environmentally-friendly business behaviour), which included ideas from BS 7750 and EMAS, in October 1992. In October 1995, the Environment Agency published the Environmental Activity Evaluation Programme. It was reported that the Agency regarded requirement of ISO 14001 as too broad and too costly for SMEs and many non-manufacturers, and therefore developed this programme as an original EMS scheme for such businesses. However, since concerns were voiced that this programme alongside ISO 14001 might create a double-standard problem, the Agency curtailed the programme.[3]

☐ Local Governments

Some local governments have local regulations or agreements that require companies to implement EMSs, although certification is not required. For example, the Kazusa Agreement of Kisarazu City, Chiba Prefecture, requires companies operating at the Kazusa Industrial Park to establish an EMS, to conduct environmental auditing and to submit reports to the city. This agreement, which has been applied since October 1994, is regarded as pioneering.

There are also local governments that support small companies to certify in accordance with EMS standards. For example, the Sumida ward of Tokyo allocated two million yen in its fiscal 1996 budget to provide seminars and consultation for small companies interested in ISO 14001 and ISO 9000 series certification.

Local governments may possibly request EMS certification as a bidding condition for government contracts. According to a survey on the cities, towns and villages of Fukuoka Prefecture, the majority expressed the possibility of taking into account the bidder's EMS implementation at tender for engineering work.

■ EMS Certification and the Training Industry

With an increase in the demand for certification, Japan is expecting a new market for the certification business together with a related training industry.

☐ The EMS Certification Industry

Both foreign and Japanese companies are participating in this market. Nikkei Mechanical's *Access Guide*, issued in May 1996, lists the following nineteen organisations (including six non-profit-making organisations, marked with an asterisk) as certification bodies. Some of them seem more focused on ISO 9000, but all of them expressed an intention to work on ISO 14001.

- ABS Quality Evaluations, Inc.
- BVQI (Bureau Veritas Quality International)
- Det Norske Veritas
- KPMG
- JTCCM (Japanese Testing Centre for Construction Materials)*
- KHK Quality Assurance Centre
- JET (Japanese Electrical Testing Laboratory)*
- JQA (Japanese Quality Assurance)*
- JACO (Japanese Audit and Certification Organisation for Environment)
- NK (Nippon Kaiji Kyokai)*
- NKKQA (Nippon Kaiji Kentei QA)
- JICQA (JIC Quality Assurance)
- JSA-Q (Japanese Standard Association)*
- JMA (Japanese Management Association)*
- TÜV Rheinland
- LRQA (Lloyd's Register Quality Assurance)
- SGS
- TÜV Product Service
- Underwriters Laboratories, Inc.

As described above, a new certification body is going to be established for the automobile industry in 1996. The chemical industry has a certification body, as does the steel industry; JACO was established by the major electronics companies. Those certification bodies that have connections with a specific industry may have a great influence on the certification market.

☐ The EMS Training Industry

Before the national programme for ISO 14001 begins, training for EMS assessors is already under way in Japan. As of August 1995, the following organisations were providing training programmes accredited by EARA (Environmental Assessors' Registration Association) of the UK: Deloitte Touche Tohmatsu (in partnership with SGS); GLOBAL TECHNO; and JSA-Q (Japanese Standards Association)

In April, 1996, thirty Japanese companies, including Nippon Steel Corporation, Canon, NEC and Toyota, established a new company, The Environmental Management Training Centre Corporation, to provide training for environmental assessors for ISO 14001. The *Nikkei Sangyo* newspaper reported on 17th April 1996 that this company is expecting 100 million yen for its first year's sales. As the national programme for ISO 14001 develops, the training market will expand and more players may come into the market.

Why do Japanese companies implement EMSs? ISO 14000 has emerged as the second 'passport to exporting' after ISO 9000. Companies that have achieved EMS certification at an early stage are often found in exporting industries such as electronics. Mitsunori Oka, Chief of the Environmental Promotion Centre of the Storage Systems Division at Hitachi, states in his book[4] that the division suffered a delay in achieving ISO 9000 certification and received pressure from its customers and therefore believes early achievement of EMS certification would be a weapon for competitive advantage.

However, many companies are implementing EMS for reasons other than the export issue. According to the survey carried out by the Japanese Environment Agency, organisations with few or no exports (e.g. utility, retail, schools) are showing interest in ISO 14001. According to the survey on SMEs, SMEs are thinking about implementing ISO 14001 on their own initiative rather than on the insistence of their customers or parent companies. Even Mitsunori Oka of Hitachi states that his motive behind EMS certification was neither export nor customer pressure, but because he thought the division needed an EMS based on an international standard. He explains that what the EMS standard requires is not only an environmental system, but also a management system; he therefore finds it beneficial to strengthen his division. It cannot be denied that concern over exports has been a motivating factor, but it now seems the EMS standards are regarded to have further implications in terms of improving management, with reasons for seeking certification also including more than merely a passport to exporting. Shinichi Iioka, Deputy Manager of the Global Environmental Department, Marubeni—one of the largest Japanese *sogoshosha* (general trading companies)—explains that, in the age of severe competition, where any piece of competitive disadvantage may take you out of the playing field, companies need consistently to demonstrate that they are environmentally aware. He states that ISO EMS certification is proof of this, in contrast to eco-labelling, which merely proves that a product is environmentally friendly.[5]

What do the EMS standards bring to the Japanese business community? 'From regulation to voluntary management' is a phrase often used in introductory books that explain ISO 14000, and now ISO 14000 and other EMS standards are regarded as among the issues that symbolise the international trends that Japanese companies have to follow. Masaru Nakagawa, Chief Programme Director at the Environmental Management Centre, Japanese

Management Association, wrote that 'ISO made us aware that as a matter of fact it is not the authorities but the market which judges values and credits.'[6] Yoshiyuki Ishi, Director of Environment Compliance at Hitachi America Ltd, describes ISO 14000 as a desirable standard for those who want the freedom to determine details, but, on the other hand, states that ISO 14000 poses a difficult task for those whose attitude is: 'If you determine the rule, I will follow it.'[7] The EMS standards can be explained in the trend towards a change in Japanese traditional reliance on the authorities or governments. Or, from another perspective, are EMS standards causing a change in Japanese management style? It seems EMS standards, as well as ISO 9000, are pushing Japanese companies towards a new management approach, part of which tendency has been observed by certification bodies. For example, Mr Ichikawa of JQA implies that the most important point of EMS standards is their systems approach, which is the most challenging part for the Japanese. Mr Shibamiya of JACO puts great emphasis on clarifying 'who is responsible for what'.[8]

Corresponding observations are made by certified companies. For example, Hisashi Sakamaki, Director for Production/Environmental Assurance at Canon, says he noted a difference in nature between the Japanese and the Europeans when Canon underwent an audit by its certification body, SGS. He explains the Europeans clarify objectives and measures before starting, whereas, on the contrary, in Japan, each employee judges and responds without a specific order.[9] It seems the Japanese business community has not been switching completely to a new Western style, but importing only useful parts. Some certified companies say that a top-down approach is needed in order to tackle environmental issues, but this does not seem to be the case. Mr Iwasaki of SGS says: 'As far as Japanese companies are concerned, they don't have to stick to top-down.' He explains that Western organisations have a tendency towards a 'don't-move-without-an-order-from-the-top' culture, whereas in Japanese organisations both management and shopfloor will act on their own initiative. He says that, as long as communication between top and bottom is healthy, strength as a system can be maintained either by a top-down or by a bottom-up approach.[10] Some 'Japanising' features can be observed in the way that the Japanese business community has been building up the ISO 14000 certification industry, with some certification bodies being formed by a group of companies in the same industry. Industry-specific certification bodies are expected to offer high-quality service with plenty of industry-specific knowledge, although critics question the impartiality of auditors. The industry-specific certification bodies are making efforts to respond to this criticism by carefully selection of the members of their judging committees. It seems the Japanese business community is leveraging the traditional culture of co-operation and 'do-as-the-neighbours-do' in order to respond to the emerging need for EMSs.

☐ *Future Prospects*

How broadly and deeply will the EMS standard influence the Japanese business community? At present, it is difficult to read the crystal ball, but the author believes the most important factor is whether the EMS standards will improve Japanese companies' environmental management. Takeshi Nishi, a SGS alumnus and independent business consultant, writes that many ISO-9000-certified companies are holding dual manuals: one for certification, the other for practice. He insists that a system is not a magic wand and that certification should not be a final goal.[11]

Many books state that certification is not an end in itself, and they emphasise the importance of 'continual improvement'. In Japan, criticism has also been levelled at the EMS standards in terms of their lack of emphasis on performance. However, many Japanese companies do not neglect performance issues when implementing EMSs. If certified companies are really aiming for continual improvement, improvement in performance seems to be expected. The difference between ISO 9000 and ISO 14000 is also important. As far as ISO 9000 is concerned, it is the customer that judges the performance and the quality. For ISO 14000, it is not only direct customers that will be affected by the certified company's performance. If not being watched, certified companies may find it difficult to maintain motivation for continual improvement. The author believes we must keep an eye on what happens in the coming years and see whether the EMS standards are really beneficial to the environment.

■ *Notes*

1. The Medium and Small Business Research Institute, *Kankyo Kanri Taisei no Kokusai Kikaku-ka to Chusho Kigyo no Taio: ISO 14000 series no donyu o maeni* (*International Standardisation of Environmental Management Systems, and Medium-Sized and Small Companies' Reaction to It: Preceding the Introduction of the ISO 14000 Series*) (Tokyo, Japan: The Medium and Small Business Research Institute, March 1996).
2. *Ibid.*
3. See Mitsuaki Tsutsui, Shinichi Iioka and Masaru Nakagawa, *Hiseizogyo no ISO 14000 Nyumon* (*Non-Manufacturers' Introduction to ISO 14000*) (Tokyo, Japan: JMA Management Centre, September 1996).
4. Mitsunori Oka and Yoshiyuki Ishii, *Surasurawakaru Kankyo ISO no Torikata* (*An Easily Understandable Guide to Achieving Environmental ISO*) (Tokyo, Japan: Chukei Shuppan, September 1996).

5. See Tsutsui *et al.*, *op. cit.*

6. *Ibid.*

7. Oka and Ishii, *op. cit.*

8. 'Kankyo Management ni okeru Daisanshaninsho no Igi' ('The Meaning of Third-Party Certification in Environmental Management'), in *Diamond Eco-line*, No. 55 (20 July 1996), pp. 1-4.

9. 'Kankyo Kanri Taisei no Seibi to BS 7750 Shutoku' ('Arranging an Environmental Management System and Achieving BS 7750 Certification'), in *Diamond Eco-line*, No. 44 (20 July 1995), pp. 1-3.

10. 'Kankyo Management . . .', in *Diamond Eco-line*, No. 55.

11. Takeshi Nishi, *Yokuwakaru ISO 14001* (*Clearly-Understandable ISO 14001*) (Tokyo, Japan: PHP, September 1996).

Environmental Management Systems
ISO 14001 Issues for Developing Countries

Aidan Davy

THE PUBLICATION of ISO's specification for an environmental management standard, ISO 14001, marks the culmination of more than three years of effort by ISO's technical committee, TC207, and the first in a series of international standards aimed at improved environmental management. It also provides an opportunity for reflection on the likely implications of the introduction of the standard in all ISO member countries and elsewhere, and on the factors that may affect the extent to which the standard becomes adopted.

The focus of this chapter is on the implications for less developed countries (LDCs), which in this context includes economies in transition. It is also mainly on environmental management systems (so far the only standards published in the series), although other standards being developed under the umbrella of ISO 14000 in key areas such as environmental labelling or lifecycle assessment clearly have wide-ranging implications also. This chapter aims to begin to provide answers to questions relating to the extent of adoption of ISO 14001, trade issues linked to its adoption, and the financial implications for LDCs. In practice, the full answers will emerge only through experience.

■ Issues for Developing Countries/Economies in Transition

While ISO 14001 will apply uniformly to both developed economies and LDCs, there are a number of issues that are of particular concern for LDCs. These include:

- Why bother to implement an EMS?
- Have LDCs had any say in the development of ISO 14001?
- How widely will ISO 14001 be adopted and in which sectors?
- What trade issues are raised by ISO 14001?
- What are the technical and financial aspects of becoming certified to ISO 14001?
- What institutional issues are raised by ISO 14001?

The latter refers both to the certification infrastructure and the regulatory framework to support the adoption of ISO 14001. Finally, a key issue for LDCs is what support is most needed, and what role might the development assistance community play in providing this support?

☐ Why Bother to Implement an EMS?

Many reasons have been cited for implementing an EMS in either developed or less developed economies. For companies making green marketing claims about products or processes, compliance with ISO 14001 reinforces their environmental commitment. Companies may implement an EMS in response to calls from shareholders or other pressure groups. Implementing an EMS is also a means of demonstrating control of environmental risks as a means of securing favourable insurance terms or perhaps project financing. For sectors where compliance with ISO 14001 may become 'mandatory', it should also help to facilitate intra-sectoral trade. Finally, some companies will implement an EMS as a basis for reducing environmental impacts for financial or ethical reasons, or to gain competitive advantage.

A limited preliminary survey of the trade implications of the ISO 9000 and ISO 14000 series undertaken by the United Nations International Development Organisation (UNIDO)[1] canvassed the views of trade associations and certification organisations in 31 LDCs. The three main reasons identified for implementing the ISO 14000 series, in order of importance, were:

- To help demonstrate conformity to environmental legislation
- To meet overseas consumer demand
- To reduce costs of implementing mandatory standards

The stated relative importance of meeting overseas consumer demands strongly indicates that pressures to comply with ISO 14001 are perceived as external rather than domestic.

Region*	No. of LDCs in ISO		No. of LDCs in TC207	No. of LDCs in SC1
	Members	*Observers*		
Sub-Saharan Africa	8	4	5	4
East Asia	8	3	8	7
South Asia	4	1	2	2
Europe and Central Asia	20	6	13	7
Latin America and Caribbean	12	8	12	8
Middle East and North Africa	5	1	2	–
Total LDCs	**57**	**23**	**42**	**28**
Total countries in ISO	**84**	**33**	**68**	**51**

* Based on World Bank Lending Regions

Figure 1: *Participation of LDCs in ISO TC207 and SC1*
(Sub-committee on EMSs)

☐ Representation of LDCs in Development of ISO 14001

Representatives of LDCs have been involved in the development of ISO 14001. Approximately two-thirds of ISO's membership is made up of LDC representatives (see Fig. 1). A similar proportion of LDCs have been involved in TC207, comprising 42 of the 68 participants or observers. Within SC1—the subcommittee responsible for drafting the EMS specification (ISO 14001) and guidelines (ISO 14004)—28 of the 51 participants or observer countries are LDCs.

While these statistics appear to indicate broad LDC representation in the development of ISO 14001, the numbers are not necessarily an accurate reflection of actual influence. First, the origins of ISO 14001 are rooted in EMS developments in Western Europe, particularly BS 7750 and the European Union's Eco-Management and Audit Scheme.[2] Secondly, the resources available to developed countries such as the US to assist in the development of standards (while promoting national interests) clearly exceed those of LDCs; in this regard it is notable that the chairpersons of all six TC207 subcommittees are from developed countries. Participation by LDCs during development of the draft international standard (DIS) was initially very weak, but later improved partly due to assistance from DEVCO—the ISO committee dealing with developing-country matters—which aims to ensure active participation by LDCs in ISO activities. For example, approximately half of the votes that led to the adoption by TC207 of the DIS in Oslo in June 1995 were cast by LDCs.

☐ *Is ISO 14001 Likely to be Widely Adopted?*

Given that ISO 14001 has just recently been published as an international standard, a degree of scepticism about the hype surrounding its introduction is appropriate. However, there is evidence from a variety of international initiatives that ISO 14001 will be widely adopted. Certification to BS 7750 is proceeding rapidly within the UK, with several European, East Asian and Brazilian companies also achieving certification. Since August 1995, almost 400 sites have been registered to EMAS within the EU. Although registration is confined to industrial sites within the EU, a number of South-East Asian companies (within Japan, South Korea and the Philippines) have been audited to EMAS standards and 'declarations' linked to EMAS have been produced, presumably as the basis for competitive advantage in supplying EU customers.

In addition, as of December 1995, at least sixteen companies had become certified internationally to Draft International Standard ISO 14001, including firms in Korea, Brazil and Argentina. In Brazil, at least three companies were certified to the DIS prior to its publication, and over 100 others were working towards certification in export sectors such as machinery and equipment manufacture, forestry and timber product manufacture (including pulp and paper), and food manufacture. This is in part supported by the Brazilian National Economic and Social Development Bank (BNDES), which has developed a line of credit offering favourable terms to borrowers looking to implement environmental controls, including an EMS. A number of governments in Latin America and East Asia are considering making certification to ISO 14001 mandatory for oil exploration and production companies.

The Japanese Ministry of International Trade and Industry (MITI) has advocated third-party registration to ISO 14001, although this is not a regulatory requirement; the Japanese electronics industry has responded by developing the necessary infrastructure. The governments of several LDCs, including South Korea, Singapore, Philippines, Malaysia and Indonesia, have demonstrated a keen interest in implementing these standards during their development by TC207—the Korean Government now requires leading Korean companies to achieve ISO 14001 certification by the end of 1997. As of the end of 1995, nineteen Korean companies were registered to the DIS or BS 7750. In the US, many major companies are reviewing their existing environmental programmes against the ISO 14001 criteria to develop implementing strategies. In January 1996, the Provincial Court in Alberta, Canada, fined Prospec Chemical Limited US$100,000 for exceeding sulphur emission limits, and ordered the company to become certified to the DIS.

Perhaps the best indication of the extent to which ISO 14001 may be adopted comes from the experience with ISO 9000, which was launched in

Region*	1993	1995
Sub-Saharan Africa*	824	1,492
East Asia	41	2,188
South Asia	9	1,038
Europe and Central Asia	82	1,708
Latin America and Caribbean	43	1,435
Middle East and North Africa	16	327
Total	1,015	8,188

*South African enterprises accounted for all 824 Sub-Saharan Africa certifications in 1993 and 1,454 in 1995. Excluding South African certifications, in January 1993 LDC certifications accounted for 0.7% of the global total; by March 1995, this proportion had increased to 5.2% of the global total.

Figure 2: *Growth in ISO 9000 Certifications in Bank Regions: January 1993 to December 1995*

1987. ISO 9000 has a similar managerial focus to ISO 14001, and most commercial organisations expressing interest in (or who have been involved in the development of) ISO 14001 are ISO 9000 certified. In addition, there is strong awareness of ISO 14001 within companies certified to ISO 9000. Initially, certification to ISO 9000 was primarily by UK companies that had prior experience with the BS 5750 quality standards, the British Standards Institution's forerunner to ISO 9000. By January 1993, of 28,000 certificates awarded worldwide, 67% were to UK companies and of the remainder, 27% were awarded to companies elsewhere in Europe, the US, or Australia and New Zealand. Excluding South Africa, LDC certifications accounted for less than 1% of the total.[3]

By December 1995, just over two years later, over 127,000 certificates had been issued and the UK's share had fallen to 41% primarily due to increases in the US and elsewhere in Europe. However, the LDC's share of total global certifications had increased to 5% (see Fig. 2), with greater than twenty-fold increases in the numbers of certifications in most regions (with the exception of Sub-Saharan Africa). In Sub-Saharan Africa, increases were largely within South Africa, although a sub-regional trend appears to be emerging, whereby neighbouring trading partners such as Namibia, Swaziland and Zimbabwe were awarded their first certificates between September 1993 and December 1995.

Some indication of the rate of certification within some LDCs can be obtained from Brazil and India. In March 1995, the number of ISO-9000-certified companies in Brazil and India were 548 and 585 respectively. By the end of

December 1995, more than 1,000 certificates had been issued in India. By March 1996, the numbers of certificates issued in Brazil had also almost doubled to more than 1,000, with at least 3,000 additional companies seeking certification.

☐ Sectoral Adoption of ISO 9000/14000 Series

Very limited data are readily available on the sectoral distribution of ISO 9000 certificates. A recent survey of certifications awarded within India by the Associated Chambers of Commerce and Industry of India (ASSOCHAM)[4] indicated that 37% of certificates had been awarded to the engineering sector, including automobile component manufacturers, 27% to the electrical and electronics sector, 12% to chemicals manufacture and 10% to textiles. Within the EU, the principal sectors achieving registration to EMAS are those where ISO 9000 has been widely adopted and include: machinery and equipment manufacture, including automobile manufacture; electrical and optical equipment manufacture, including telecoms and electronics; food, beverages and tobacco manufacture; chemicals and chemical product manufacture; plastics and rubber manufacture; and pulp, paper, publishing and printing. Other sectors for which sites have achieved registration include manufacture of metal products, textiles and wood products. Given the correlation between ISO 9000 and EMAS in Europe, and the emerging US interest in ISO 14001 in sectors where ISO 9000 has been implemented, it is reasonable to assume that ISO 14001 will have the greatest uptake in similar sectors.

☐ Trade Issues Related to ISO 14001

Likely extent of supply-chain pressures. One of the objectives of the ISO 14000 series is to prevent divergent national EMS requirements from becoming 'technical barriers to trade'. Concerns have emerged, however, that ISO 14001 could help to reinforce the trade barriers that it aims to prevent, particularly for international trade between developed and less developed countries.[5] A key issue for companies involved in international trade is whether ISO 14001 will be a basis for marketplace distinction or a prerequisite for survival (and in which sectors and regions). In this regard, the provisions of ISO 14001 requiring consideration of the environmental performance of suppliers are most relevant. A related issue of concern to governments is whether ISO 14001 represents a non-tariff trade barrier under World Trade Organisation (WTO) agreements.

Early drafts of ISO 14001 contained an explicit requirement for evaluation of suppliers' adherence to environmental management standards, which was analogous to the ISO 9000 requirement to ensure consistency of quality of raw materials. Although this explicit requirement is not part of the published standard, the section that deals with operational controls (4.3.6[c]) requires establishment of 'procedures related to the identifiable significant environmental aspects of goods and services used by the organisation and communicating relevant procedures to suppliers and contractors'. While somewhat ambiguous, this clause is being widely interpreted as the means by which suppliers may be required to adopt ISO 14001.

In determining the environmental aspects of its activities, products or services that 'it can control or which it can be expected to have an influence' (Clause 4.2.1), many organisations placed closest to the consumer in developed countries are adopting a lifecycle assessment (LCA) approach. LCA is the subject of the deliberations of ISO TC207 SC5, and the draft standards they have produced (ISO 14040–14043). An LCA approach to determining the environmental aspects of products is likely to place the most significant environmental burdens at various points within the supply chain, which further increases the likelihood of suppliers being required to adopt ISO 14001 or equivalent. In addition, the introduction to the ISO 14004 guidance document identifies one of the key principles for organisations implementing an EMS as 'to encourage contractors and suppliers to establish an EMS'.

Evidence from the adoption of ISO 9000 indicates that, in certain sectors, the adoption of ISO 14001 is likely to be a prerequisite for economic survival. In the US, for example, the Food and Drug Administration has adopted ISO 9000 as 'good manufacturing practice' in the manufacture of medical devices. This has had a knock-on effect which, in practice, extends the requirement to comply with ISO 9000 to businesses at every level in the global supply chain that serves the medical devices industry. The only alternative to manufacturing companies closest to the end-user would be to audit extensively the quality assurance systems of their suppliers. In practice, it is much easier to require them to comply with ISO 9000. The sectors previously identified as most likely to certify to ISO 14001 are also most likely to be the subject of supply-chain pressures to adopt an EMS.

Validity of ISO 14001 under the Uruguay Round Agreements. A related trade issue is the conformity of ISO 14001 under the Technical Barriers to Trade (TBT) Agreement and the Sanitary and Phytosanitary (SPS) Agreements under WTO (the latter applies to the protection of people, plants and animals). Both agreements are concerned with national laws on the environment or public health becoming non-tariff barriers to trade. As a means of preventing TBTs, both require the use of available international standards as the basis for national standards; in this regard, the adoption of ISO 14001 might be viewed as an obstacle to trade barriers.

The TBT agreement states that members must not use voluntary product standards (such as those relating to eco-labelling; for example, the ISO 14020 series or Germany's 'Blue Angel') in a discriminatory way, or as 'unnecessary obstacles to trade' in order to meet a 'legitimate objective', which might include environmental protection. The SPS agreement also prohibits discriminatory measures based on risks to the health of people or the environment. Within this framework, the issue is whether ISO 14001 can be considered an international standard or whether it is consistent with the principles of non-discrimination, legitimate purpose and necessity. In this regard, it is important to consider the privately-sponsored voluntary nature of the standard. ISO 14001 is unlikely to form the basis for environmental regulation by governments; however, where governments promote its implementation by private companies, or require supplier compliance as a condition of procurement, ISO 14001 may fall under the framework of the agreements.[6]

Irrespective of possible future challenges to the legitimacy of measures adopted by governments, individual companies will not be subject to such constraints. Unilateral measures that relate to the purchasing policy of private companies are not subject to WTO regulation. Consequently, if a private company chooses to include ISO 14001 as a condition of its purchasing policy, compliance with the standard becomes essential to doing business and a *de facto* trade barrier. Consequently, the issue of whether compliance with ISO 14001 is 'appropriate' by any objective criteria becomes irrelevant. Similarly, the issue of whether ISO 14001 is in the best interests of LDCs also becomes academic.

The majority of respondents (trade associations and standardisation organisations) in the limited UNIDO survey expect the ISO 14000 series to strengthen non-tariff trade barriers. This fear is particularly widespread in Asia, even though most respondents also feared a loss of market if the ISO 14000 series were not introduced.

☐ *Technical and Financial Aspects of Certification*

There is an inextricable link between the technical and financial issues for LDCs relating to ISO 14001. Technically, there may be a lack of knowledge regarding:

- The definition and application of management systems
- The aspects of an organisation's operations that significantly affect the environment
- The regulatory requirements in relation to these environmental effects
- The technological or procedural approaches to implementing environmental control

For individual companies, access to internal (or external) human resources with appropriate skills or knowledge can be problematic. This is particularly valid for small and medium-sized enterprises (SMEs) which are typically more wary of taking on additional bureaucracy and associated costs.

The costs of implementation and certification are likely to be high, particularly for companies without a well-defined business or quality management system or who have previously given little attention to environmental management issues. Other factors affecting costs include:

- The complexity of an organisation's processes
- The extent to which local or international consultants are involved in development and implementation of the EMS (which is linked to availability of internal resources)
- Existence of an environmental programme within, and current environmental performance of, the organisation
- The possible need for expenditure on clean technologies or control technologies to achieve regulatory compliance
- The availability and credibility of local certification infrastructure

The implementation costs should mainly be internal and arise from employees' time for activities such as awareness training, documenting procedures to demonstrate environmental control, training for procedural controls, maintenance of the system, and internal auditing to check for compliance to the requirements of ISO 14001. However, many companies may use consultants to assist in the development of the management system or drafting of procedures which adds to the costs of implementation. In addition, if, as a consequence of implementing an EMS, companies discover that they are unwittingly in breach of environmental regulations, unplanned investment may be needed to ensure regulatory compliance.

Given that ISO 14001 provides for self-declaration, the costs of 'certification' could theoretically be internalised. However, in practice, third-party registration is likely to be essential to gain marketplace credibility. Most experts agree that self-declaration will be little used in practice, which raises an important question regarding the existence of a certification infrastructure in LDCs and the associated costs. Estimates cited[7] vary between US$10,000 and US$30,000, depending on the size and nature of the facility for certification audits only. However, such estimates are typically based on the use of US or European consultants. A small informal sample of ISO 9000 registrars in India for this Chapter indicated average costs in the range of US$1,800 to US$5,200 (3–10 days' effort) for SMEs with between 20 and 500 employees using internationally-accredited local certifiers. Where local credible certification bodies do not exist, the costs of registration are likely to increase due to the need to resort to international certifiers. This aspect is discussed in more detail below.

☐ *Presence and Credibility of Certification Infrastructure*

Given that self-declaration to ISO 14001 by organisations in developing countries is unlikely to be acceptable in the international marketplace, access to certification bodies is an important issue. A secondary related concern is the acceptability of locally-issued certificates from LDCs to organisations in the targeted market.

Clearly, all member countries of ISO have a national standards body, although some LDCs have no such organisation. However, few of these countries have a national accreditation body responsible for 'licensing' registration bodies, and only fourteen LDCs had third-party organisations offering registration programmes in July of 1992.[8] Some LDCs are in the process of establishing national certification schemes for ISO 14001, although many others are unlikely to, at least in the short term.

Of related concern is the credibility of certificates issued by local registration service providers. For example, in India, approximately 60% of ISO 9000 certificates have been awarded by international certification organisations (albeit from local well-established offices). This partly reflects the desire of companies to enhance the international credibility of their certificate. In many less developed markets, international certifiers may provide services from neighbouring or distant regional offices. One anecdotal report from South Korea mentioned a recently-certified company publicising the passports of their international certifiers!

The issue of adequacy of certification infrastructure provides a window to the much broader consideration of the adequacy of national systems of metrology, standards, testing and quality (MSTQ).[9] With respect to environment, **metrology** standards and calibration services are critical for assessing pollutant levels accurately. International **environmental standards** (of which ISO has published more than 300) encompass standard methods of measuring pollutants or performance standards for pollution control technologies. **Testing** in accordance with such standard methods demonstrates conformity with regulatory limits. Finally, **quality** systems (analogous to EMSs) are concerned with pollution prevention, for example.

There is currently no comprehensive data source on the adequacy of national certification or MSTQ capacities. However, the US National Institute of Standards and Testing (NIST) is sponsoring a survey of international MSTQ infrastructure on a country-by-country basis. This should serve as a MSTQ needs assessment for LDCs.

MANY LDCs price energy, particularly electricity, below the costs of production and distribution, whereas developed countries tend to recover full costs (at least). In 1992, electricity subsidies for developing countries were estimated at US$100 billion.[1] This has a double negative environmental impact: first, low prices cause excessive demands, making demand-side energy-efficiency measures unlikely or unprofitable; and secondly, the power sector is financially 'strapped', leading to under-investment in new or cleaner technologies and inefficiencies on the supply side. Sound pricing policies are therefore essential for achieving energy efficiency.

Energy-efficiency measures become increasingly important as global electricity demands continue to rise. Current estimates[2] predict the following trends in CO_2 emissions from energy production:

Global CO_2 Emissions (billion tonnes)

	1990	2000	2010
World	21.1	24.7	31.5
Asia	4.0	6.3	9.6
China	2.4	3.5	5.1

1. World Bank, *World Development Report* (Washington, DC: World Bank, 1992).
2. International Energy Agency and Organisation for Economic Co-operation and Development, *Global Energy Outlook* (Paris, France: IEA/OECD, 1996).

Figure 3: *Energy Usage in Developing Countries*

☐ *Economic Incentives, Institutional Framework and Global Environmental Benefits*

The costs of implementing an EMS and registration are discussed above without reference to the benefits. Beyond reducing the incidence of regulatory non-compliances and the marketplace benefits of gaining or maintaining access to markets or market share, the strongest incentive for implementing an EMS relates to the savings arising from internal efficiencies relating to:

- Controlling costs of waste, effluent or emissions disposal/abatement
- Reducing costs of energy or raw materials

Many organisations have demonstrated clear cost-savings as a result of instituting environmental programmes. However, the first of three important factors influencing this 'win–win' situation is a moderate-to-high degree of inefficiency (at least initially), which is common. The second is that the institutional/regulatory framework supports true-cost pricing of energy (see Fig. 3), waste disposal and raw materials consumption, such as petroleum products and metals. Lastly, the institutional framework should adequately promote environmental protection (with respect to environmental regulation, policing and imposition of fines for non-compliances). In many LDCs, not all of these factors are in place.

Proponents of ISO 14001 point to the global environmental benefits that could result from its widespread adoption. These include not only the direct benefits of, for example, reduced emissions to air or water, but also indirect benefits such as reduced incidences of disease or enhanced economic potential of areas subject to environmental improvement. These potential global benefits are critically dependent, not only on the widespread adoption of ISO 14001, but also on the factors influencing the 'win–win' situation described above. These factors are likely to dictate the extent to which an organisation improves its environmental performance as opposed to its marketing potential.

The last factor raises the issue of whether ISO 14001 or EMSs could have a role in reducing the regulatory burden associated with command-and-control approaches to environmental protection. Clearly the EU thinks so, with the introduction of EMAS. However, EMAS operates in the context of disclosure of environmental effects, and of third-party verification of corporate self-assessment of compliance with stringent environmental regulations. Where the basic regulatory framework is less well defined and disclosure is voluntary, the value of self-assessment is obviously diminished.

■ *Support from the Development Assistance Community*

One possible source of assistance for LDCs in coming to terms with ISO 14001 is the development assistance community, including multilateral financial institutions such as the World Bank and the European Bank for Reconstruction and Development, or bilateral aid agencies such as the US Agency for International Development (USAID) or the UK Overseas Development Administration (ODA). Target areas for this assistance might include:

- Assistance with establishing or reinforcing the necessary certification infrastructure
- Institutional support for promotion of ISO 14001 within the industrial sector
- Development of sectoral guidance on implementing EMSs, particularly in relation to SMEs
- Financing implementation of EMSs or registration to ISO 14001

These target areas are briefly discussed below.

☐ *Establishment or Reinforcement of Certification Infrastructure*

Where no credible certification infrastructure exists, support could be directed to its establishment or reinforcement. In practice, this could extend to either accreditation bodies (typically public sector entities) or certification bodies (either public or private sector entities), with the objective of improving access to affordable credible registration bodies. Issues to consider in determining such support might include: the broader MSTQ standard-setting capacity within the country/region; the regulatory framework and pricing regimes; and the representation of key sectors likely to be affected by ISO 14001.

☐ *Promotion of ISO 14001*

The issue of apportioning costs between the public and private sector for promotion of ISO 14001 was raised in the UNIDO survey of trade associations and certification bodies in LDCs. The overwhelming majority of respondents thought that governments should bear the cost of awareness campaigns, whereas companies should pay for implementation and registration costs. Ministries of trade, industry or environment are the most probable targets of such support, which might involve training, workshops, or other aspects of institutional development. This might best be achieved in the broader context of developing certification or MSTQ infrastructure or appropriate economic incentives.

☐ *Development of Sectoral Guidance*

Development of sectoral guidance on how to implement environmental management systems would be of particular value for SMEs. This should take account of those sectors where compliance with ISO 14001 is most likely to become 'mandatory' (such as electronics and engineering), in addition to those sectors that would most benefit from implementing an EMS based on potential environmental impacts, e.g. oil and gas, power generation, and mining. Obvious targets for sectoral guidance include an overview of legal requirements and international standards, an assessment of significant environmental effects, or guidance on procedural or technological controls.

☐ *Financing Implementation of ISO 14001*

This might be achieved through lending instruments such as Pollution Control and Environmental Management Funds, targeted at SMEs in the sectors likely to be affected by ISO 14001. This would be an appropriate mechanism for providing small-scale loans (US$5,000–US$5m) to companies intending to implement an EMS or become registered to ISO 14001. The funds might be managed by public or private financial intermediaries.

■ Conclusions

In conclusion, it appears likely that ISO 14001 will have far-reaching effects within certain sectors or regions that will encompass LDCs. In particular, supply-chain pressures may lead to *de facto* trade barriers in some instances.

Undoubtedly, LDCs will require assistance in coming to terms with these effects, either for establishing the necessary infrastructure in support of ISO 14001, in having access to sectoral guidance in relation to EMSs, or in financing implementation of EMSs (including ISO 14001).

Finally, the development assistance community has an important role to play in assisting LDCs in coming to terms with these effects

■ Notes

1. United Nations Industrial Development Organisation, *Trade Implications of International Standards for Quality and Environmental Management Systems* (Geneva, Switzerland: UNIDO, 1996).
2. See European Commission, *The Community Eco-Management and Audit Scheme: An Overview, Progress to Date and Current Issues* (Directorate General XI; Brussels, Belgium, 1996).
3. See Mobil, *The Mobil Survey of ISO 9000 Certificates Awarded Worldwide* (5th edn; London, UK: Mobil Europe Ltd, 1996).
4. Associated Chambers of Commerce and Industry of India, *Directory of ISO 9000 Certified Companies in India* (New Delhi, India: ASSOCHAM, 1996).
5. See United Nations Conference on Trade and Development, *Newly Emerging Environmental Policies with a Possible Trade Impact: A Preliminary Discussion* (Geneva, Switzerland: UNCTAD, 1995).
6. See The World Bank, *ISO 14001: Environmental Management Systems* (LEGLR Advisory Note, No. 10; Washington, DC: World Bank, 1996).
7. See T. Feldman *et al.*, *ISO 14000: A Guide to the New Environmental Management Standards* (Chicago, IL: Irwin Professional Publishing, 1996); United Nations Development Programme, *Environmental Management Standards and Implications for Exporters to Developed Markets* (New York, NY: UNDP, 1996).
8. International Organisation for Standardisation, *Directory of Quality System Registration Bodies* (Geneva, Switzerland: ISO, 1992).
9. See R.B. Toth, *The Role of MSTQ in Sustainable Industrial Development* (unpublished; R.B. Toth Associates, 1995).

● The views expressed in this chapter are those of the author and should not be attributed to The World Bank.

10 Training and Environmental Management Systems

Andy Wells

ENVIRONMENTAL management training has been found to be an area of deficiency in commercial training provision. For comparison and reference purposes, the development of environmental management systems (EMSs) for the international arena can be viewed in microcosm by reference to their historical development in the UK, where standardised EMSs have been formally developed since 1990. This training shortfall was particularly highlighted in the UK by a company survey undertaken by the Centre for Environmental Education (CEE) as early as 1987[1] well in advance of the initial developmental stages of ISO 14001, the early drafts of BS 7750 and the European Union's Eco-Management and Audit Scheme (EMAS). This was further substantiated by a subsequent survey six years later identifying the low level of financial resources allocated by business to this area of training.[2]

Despite many companies affirming the provision of some type of environmental training, this could often be considered as inappropriate or perceived as mere lip service, as, in practice, its provision was often patchy, inconsistent and inaccurately targeted.[3] The potential consequences of this accorded well with spurious, confused and often misleading claims made, mostly for product marketing purposes, by many international companies. As an example, these claims included the incorrect statements that '... now being able to run a vehicle on unleaded petrol would help prevent depletion of the ozone layer...', and that adding a catalytic converter to a vehicle would cause it to become 'an air freshener'.

The CEE survey[4] found that formal training was available irregularly and only made available to a small percentage of employees: the limited resources allocated to the area of environmental management training resulted in it being a very minor component of organisations' total training provision, although the level of this varied between industrial sectors.

The amount of work time dedicated to any training can vary considerably between organisations. In a further commercial survey,[5] it was found that 62% of the companies sampled undertook 'some form of training', although this was described as being of 'a general nature' and was in total allocated only approximately 2.5% of the time. The phrasing of this survey question permitted a wide interpretation, and it must be stated that, even if these figures apply only to environmental personnel, companies responding under a questionnaire heading of 'General Training' would also be likely to include induction training for new employees. Hence it is also reasonable to assume that the 62% figure quoted by the survey would therefore not have been truly representative of any workforce upskilling in non-traditional procedural techniques such as environmental management systems development.

In fact this survey found that environmental training represented 'only a small proportion' of total training, and, not surprisingly, concluded that environmental training will 'be taking an increasingly prominent role in the training provision of companies'.[6]

■ Training within an Environmental Management System

Effective communication is an essential element of good training, where concepts and actions are clearly communicated from the trainer to the trainees and are ultimately acted upon in the workplace creating a practical change.

ISO 14001 Environmental Management Systems: Specification with Guidance for Use is the unifying international standard for environmental management systems. However, it is likely that there will be differing interpretations of both the letter and spirit of the standard, both in its implementation and auditing by organisations on a company- to country-wide level. In the absence of universal guidance and control, this could result in differing interpretations of competency, the need for training and for training provision.

ISO 14001 categorises both training (Clause 4.4.2) and communication (Clause 4.4.3) as essential components in the implementation and operation of an environmental management system. It further requires commitment from the top management of implementing organisations, with regard to their environmental performance; practical commitment as a motivating force for the development of an environmental management system.

☐ Commitment

Levels of commitment throughout the organisation are necessary to implement fully an environmental management system. At the director level, practical experience has shown that commitment can be essential, for example to initiate strategy, outline policy and allocate appropriate financial and personnel resources. Directors are often responsible for policy decisions within many organisational structures; however, the origination and implementation of procedures to achieve those policies are often performed at the managerial level through the communication of work instructions to, and supervision of, the organisation's workforce. Environmental policies need to be fully understood at the managerial level and effectively communicated to the organisation's workforce and contractors through training programmes supported by other methods of internal communication.

☐ Internal Communication

An effective internal communication procedure is necessary for the implementation of an environmental management system. However, the communication of a policy cannot guarantee the translation of that policy into action. There can be various stages in this communication process from concept to reality as exemplified by Figure 1. All of the stages in Figure 1 regarding the internal communication of ideas into action can be dependent upon the knowledge levels, competency and motivation of individuals within this chain of translation. Additionally, and often of equal importance in effective communication within organisations, there is the need for good upward as well as downward communication. This can influence the development of environmental performance objectives through changes suggested in actual practice. Environmental management can often be most effectively improved

1. Policy	The strategy
2. Practice (perceived)	That which is believed to be done, through the production of procedures
3. Practice (actual)	That which is actually done through the application, understanding or misunderstanding of the work Instruction

Figure 1: *The Staged Process of Internal Communication*

through achieving staff involvement, ownership and meaningful involvement in an environmental management system.

☐ The Training Need

The process flow from concept to reality also requires the necessary competency to achieve the desired end point. Unlike the implementation of a quality management system, where it would be reasonable to assume that individuals within the organisation could confidently define 'quality' or 'replicability' specifications for their products or services, some of the elements of an EMS require a degree of environmental knowledge or a practical subject-specific ability, in addition to requiring the provision of systematised management expertise.

It is less likely that an organisation's personnel resources will be able to respond as readily to provide the subject-specific environmental skills required to implement an environmental management system. For many companies, these subject-specific skills have often been acquired, at least in the first instance, by the appointment of external specialists. In the UK, many organisations have utilised external consultants to a greater or lesser extent in the development of their environmental management systems.

However, in the medium to long term, many such organisations have further developed their strategies, often in line with their business requirements, and are upskilling their staff through environmental training for both general and specific purposes.

Environmental training for an organisation's staff is a requirement of the environmental management system and complementary standards (ISO 14001, ISO 14004, ISO 14010, ISO 14011 and ISO 14012; see page 43). For example, ISO 14001 Clause 4.4.2. defines the requirement for staff training:

> The organization shall identify training needs. It shall require that all personnel whose work may create a significant impact upon the environment, have received appropriate training.
>
> It shall establish and maintain procedures to make its employees or members at each relevant function and level aware of
>
> a) the importance of conformance with the environmental policy and procedures and with the requirements of the environmental management system;
>
> b) the significant environmental impacts, actual or potential, of their work activities and the environmental benefits of improved personal performance;
>
> c) their roles and responsibilities in achieving conformance with the environmental policy and procedures and with the requirements of the

environmental management system including emergency preparedness and response requirements;

d) the potential consequences of departure from specified operating procedures.

Personnel performing the tasks which can cause significant environmental impacts shall be competent on the basis of appropriate education, training and/or experience.

This ISO 14001 requirement places a greater emphasis on the staff training need than BS 7750 previously did, although the required areas of awareness demonstrate a remarkable similarity.

Many companies or organisations do not have an exclusively self-contained workforce, but rely to a greater or lesser extent on contracted-in expertise or labour to perform some of their operational functions. Clause 4.4.6c of ISO 14001 requires that contractors and suppliers be also made aware of the organisation's environmental management system and be communicated the relevant operational procedures and requirements that relate to the organisation's identifiable significant environmental aspects.

Many environmental management systems have initially received criticism for shortfalls in the awareness, not only of staff, but also of contractors, with regard to the organisation's environmental management system requirements. So, whereas some type of environmental management system awareness training of contractors was a specified requirement of BS 7750, it would also appear to be an implicit need within ISO 14001, as the control of contractors' activities can be a key issue for the success of an organisation's environmental management system.

While the complementary international standards ISO 14010, ISO 14011 and ISO 14012 outline the training need to achieve and maintain necessary competencies in the fulfilment of environmental management system roles, in Europe EMAS[7] is less specific with regard to training. EMAS requires that management, in co-operation with all levels of its workforce, should foster a sense of responsibility for the environment. This is the only approach to specify workforce collaboration as a need, an issue arguably lacking from the environmental management system standard approaches (ISO 14001: 1996; BS 7750: 1994).

The active participation of adults in their own learning experience can often be a prerequisite for their ownership of, and involvement in, concepts such as environmental management systems. This European approach was not reflected in the Environmental Management Systems International Standard ISO 14001, despite an unsuccessful request by Germany to include a requirement that training needs should be identified in consultation with employees.

■ Training Evaluation

The evaluation of training and relating it to environmental performance is of key importance. Although both ISO 14001, and BS 7750 before it, specified training requirements, in practical terms there can be seen to be a need for:

1. Training in the procedural requirements of the standards
2. Practical training of real value in terms of environmental performance and environmental merit

Real value in terms of improvement in environmental performance can be automatically self-referential within a company, and can be audited through the EMS. Practical training in terms of environmental merit can be wide-ranging, but should expand the environmental competency level and knowledge base of the company's personnel, and in many cases can be of direct environmental benefit in terms of increasing understanding and awareness. Evaluation of training can help identify the outcomes of training programmes and determine if practical meaningful actions are being taken as a result.

There are process as well as subject issues here. The quality of the training and the effectiveness of the communication is important, not just the subject area covered. Inappropriate evaluation of training outcomes could well corrupt the process by allowing inaccurate conclusions to be drawn from incorrect or incomplete data on competency. Evaluation should fundamentally pose the question: 'Of what value is the training to the environmental performance of the business as well as to the environment *per se*?'

In considering the above, there are many different ways in which the environmental value of the training may be assessed: for example, by the organisation's stakeholders and interested parties, who may well have different perceptions or points of origin. 'Stakeholders' and 'interested parties' in this sense are those who have an interest in the operation of an organisation or are affected by its environmental performance, and may include an organisation's customers, regulators, bankers, shareholders, local community and employees, among others.

■ Types of Training

In developing an environmental management system, there is a need for environmental awareness training for all staff, a point also relevant to the standardised methodologies (ISO 14001: 1996; BS 7750: 1994). This need has been broadly interpreted and has been fully translated into action in some organisations. In addition to general awareness training, procedural and specific training will also be required for specialist activities (ISO 14012) in the process of EMS implementation and monitoring. Additionally, specialist training may be required to develop the competence of third-party auditors and certification body assessors.[8] These different areas and types of training need, are summarised in Figure 2.

☐ General Environmental Awareness Training

Staff should be trained in general environmental awareness, awareness of the general concept of EMS, the role and function of an environmental management system within their organisation, and their individual and collective contribution to it.

The general awareness training should encompass a wide range of environmental issues, from global to local in significance, eventually focusing down to the individual level. As with most effective communication, the vocabulary, terminology and concepts should be presented in a form accessible to the participants. Environmental management systems have been represented on one such course as a woolly lion self-assembly figure to represent, demystify and dejargonise the communication of an organisation's environmental policy to its workforce. Far from being gimmicky, such innovative approaches can achieve high success in inspiring interest when communicating such non-traditional operational concepts as environmental management systems.

☐ Procedural Training

Procedures to maintain and monitor an environmental management system, such as activities that minimise site environmental impacts, and methodologies for the undertaking of work instructions, may require specific training programmes. The exact nature and type of training for the implementation of procedures and work instructions will vary from organisation to organisation dependent upon its environmental management system policy and objectives and any standard or external specification to which it is operating. Such a training need may be process-related, such as a practice for minimising waste production or energy use when undertaking a specific activity; or may be more directly related to environmental management, such as a water sampling procedure for monitoring purposes.

This chapter does not attempt to cover the detail of this area of training due to its bespoke nature to each organisation. However, the general concepts enshrined in the environmental management system standard (ISO 14001,

General	Awareness of: • Global, regional and local environmental issues • Legislation requirements • Industry and the environment • Pollution • The role of the company • What an individual can do • Awareness of an environmental management system
Procedural	Specific to skills in the undertaking of work instructions. May be process-related: such as a practice for minimising waste production or energy use when undertaking a specific activity; or may be more directly related to environmental management, such as a water-sampling procedure for monitoring purposes.
Specialist	This is defined as specialist to the environmental management system. Such activities include site environmental skills, knowledge of environmental legislation, management and system development skills, environmental management system auditing skills.
Assessment	Particular to the environmental management system assessment procedure, as undertaken by third-party auditors or certification assessors. An extension of the skills and experience in this category may also include verification/validation skills for environmental statements, etc.

Figure 2: *Environmental Training within an Environmental Management System*

Clause 4.4.2[d]; and previously BS 7750, Clauses 4.3.4[d]) require procedures to be in place to ensure awareness of the potential environmental consequences of departure from specified operating procedures. By implication, understanding of, and competence in, the performance of work instructions is expected. Training may therefore be necessary on an ad hoc basis to ensure an adequate level of staff performance in line with the environmental management system requirements. ISO 14001 (Clause 4.4.2) further specifies that personnel performing tasks which can cause significant environmental impacts shall be competent on the basis of appropriate education, training and/or experience. As well as practical and manual activities these procedural areas may also cover specific managerial functions such as the project management responsibility for overseeing the implementation of an environmental management system.

Those delegated responsibility for environmental management within organisations are often drawn from, or also perform, other functions such as quality and health and safety management, and as such may require subject-specific environmental management, or at least an updating, training input. This point is further substantiated by a 1993 survey of 210 environmental managers,[9] which found that only 39% of the sample surveyed had had any prior environmental management experience.

The role of environmental manager was also identified as a recent phenomenon, with 66% of the respondents appointed since 1990, a fact justified as being influenced in the UK by legislation, in the form of the Environmental Protection Act (1990). The survey identified a trend of change in activity away from a more traditional legislative compliance function towards incorporating a more proactive role in the implementing of environmental management systems, the environmental auditing of suppliers and waste production, and the co-ordination of organisational training. However, a significant shortfall in training and education of environmental managers was also illustrated, with 61% of the sample having had no environmental content in their basic education and yet receiving little in the way of subsequent practical training.

This apparent lack of an academic environmental background would seem to illustrate a disharmony between industrial expectation and educational provision as highlighted in the Toyne Report,[10] which recommended more emphasis on cross-discipline environmental education and the provision of courses better targeted to the needs of industry. Procedural training within an EMS should include an awareness of the environmental benefits of improved personnel performance as required by ISO 14001 (Clause 4.4.2[b]). This clause accords well with comments made in the Toyne Report:

> Everybody has some scope for doing his or her job in a more environmentally responsible way, and needs to understand the importance of this.[11]

and is further supported by a Confederation of British Industry Guide[12] which translates to the practical situation in asserting that employees should be able to perform their work activities with the least level of environmental risk.

Procedural training for performance in the normal operational scenario is extended to cover the need for training for employee awareness of roles and responsibilities, including emergency preparedness and response requirements (ISO 14001, Clause 4.4.2[c]). Although some of the environment components of ISO 14001 are perceived as considerably weaker than those of BS 7750,[13] with regard to environmental training ISO 14001 displays an improvement on BS 7750. Within BS 7750 (Clause 4.4.2) there were the requirements to consider environmental impacts as a consequence of normal, abnormal and emergency conditions, but without a necessarily complementary requirement for training on how environmental performance may change in these situations.

■ *Specialist and Assessment Training*

The UK EMS standards model is acknowledged as having a successful history regarding development, pilot programming, implementation, assessment and certification that can also provide a meaningful learning experience due to the historical duration of the process. There is currently no universally accepted formal international scheme for the recognition of training, background and experience for the registration of environmental management system certification assessors and/or for professional environmental management system auditors.

Internationally, environmental management system certification is supervised by the appropriate accreditation bodies. In the UK, the United Kingdom Accreditation Service (UKAS), formerly the NACCB, has a role that includes the accreditation of the range of certification bodies assessing environmental management systems to ISO 14001 and/or EMAS. The role of UKAS is in turn viewed by its European peer group organisation the European Accreditation Council (EAC). The key professional bodies, including the Environmental Auditors Registration Association (EARA), the International Register of Certified Auditors (IRCA)—formerly the Registration Board of Assessors (RBA)—and the Institute of Environmental Management (IEM), are all involved in the UK, but no specific outcome is yet clear regarding this formal recognition and registration. Indeed, whether the accreditation bodies will be the maker, or merely the conveyor, of any decision may well also be a matter for current debate, as it was previously regarding the UKAS qualifications to assess the environmental expertise levels for BS 7750 certification and EMAS verifications.[13]

UKAS are currently leaving the professional bodies in the marketplace to devise appropriate registration schemes, and this process is reflected in many other countries. EARA have produced an international Advanced EMS Auditing training syllabus to cover third-party environmental management systems audits, with differing requirements to permit approval for specified countries. IRCA have also responded by producing a syllabus for training in the area of EMS assessment, whereas the IEM training scheme is perhaps more focused on the competency requirements of its membership, and, as such, does not present a registration specifically for environmental management system auditors or certification assessors. There is therefore some level of competition between the professional bodies for the recognition to register environmental management system auditors or assessors.

However, the publication of the international standard ISO 14001 affects the internationalisation of environmental management system recognition. It is likely that any organisation approved for the registration of environmental management system auditors or certification assessors may also itself require an international presence and international recognition to be seen as globally credible, as well as to help allay some of the suspicion with which environmental management system standards have been viewed by many developing countries.

■ Conclusion

If the assessment of an environmental management system standard is not sufficiently rigorous, and as it may be reasonable to assume that organisations will generally wish to minimise resource and financial inputs due to business pressures, then a trend for environmental management implementation to be demand-led by the competency and performance levels to which accreditation, certification and verification bodies operate, may well develop. Therefore there is a need for the setting of standards of expectation through the thoroughness of practice operated by these bodies. Internationally, the ultimate controller of this would be the accreditation bodies for ISO 14001, who would also need to operate some level of regulation to ensure international consistency in their levels of performance in this field.

Commercial and political pressures could well run contrary to this argument due to a desire initially to certify a number of organisations to the standards or specification requirements (similar to the marketing 'minimum stock' theory), and to create a momentum and establish the standards within business considerations. This could well prove counter productive to the setting of best practice standards in auditing and implementation. In the UK, the variable quality of the early-validated environmental statements for EMAS can be suggested to provide testimony to this. It is difficult to identify a similar trend in certification to environmental system standards, as the public disclosure requirements are for a much less revealing level of documentation.

In the main, training provision at the assessment/specialist level for the setting of training and performance standards in the auditing of environmental management systems is currently left to the competing professional bodies, operating under market-forces pressure. Although this may offer a comprehensive coverage of subject matter and approach, it also presents a confusing fragmentation of methodologies and could well prove counter-productive to the environment element, due to the competitive nature of the marketplace.

There is a need for all staff and an organisation's contractors to receive environmental awareness training, at all corporate levels, and a corresponding need for standardised guidance to ensure the content encompasses a wide range of environmental issues from global to local in significance, and eventually concentrating down to the actions of the individual. Procedures for the provision of environmental awareness training (although no major guidance for its content, quality or duration) are specified in standardised approaches to environmental management systems (ISO 14001: 1996; BS 7750: 1994).

Co-ordinated assistance with the procedural and environmental awareness levels of training need is virtually non-existent and is left to individual company attitudes and consultants to fill the market gap. Procedural training need can vary considerably from company to company, even within the same industrial sector; however, environmental-awareness-level training provision could be standardised to some degree and receive recognition through, for example, international commercial bodies. The provision of both procedural and environmental awareness training could then be monitored and tailored to meet specific sectoral needs through appropriate industry trade bodies.

Also of key importance is environmental management system training, which could cover areas of practical system implementation, site review and internal auditing. To some extent, these areas are covered by the professional bodies, such as by the EARA Approved EMS Implementation training course syllabus. However, there is a need for greater co-ordination and consistency of international provision by the universal recognition of schemes for the standardisation of these courses. Hence the evaluation of training and relating it to environmental performance improvement is of key importance.

Although the accreditation bodies have a remit to monitor, there is a lack of opportunity for moderation. Moderation of the quality of delivery of the environmental management system certification process and of the consistency in the practical manifestation of this assessment process between certification bodies would help ensure perceived, observed and actual equity of the EMS assessment process. This shortfall also serves to strengthen the recommendation that there is a need for criteria to be established outlining the competency requirements at all levels of environmental management system implementation and auditing.

A standardised recognition of training requirements, including a focus on the environmental subject-specific elements and appropriate continuing professional development, is needed to support the essential establishment of a single international registration scheme for environmental management system auditors and assessors. Such a register, with stated training and environmental competency requirements, is greatly needed to prevent competitive fragmentation, and national or international devaluation of the benefits of the international environmental management system standards.

■ *Notes*

1. S. Mathrani and the Council for Environmental Education, *Environmental Education and Training in Industry* (Reading, UK: CEE, 1987).
2. David Bellamy Associates, *Environmental Training Needs* (Durham, UK: DBA, 1993).
3. See Mathrani and CEE, *op. cit.*
4. *Op. cit.*
5. DBA, *op. cit.*
6. *Ibid.*
7. Commission of the European Communities, *Community Eco-Management and Audit Scheme, Official Journal* L168 Vol. 36, 10 July 1993 (Council Regulation [EEC] No. 1836/93; Brussels, Belgium: EEC, 1993).
8. UK Accreditation Service (formerly NACCB), *Environmental Accreditation Criteria* (London, UK: UKAS, 1995); UK Accreditation Service, *Addition of ISO/DIS 14001 Certificates to Accredited BS 7750 Certificates* (London, UK: UKAS, 1995).
9. Environmental Data Services, 'Environmental Managers in Business', in *ENDS Report*, No. 225 (October 1993).
10. Department of Education, *Environmental Responsibility: An Agenda for Further and Higher Education* ('The Toyne Report'; London, UK: HMSO, 1993).
11. *Ibid.*
12. Confederation of British Industry, *Environmental Education and Training: Guidelines for Business* (London, UK: CBI, 1993).
13. See Environmental Data Services, 'Weak ISO Draft Threatens Europe's Environmental Management Standard', in *ENDS Report*, No. 240 (January 1995).
14. See R. Hillary, *The Eco-Management and Audit Scheme: A Practical Guide* (Letchworth, UK: Technical Communications Publishing, 1993).

11 Environmental Management Standards: Who Cares?

Andrew Blaza and Nicky Chambers

IN THE EARLY 1990s in the UK, there was a rapid uptake—by pioneering companies at least—of strategic approaches to the environment, which led to the development of systematic management processes and hence standards such as BS 7750, ISO 14001 and EMAS. While many of the same pioneering companies are implementing these environmental management systems (EMS) standards, the uptake has still not been widespread, particularly among smaller companies (SMEs). There has been a move on the part of some larger businesses to extend environmental responsibility through their supplier chains, but in general there has been no great influence on the purchasing behaviour of the final consumer.

The consumer pressure that initiated the process in the late 1980s is now less apparent. Public concern about environmental issues is still evident, as shown by the 80% of respondents who expressed concern in a survey conducted by Harris in 1995.[1] However, public confidence in corporate environmental programmes does not appear to be providing sufficient consumer response to encourage more companies to move further. Even the so-called pioneers are beginning to ask themselves why they are doing it, who really cares and where the competitive advantage has gone that these programmes offered at the outset.

The apparent lack of positive public response in choosing products and services from environmentally-responsible suppliers could be the result of three processes:

- The recognition by consumers that, in the past, some companies have indulged in 'greenwashing'—making superficial claims about their environmental programmes and products—and consequently assuming that all companies operate that way.

- Failure, even by some progressive companies, to communicate effectively their achievements and thereby to reap the rewards of their endeavours through increased consumer support. Environmental reporting, where practised at present, tends to be one-way and not designed to achieve feedback from the target audience. Most reports are not even aimed at the general public or final consumer.

- The lack of a mechanism to provide access to really credible, balanced information on total company performance, i.e. beyond the current trend of simply product-based 'eco-labels'. In the absence of such information, most consumers are unable to make informed purchasing decisions that positively favour the more environmentally-responsible companies and their products. At best, action is restricted to boycotting companies known to have a poor record on the environment.

In this chapter we trace the recent development of environmental management systems and standards in the UK. We summarise past consumer opinion surveys on business activity on environmental issues. We suggest ways in which companies can mobilise their ultimate customers to influence changes in lifestyles and achieve future patterns of production and consumption which will be the basis of 'sustainable development'.

■ The Battle Lines are Drawn

Throughout the 1980s, and even before, much of industry was seen as the 'villain' in matters concerning pollution and the environment. Campaigning environmental pressure groups regarded industry as the 'enemy' and there were calls for increased environmental legislation to force companies into action in the mistaken belief that this was the only way forward. In return, many companies that responded did so with defensive, reactive programmes designed merely to comply with such legislation. At the same time, they lobbied regulators hard, emphasising the costs and burdens of this increasing legislation and the effect it would have on jobs and on the national economy.

There is no denying that change was necessary; however, confrontation was hardly the best way to encourage it. Industry, while seen as the cause of the problem, was rarely regarded by those outside it as part of the solution. The battle lines were drawn, and communication between industry and environmentalists amounted to little more than shouting at each other across a vast chasm with little obvious benefit to the environment. In this atmosphere, few companies became engaged in creative programmes and fewer still were willing to discuss their production processes and operations, even with their own staff, let alone the communities surrounding their plants, NGOs or the public at large.

■ *And Then There Was Light!*

In early 1990 in the UK, there was an atmosphere of change. Positive-minded factions within certain environmental groups, including Greenpeace and WWF-UK, as well as representatives from some parts of industry, including the Confederation of British Industry (CBI), independently realised that this state of affairs would lead nowhere. It was recognised that no amount of campaigning in isolation, even by dedicated and well-meaning environmentalists, would bring about environmental change. It was not realistic to expect legislation alone to achieve the kind of changes needed in the time available, because legislation on its own does little, if anything, to change corporate cultures or personal lifestyles. What was needed was a new form of partnership between all concerned, in a spirit of true collaboration, not confrontation, which would encourage business through 'voluntary' actions that went way beyond mere compliance with legislation, to develop innovative action programmes aimed at improving and enhancing the environment for the good of all.

Out of this new thinking was born, among other things, an initiative by the CBI under the slogan 'Environment means Business', with the aim of changing the way UK businesses addressed environmental issues. Far from a reactive, defensive response, the new approach sought to identify opportunities to bring benefits to the business from improved environmental performance. It aimed to encourage companies to adopt a strategic, 'proactive' approach, without waiting for legislation or any other external agency to force action.

So what were the incentives to follow these recommendations? Principally, there would be enormous short-term cost savings as well as longer-term benefits that would contribute to the future viability of the business. The underlying need was to establish confidence among a constituency much wider than the traditional target audiences of customers and shareholders—in fact with anyone who had an interest in the business. The term 'business stakeholders' had arrived and there were now three major reasons for taking action on the environment: the law; cost reductions; and stakeholder demands. It soon became apparent that there is nothing quite like enlightened self-interest to encourage business people to act.

When the CBI published its simple practical guide to environmental management, *Narrowing the Gap: Environmental Auditing Guidelines for Business* in

June 1990, there was a call from the business membership for a systematic management process that companies of all types and sizes could adopt, safe in the knowledge that the system, and, more particularly, the outcome from it, would be accepted by the world at large.

Discussions followed immediately with British Standards Institution (BSI) which led to the publication of the world's first standard for environmental management systems, BS 7750, and subsequently to its replacement on the international stage by ISO 14001. The rest, as they say, is history—or is it?

■ *Management System or Performance Target?*

From the start of the two-year development of the proposed 'standard', some people from within the business community and without believed that the acquisition of certification to the published standard would be a quick way to establish environmental credentials that would persuade regulators and customers alike somehow to trust the company with all matters environmental. Still to this day, there are many eminent people who confuse environmental management systems standards with environmental performance targets, often unfortunately called 'environmental standards'.

Admittedly, written into BS 7750 from the beginning was the need for 'continual environmental improvement'. But the existence of a certification to the standard is not in itself a guarantee of outstanding environmental performance. What it does indicate is the existence within the organisation of a process that can be geared to achieve stated, and preferably published, environmental improvement targets. It is the setting of those targets themselves that ultimately determines the environmental performance of the overall operation. If the targets are too low, little will be achieved, either for the business or for the environment, even though the organisation could now boast ISO 14001 on all its promotional literature.

While there are some who would still argue for much wider representation in the technical process for developing environmental management systems standards, it is surely much more important for there to be wide-scale consultation and involvement in the establishment of the actual *performance targets* which a company sets itself to be 'managed' within the system. Furthermore, far more people outside the company should be concerned about whether or not these targets are actually achieved rather than necessarily how they are achieved.

■ *Now That I've Got It, What Do I Do With It?*

Many companies are not ready for the somewhat formal rigours of the management systems proposed in the currently available standards, whether ISO 14001 or the European Eco Management and Audit Scheme, EMAS. Often a more ad hoc approach is adopted as a way of getting started. This is fine, if the process is compatible with, and provides a means of preparing for, the more formal standards approach at some future point. Maximum credibility will be achieved, however, if the management system can be externally verified, or certified, which is certainly the case with the established standards.

A company that is committed to achieving high environmental ideals needs to be rewarded for its past efforts and encouraged to go further in the future; otherwise, why do it at all? Beyond the internal benefits that arise from issues such as energy efficiency and waste minimisation, which bring significant cost savings, what benefits are there if your competitors act like proverbial cowboys and suffer little at the hands of the market? Yet credit cannot be given unless progress in implementing the environmental management system, and the results it achieves, are communicated to all stakeholders, their comments invited and the feedback used to modify or improve future management plans. Gaining a certification to ISO 14001 or EMAS is only part of the ultimate goal, and although EMAS does go some way to encouraging reporting of performance on the targets set, the emphasis is on disclosure rather than positive two-way communication. It is not so much a case of what you do within your established environmental management system—important as that may be—but rather how you communicate the results of the process to extract maximum benefit both inside and outside the organisation.

The complete process of preparation, operation and communication within the context of an environmental management system is represented in Figure 1.

■ *Perception is Everything*

It is no longer sufficient to do as the agrochemical industry did in the UK, for example, in the 1970s. For the average member of the public at that time, agrochemicals were the 'number one polluters'; the perception was that they polluted when they were manufactured, polluted when they were applied to farmers' fields, and polluted through the residues left in the food on the supermarket shelves. What the industry had failed to do was to communicate the facts adequately and invite feedback as to whether this information was sufficient to allay genuine fears that an otherwise uninformed, even ill-informed, public might have had.

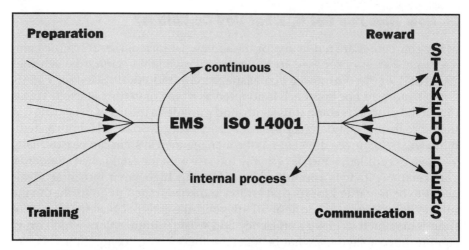

Figure 1: *Preparation, Operation and Communication within an EMS*

More recently, in the absence of widely perceived environmental performance improvements from a number of industries, the regulators, including the European Commission, have stated quite clearly that they will no longer be able to resist the demands for more environmental legislation by the critics of industrial voluntary initiatives. This is especially true where there is little *visible* evidence that these initiatives are being taken up by the vast majority of companies. 'If they will not do it voluntarily, then force them to do so through legislation' is the quite understandable cry. Yet legislative measures—the 'sticks'—apply to everyone. There is usually no attempt to differentiate between those who have taken action voluntarily and those who have been content to sit back and watch in the wings. There is a case perhaps for extending the argument about creating the 'level playing field' so that it is very much tilted in favour of those who are prepared to take action beyond legislation.

The pioneers of business 'voluntary' environmental action, far from wanting to eradicate legislation altogether, are seeking to encourage regulators to introduce a supportive framework of realistic, achievable and, more importantly, enforceable legislation. This would set the minimum standards that even the cowboys must adhere to, but it should then be reinforced by creative market instruments—the 'carrots'—to encourage innovation and action beyond the minimum, rather than simply penalising apparent transgressors.

However, apart from signals from the regulators, the pioneers are also anxiously awaiting positive signals from the market which, as we shall see below, are hardly there in profusion. Here the argument begins as to whether companies should be leading or following their customers. There is no doubt in the minds of the authors that strategic companies should be leading their customers, even if this temporarily puts them out of step—dare we say, ahead—of their competitors. This is the beginning of the path towards true sustainability.

■ *Sustainable Development: What's In It for My Company?*

The real turning point for many in industry was the Shell debacle over its oil operations in Nigeria and then the proposed deep-ocean disposal of its oil platform, Brent Spar. It seemed that even a company with a proven track record on environmental management somehow could not get its communications strategy right.

For some time now, the debate has been moving on from the relatively low-level one about the 'greening' of business, where many companies choose their particular shade of green, sometimes unfortunately based on how little they think they can get away with, or more particularly, what the competition is currently doing! The emphasis has progressed towards the more strategic concept of 'sustainable development'. Here the debate enters a new realm, for there really are no 'shades' of sustainability. What is variable, if anything, is the speed with which we, whether as a society or as an individual business, decide to try to achieve sustainability.

As the environmentalist Jonathan Porritt put it recently in his regular feature in a popular journal: 'If something is sustainable, it means that you can go on doing it indefinitely. If it isn't you can't.'[2] Perhaps it would have been more appropriate to reverse the second phrase as a test for deciding whether or not a particular activity is sustainable and say, 'If you can't, it isn't'!

In the formula for summarising the process of environmental reviewing included in the CBI publication *Narrowing the Gap*, we posed the following questions to be asked by any business of its operations:

> What are we doing?—Can we do it differently?—Can we do it more efficiently?
>
> Can we do it more cheaply? Can we do it with less impact on the environment?

To these should now perhaps be added the question:

> **Can we do it indefinitely?**

In other words, is our business *sustainable*? Is it environmentally as well as economically *viable* in the long term?

Half the battle, of course, is defining the concept of sustainability in terms that business can understand and act upon. At the time of the UK Government's 'Partnerships for Change' conference in September 1993, there was no really practical definition by which business could operate. Consequently, even the representatives of companies that were trying to adopt meaningful strategies for environmental issues were declaring that social and economic issues affecting, for example, local communities in developing countries where they operated, were the responsibility of the local governments. Their companies were, after all, operating according to 'licenses' granted by those governments.

As the discussions within the conference progressed, they soon realised that this was too unrealistic and simplistic an approach to the question. A possible definition of the 'sustainable business' was offered by the author and it is repeated again here by way of re-emphasising the importance of communication:

> The 'sustainable' enterprise conducts its business in such a way as to meet the needs and demands of all its stakeholders (including customers, shareholders, bankers, insurers, employees, suppliers, regulators, media, environmental groups and local communities).
>
> One of the keys to success will lie in establishing, through effective communication, the confidence of all these stakeholders in the management's ability to address the potential environmental impacts of the total business operations in a positive, strategic way, seeking to minimise or preferably eliminate the negative impacts.

■ *Communicating beyond the World of Business*

Environmental reporting began in the UK in the latter part of 1990 with the publication of the Norsk Hydro UK report, soon to be followed by a host of other pioneers, many of whom won awards for their efforts. But reporting internally to staff, even to some of the external stakeholders, including business customers, financial institutions and the 'media', is relatively straightforward, given that they probably recognise and understand much of the business language employed. However, the response needs to come from the market itself, if real rewards are to be reaped. Not only is much of the reporting hitherto not aimed at the final consumers, nor written in language that they might understand, but there has been little evidence of two-way communication with final consumers that attempts to ensure the message has been received, understood and feedback from it invited. Little wonder then that there is such scarce direct response from general consumers to company environmental activity.

■ *What Evidence is there that Consumers are Responding?*

There have been numerous market research surveys conducted over the years into so-called 'green consumerism'. Many are based on such facile questions as 'Are you concerned about the environment?' and 'Would you buy environmentally-friendly products?' Not surprisingly, the answers are usually 'yes' on both counts. But do those same consumers actually buy more environmentally-benign products and services in practice? The answer to this is not always the same, especially where those products cost more than the traditional alternatives. The situation becomes even more confused when, as happened recently in the UK, a major supermarket chain, Sainsbury, withdrew its so-called 'environmentally-friendly' range of products on the grounds that the 'regular' products were no more environmentally damaging and they were cheaper!

In one of the more comprehensive surveys of recent years, the Harris Research Centre reported in March 1995[3] the following main findings:

1. People are concerned about environmental protection but do not know what to do to help. In support of this they quoted the following data:

- 80% of respondents expressed concern, 23% of which were *very* concerned.
- 63% would like to be able to do more to protect the environment. Of those, 28% did not know what sort of things to do.

2. People are not taking action—why not?

- 46% said they had no spare time to do so.
- 35% were not sure how to go about it.
- 16% needed someone to show them what to do.

It was generally reported that cynicism and frustration were seen to be running high as a result of a perceived lack of effective and reliable information, support and encouragement for action.

3. There is a lack of environmental understanding.

- 48% of respondents expressed an interest in receiving more information.
- Of those, 59% thought requesting information through the post would be best.

4. As to where the public gets its information, the replies included the media, local authorities, the national curriculum, pressure groups, science programmes, work activities and trade unions. It is notable that communications from business were not included in this list!

In the MORI Business and Environment survey conducted in August 1995:[4]

- 74% of respondents expressed disagreement with the statement, 'Too much fuss is made about the environment these days'.
- 49% stated they bought products that came in recycled packaging.
- 53% bought products made from recycled materials.
- 69% agreed that 'British companies do not pay enough attention to their treatment of the environment'.
- Only 23% said they avoided using the products or services of a company that they considered had a poor environmental record.

The most recent information comes from a consumer investigation into marketing claims about the environment from the UK National Consumer Council in March 1996.[5] The report states:

> Overall we were struck by the low level of knowledge about environmental issues in general. We believe there is a need for a co-ordinated, coherent government information to remedy this situation.

> We found very few examples of environmental claims in the context of services, other than financial services, which are subject to separate regulation outside the terms of our brief. We have therefore concentrated on products.

The following data were quoted in response to the question, 'Which if any of the following have you done in the last twelve months through concern for the environment?'

- 45% had bought products in recycled packaging.
- 52% had bought products made from recycled materials.
- 33% had bought products in biodegradable packaging.
- Overall, 42% of respondents said they had selected one product over another because of its environmentally-friendly packaging, formulation or advertising.

In all of these surveys, what is generally missing is anything that seeks to understand the willingness of the general public to make purchasing decisions on the basis of corporate, rather than product, performance: in other words, favouring the products and services of one company against another through the perceived better overall environmental commitment and performance of the one against the other. The MORI survey came closest, but even then there was only a rather negative question about British companies in general and an equally negative one about *not* using the products of a 'poor performer'. We conclude that this is because there is no differentiation between 'green consumerism' and 'sustainable consumption'. Surveys on green consumerism tend to focus on willingness to pay for so-called 'green products' and hence the interminable debate about eco-labels. Sustainable consumption takes the issue further up the supply chain to sustainable production and service. It is a much more strategic action that requires more sophisticated decisions and the additional information about corporate performance to support them.

The general public is apparently not taking large-scale action on the environment. In fact, some consumer groups believe that the environment is not an issue for consumers at all. This is perhaps too extreme a conclusion. Lack of consumer action may be due to:

- Not knowing what to do
- Leaving it to others (environmental groups, governments and business organisations) to act on their behalf
- Believing that whatever they do as individuals will not make a difference
- Past mistrust of any business claims and therefore a reluctance to believe one company's claims above another's

■ *The Missing Link*

We now have the environmental management systems in place for companies to identify and manage all the potential environmental impacts of their business operations. We have many examples of company willingness to pioneer and champion new environmental initiatives ahead of legislation, with support among members of staff from the boardroom to the shopfloor. We have evidence that consumers are generally aware and concerned about environmental issues but that they are not sure what they as individuals can do beyond the obvious acts of buying recycled products. This enthusiasm needs to be nurtured and extended into every purchasing decision.

While we have all the means for companies to respond to the challenges that a better-informed market would bring, what appears to be missing is the mechanism to bridge the gap between the two—a positive mechanism for developing consumer response through informed purchasing decisions, based on access to information not just about the products, as with current eco-labels, but also about the overall environmental performance of the manufacturers and suppliers. By bridging the gap between customer and supplier in this way, we believe we can not only overcome the obvious frustration expressed by participants in recent consumer surveys, we can also provide a vital spur to encourage companies to progress and take further advantage over their competitors in response to positive consumer action.

For the future we believe the general public should be encouraged to look beyond the mere existence of an environmental 'badge', even one as apparently prestigious as ISO 14001 or EMAS. Positive environmental communication of the *outputs* from the management system should help provide information about exactly what environmental performance is being achieved by the organisation, its relevance to the overall protection and enhancement of the global environment, and how the individual consumer can participate to encourage and reward further action.

Three primary things need to happen if this missing link is to be activated:

1. Companies can be helped to develop more sustainable production methods and to generate innovative ways of communicating their environmental intentions and achievements through the process of the '3 **Rs**': Report – Response – Reward. This process involves determining what to communicate; how best to communicate it; and to whom; finally, how to elicit positive feedback and response from customers and thereby reap a justifiable reward in the marketplace.

2. Those same companies could also be encouraged to exert a positive influence, through their advertising and promotion, on their customers and potential customers to adopt more sustainable lifestyles.

3. A neutral, credible and balanced source of information needs to be established to provide ready access to everything a potential customer would need in making a well-informed purchasing decision.

These three activities must be linked together. Without this linkage, and the partnership approach it seeks to facilitate, we do not believe that the current discussion, where it occurs, about sustainable lifestyles or sustainable consumption and production can be much more than simple rhetoric. On the other hand, we are convinced that, provided access to appropriate information is created, companies and customers can unite to work positively towards sustainability and thereby provide a worthwhile future for arguably the most important stakeholder group of all: our children and their children's children!

■ Notes

1. Harris Research Centre, *The Harris Report* (Prepared for the UK Department of the Environment; Richmond, UK: Harris Research Centre, March 1995).
2. See J. Porritt, 'The Sustainability Story', in *Country Living*, July 1996, p. 27.
3. *Op. cit.*
4. MORI, *Business and the Environment: British Green Consumers* (MORI Survey of UK general public; London, UK: MORI, September 1995).
5. National Consumer Council, *Green Claims: A Consumer Investigation into Marketing Claims about the Environment* (London, UK: National Consumer Council, March 1996).

12 Targeting Sustainability
The Positive Application of ISO 14001

Philip Sutton

HAS YOUR FIRM adopted an ISO-14001-conforming corporate environmental management system (EMS) to help minimise *its own* environmental impact? Or is it using the EMS to help it contribute to *society's* achievement of ecological sustainability? If your answer is 'yes' to the former and 'no' to the latter, then you are missing most of the social and commercial value of ISO 14001.

ISO 14001 is a major step forward for environmental management in many countries. However, a 'default' approach to EMSs has arisen because the first firms to introduce EMSs had major direct environmental impacts, often caused by pollution, and were subject to tough regulatory requirements, high materials costs, and critical public scrutiny. This 'default' approach, however, is not suitable for firms that want to contribute most effectively to the achievement of sustainability. Alternative approaches are possible, and this chapter develops the framework of a sustainability-seeking interpretation of ISO 14001. If this framework for applying ISO 14001 is adopted widely, it will dramatically improve the chances of achieving sustainability.

This chapter is written for those who want to know how their firms could become sustainability-seeking or those who need to know what drives sustainability-seeking firms and how they can use ISO 14001. For example, it is written for:

- Firms that are 'greening up' or are considering it
- People who want to help their firm become sustainability-seeking
- Firms and other organisations that want to service the 'green' market[1]
- Firms whose customers are demanding 'green' products

- Firms who are worried that their customers might demand 'green' products
- Government agencies and advocacy organisations that want to encourage firms to become sustainability-seeking

■ Changing Times, Changing Needs

Over the last hundred years, three changes have given rise to a major increase in environmental impacts around the globe. There has been a consistent long-term trend for: (a) resource-intensive goods to get cheaper; (b) populations to grow; and (c) the scale of economic activity per head in the rich countries to rise. The use of resources has therefore risen dramatically. This in turn has generally translated into greater levels of environmental impact with a consequent need for stronger environmental laws, first at the local level, and then progressively for higher and higher jurisdictions, culminating in international law. It is now widely recognised that the historic trend of rising environmental damage and resource use is unsustainable and must be reversed.

■ Sustainability: What Does it Mean?

So there is much talk about the need for communities, governments and firms to pursue sustainable development. While it is generally agreed that sustainable development is a good thing, there is a lot of uncertainty about what it means and what its implications are. The definition of sustainable development is: 'development that brings about ecological, social and economic sustainability while contributing to the achievement of society's other goals'. Ecological sustainability involves both:

- The protection, in perpetuity, of **life support systems** (e.g. ecological processes for nutrient and water cycling, soils, the protective and climate controlling functions of the atmosphere) and **biodiversity** (i.e. species, genetic variety and ecological communities)
- The conservation of **material and energy resources**

Social sustainability involves, at a minimum, maintaining society's ability to solve major social problems and, more generally, it means maintaining at least a given level of community caring, vibrancy, tolerance and stability. Economic sustainability means that at least a minimum level of economic welfare is maintained in perpetuity. The rest of this paper will largely focus on *ecological* sustainability.

Sustainability is a *system* characteristic. An ecosystem, human community or an economy together with the associated environment can be ecologically sustainable whereas a stand-alone product, material, technology or factory cannot. A firm might be *economically* sustainable (i.e. last indefinitely), but it cannot be *ecologically* sustainable by itself since it is connected to the environment. Given the global interconnections of ecological systems, societies and economies, it is not possible to have local sustainability without global sustainability.

Since it is only *societies* with their environment that can be judged to be ecologically sustainable, an EMS can only be judged to be consciously contributing to ecological sustainability if it is focused on effectively helping society to become sustainable rather than just helping the firm to reduce its own environment impact.

■ *The Emergence of the 'Sustainability' Market*

A gradual but intensifying process of 'greening' has been under way in many markets around the world for the last few decades. This evolution has been punctuated by two major surges of growth in environmental concern. The first was in the late 1960s and early '70s, and the second occurred in the late 1980s and early '90s. The first surge alerted the community to the extent of bad environmental practice throughout industry and the community. The second surge, triggered by a deepened realisation of the global dimensions of environmental decline, alerted people to the need for sustainable development. But it will take a third surge to push societies over the threshold so that they make sustainable development as defined above actually happen.[2]

What might trigger such a third surge? The next decisive surge is likely to occur as:

- The scientific basis for major environmental threats becomes more firmly established.
- Transformative technologies spread.
- Community-based lifestyle programmes with an educational feedback-loop to business proliferate.
- Win–win macro-economic policies are developed and begin to be applied.
- Management tools for sustainability-seeking businesses become readily available.

- Key corporations break away from the pack and go for sustainability to compete.
- Media interest in these changes grows.

☐ Scientific Evidence

People are unlikely to make difficult adjustments unless they know they have to change. So, for example, the recent scientific consensus conclusion of the Intergovernmental Panel on Climate Change that 'the balance of evidence suggests a discernible human influence on global climate'[3] is an important motivator.

☐ Transformative Technologies

These have the power to change the environmental impact of industry dramatically. But, even more importantly, by altering the industrial structure, they will change the relative influence of different industry sectors and they will change public perceptions about what is possible.

For example, Amory Lovins and co-workers at the Rocky Mountain Institute have been promoting an new car design, based on a hybrid electric drive and a carbon fibre shell, that could reduce energy use by 75%–90% for the same performance level as current cars. They have attracted about 25 significant companies to invest around US$1,000 million in this development and expect to see the new cars rolling out of factories by about 2000. This technology will have very serious implications for the oil and steel industries whose products will be virtually displaced in the automobile market.[4]

Another transformative technology that is expected to come on stream by about 2000 is grid-connected solar photovoltaic electricity generation that will be price-competitive with coal-based electricity. This technology, pioneered by Pacific Solar in Australia, will fundamentally shake up the coal and electricity generation industries.[5]

☐ Community Lifestyle Programmes

Already programmes such as the Global Action Plan are spreading through Europe, North America and beyond. They involve householders in reviewing their environmental impact and their consumption patterns. Once an educational feedback-loop back to business is added to these programmes they will become a very powerful influence on 'green' product development and uptake.

☐ *Win–Win Macro-Economic Policies*

There will not be much corporate enthusiasm for the pursuit of ecological sustainability while most businesses think the economy will suffer as a result. But there is increasing awareness that win–win economic strategies can be developed to deliver superior environmental, employment and GDP results.[6] The European Union is leading this development, but interest is being generated elsewhere in the world.

☐ *Green Business Tools for Sustainability*

Businesses will not be able to become proactive in their pursuit of sustainability until tools are readily available to help them manage themselves as sustainability-seeking firms. A number of organisations are developing just such tools.[7]

☐ *Key Corporations Go for Sustainability*

With the globalisation of the economy, large multinational corporations have become the dominant players in business. Until a number of these firms break away from the pack to pursue ecological sustainability as a serious source of competitive advantage, the bulk of firms will not see sustainability as having more than public relations significance. A small but influential band of large multinationals are now moving tentatively in the direction of such a transformation.[8]

☐ *Media Interest*

It is rare for major shifts in public, corporate or government thinking to occur until the issues are picked up by the media. The globalisation of the media has gone so far that local media outlets usually only treat an issue as newsworthy after the international news networks have picked it up in what they see as the trend-setting countries.

■ *Ecological Sustainability and the Role of Firms*

To achieve ecological sustainability, society will most probably have to do the following:

- Reduce CO_2 emissions globally by at least 60% of the 1988 levels,[9] rather than the reductions of 20% or less that governments have been considering.

- Prevent the accumulation of toxic and artificial substances in the natural environment.
- Ensure that the chemical and physical composition of the atmosphere and of water stays within natural levels.
- Ensure that there is virtually no further destruction of natural environments globally.
- Restore natural habitats to the extent needed to safeguard threatened species.
- Adopt a closed-cycle economy in which virtually all wastes are recycled.[10]

What are the consequences of objectives such as these? Let us take the example of CO_2 reductions and assume that we will accommodate expected levels of population increase and economic growth and avoid the widespread adoption of nuclear power. Under these circumstances, it would be possible to achieve a 60% reduction in CO_2 from the 1988 levels if the global efficiency of resources use was improved by about 5% per annum.[11] Improvement at this rate is technically feasible, but the implied rate of social and technical change is extremely testing. Certainly, this magnitude and pace of change will not be achieved if significant forces in society are opposed to the changes, causing the reform process to become deadlocked.

So the rate and scale of change needed cannot be achieved without the support of strategic sections of industry. But, if this support is to be worth having, it has to be won without compromising the commitment to the achievement of ecological sustainability. To achieve ecological sustainability, an increasing number of firms, large and small, will have to take on an internal commitment to help catalyse the change. These firms need to become 'sustainability-seeking'. The most effective way to do this is to adopt a sustainability-seeking environmental management system. However, this is easier said than done. Almost all EMS activity at present falls into a very predictable 'default' pattern which unfortunately is particularly unsuited to the pursuit of sustainability because of its restricted focus, as is explained below.

■ A Typology of Environmental Action

Before describing the 'default' approach, a number of concepts used in the following discussion need to be explained. In this chapter, a distinction is made between three contrasting pairs of terms.

- *Negative* **environmental impacts** result in environmental *damage* or resource *wastage*.

- *Positive* **environmental impacts** result in improvements in environmental *conditions* or increased resource *stocks*.

- **'Product-and-organisation improvement' environmental management programmes** involve the reversal or prevention of negative environmental impacts caused *directly* or *indirectly* by a firm. Broadly, an 'improvement' environmental management programme could be characterised as involving 'a flight from bad practice'.

- **'Sustainability empowerment' environmental management programmes** are aimed at helping others fix their environmental problems or create environmental opportunities—through new products (including services) or through the appropriate use of influence.[12] Broadly, a sustainability-empowerment environmental management programme could be characterised as involving 'a race to sustainability'.

- *Direct* **environmental impacts** are those caused by the *physical activities* of the firm in question.

- *Indirect* **environmental impacts** are caused by others as a consequence of *decisions* taken or *influences* exerted by a firm, e.g. the indirect effects of a manufacturing firm would include consequential environmental impacts caused by its materials suppliers and those who use and dispose of its products.

■ The 'Default' Implementation of EMSs and its Limitations

Environment management systems were first developed to help firms that:

- Had large, direct and negative environmental impacts with a significant risk of public exposure
- Were subject to tough legal controls
- Were sensitive to materials costs

These firms were particularly concentrated in sectors such as chemicals, heavy manufacturing and mining. Their EMSs were introduced to facilitate major improvements in past bad practice and to put in place controls to prevent mistakes. The key issues covered were pollution, waste minimisation and energy conservation. These were the environmental issues that were most easily kept in front of operations managers on a routine basis.[13] Firms were often motivated to act because of concerns about legal compliance, risk management and cost saving: all issues that had a very direct bearing on the welfare of the firm.[14]

The 'default' approach, by focusing on *direct*, negative environmental impacts, caused firms to concentrate on their own production processes first. Some firms' EMSs eventually led them to manage the *indirect* negative impacts of their products on a lifecycle basis (product stewardship). Only rarely did the EMSs lead to the introduction of new classes of products to help customers contribute to the 'greening' of society. The 'default' approach to EMSs made sense for the heavy industry sectors, especially while firms were actively bringing their performance up to a high environmental standard. But the 'default' approach is not very relevant to a significant majority of businesses whose indirect impacts are greater than their direct impacts.[15] This helps explain why the majority of firms is taking a long time to become interested in adopting an environmental management system. Firms with low direct impacts are either unaware that they could or should be managing their *indirect* impacts, or they feel that conventional EMS approaches do not provide an appropriate framework for them.

Unfortunately, what made sense for *certain* industries for a *certain* period of time is now being generalised by many environmental practitioners as *the* way to implement an EMS in all industry in all circumstances. Indeed, many practitioners find it hard to imagine how ISO 14001 could be implemented in a non-'default' way. But there are major limitations to the 'default' approach. These are summarised in Figure 1.

■ *Opening up Possibilities beyond the 'Default'*

Far from the 'default' approach being the only way to frame an EMS, it merely represents one of four possible focuses for action (see Fig. 2):

- Production process improvement: the major 'default' focus
- Product stewardship
- New product development
- The use of influence

Also, it draws upon only one of two major sets of motivations:

- Private interest (firm's and direct customer interest): the major 'default' focus
- Society's interest

These wider possibilities can be grouped into two EMS models or paradigms:

- The current or 'default' model, driven primarily by private interest
- A preferred sustainability-seeking model, driven by both society's interest *and* private interest[16]

If EMS is driven *only* by:	Then:	So:
Legal compliance	Firms will only be sensitive to those issues that are the subject of strong, well-enforced laws.	Issues such as biodiversity, soil, and heritage conservation and urban planning will be missed.
Issues caused directly by the firm	Indirectly caused impacts will be missed (e.g. biodiversity loss, poor land-use planning).	Firms will not contribute actively to 'whole-system' solutions.
Negative impacts	Firms will not try to provide solutions to problems caused by others.	Opportunities for developing new products or new areas of business will be missed. Opportunities to advance significantly *society's* capacity to be ecologically sustainable will be lost.
Issues that generate politically-significant stakeholder reactions	Issues will be ignored until politically-significant stakeholder reactions build up.	A lot of avoidable damage will be done before action is taken. Problems that are not immediate or acute in their impact on humans or that affect a small percentage of any given population are likely to be ignored.
Significant materials costs	Firms will not take action if money cannot be saved, and they are also unlikely to take action if materials costs are a small percentage of total costs.	Large sections of industry will fail to conserve resources (even when doing so is cost-effective).
Legal compliance, risk reduction and cost-saving	Firms will not explore all opportunities for competitive advantage (e.g. non-price or opportunity rather than constraint-based advantages).	Opportunities for improving the quality of products environmentally or for developing new products or new areas of business will be missed.
Short-sighted self-interest	Firms will not exhibit corporate responsibility or enlightened self-interest and thus they will not tune in to the customers' needs as a whole in the context of their families (for people) or strategic alliance network (for firms), the local and global community and the wider environment.	Opportunities for a positive relationship between the firm on the one hand, and its customers and the community on the other, will be lost. Opportunities for product improvement or new product development based on a wider perspective of customer need will not be identified.

Figure 1: *The Consequences of Locking in on the ISO 14001 'Default'*
© Green Innovations, Inc.

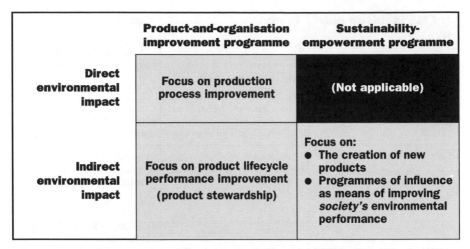

Figure 2: *The Four Potential Focuses for Environmental Management Programmes*
© Green Innovations, Inc.

The 'default' model starts with production process improvement and might eventually progress to new product development if the incentives and opportunities are strong enough. The preferred model explores opportunities first in the area of new product development and the use of the firm's influence and then progresses through product stewardship to process improvement (see Fig. 3).

New product development is needed to achieve win–win results for the environment and for economic performance. Incremental improvements in products will not achieve sufficiently large and rapid improvements in environmental performance, nor will they achieve this while maintaining the community's standard of living. The transformative technologies discussed earlier show what is needed and what is often possible when sufficient creativity is applied to the task. Under a preferred sustainability-seeking model, EMSs should not only help firms to 'do things right' (i.e. minimise the environmental impact of existing processes or products), but, more importantly, they should ensure that firms are 'doing the right thing' (i.e. producing appropriate products and exercising influence in an appropriate fashion).

So firms should design their EMSs to help them to simultaneously create maximum environmental value for society (i.e. ecological sustainability) and gain maximum competitive and strategic benefit for themselves. Their EMSs should help them develop green markets and position themselves so that they are ideally suited to satisfy these markets. Developing 'green' markets involves more than marketing and product development. Firms will also

	The 'default' model or paradigm: 'Flight from bad practice'	The preferred model or paradigm: 'Race to sustainability'
Main motivation	Primarily private interest (the firm's and the direct customer's)	Both society's need and private interest
Main drivers	Pursuit of *private opportunity* within *social constraints* set by laws and economic incentives	Pursuit of *social needs* and *private opportunity*
Typical action sequence	Production process improvement > product stewardship > new product development	New product development > product stewardship > production process improvement

Figure 3: *The 'Default' and Preferred EMS Paradigms*
© Green Innovations, Inc.

need to be active in encouraging governments to create appropriate incentive structures to help bring forth such markets.

Although ISO 14001 does not *require* an EMS to be framed in the 'default' style, the specification standard and the 14004 implementation guidelines are written in a way that powerfully, if unintentionally, reinforces the 'default' rather than a sustainability-seeking approach. Unless there is something in the ISO 14001 or 14004 procedures that can be relied upon to lead firms 'out of the box' and to a sustainability-seeking orientation, their original preoccupations (with pollution, compliance, risk and liability reduction, cost saving, and reducing the firm's direct impacts rather than helping to achieve sustainability for society) will create a lasting culture that defines and limits what environmental management means for them.

ISO 14001 builds in four mechanisms that some might argue have the potential, over time, to draw a firm across to a sustainability-seeking orientation. These are:

- A suggestion that lifecycle thinking could be used
- A requirement to comply with relevant laws
- A requirement to engage in continual improvement in line with the firm's environmental policy.
- A requirement to consider the views of interested parties

The adoption of lifecycle assessment methodologies, encouraged by ISO 14001, will widen firms' focus from direct effects to include the consideration of indirect effects. But the existing methods for lifecycle assessment cannot handle the explicit consideration of indirect impacts on biodiversity, so this issue will still be overlooked by the majority of businesses. Even if indirect effects are considered, this does not mean that their management has to become a major priority.

The ISO 14001 requirement to comply with relevant legislation and regulations means that, if society legislates to control new issues or manage existing issues differently in order to achieve ecological sustainability, firms will have to respond. However, if firms are not internally committed to promoting sustainability, they are just as likely to lobby to prevent the legislative change, while maintaining an ISO-14001-conforming EMS.

ISO 14001 requires firms to institute a programme for continual improvement in their environmental management systems *in line with their own environmental policy*. So long as firms commit themselves to legal compliance and pollution prevention, and as long as their certifiers think that their EMS policy is appropriate to the nature, scale and environmental impact of their activities, products or services, they are not obliged to take on a sustainability-seeking approach.[17] So a firm that was committed to the 'default' approach would simply have to get better at that approach to remain ISO-conforming.

ISO 14001 requires firms to determine their issues-scope and environmental directions in their environmental policy. The views of interested parties are only required to be considered when the already-determined policy is translated into objectives and targets. So, for example, an oil company that committed itself in its policy to reducing the environmental impacts *of oil production*, could ignore arguments from interested parties that the firm should shift from oil to renewable energy production when the objectives and targets of the EMS are set. Given the lack of effective drivers for the adoption of a sustainability-seeking approach in the ISO 14001 process, a firm wanting to avoid the reflex adoption of the 'default' will have to take its own deliberate and self-motivated steps to ensure that it moves in a different direction. Simply following the ISO 14001 requirements and the ISO 14004 guidelines will not be enough to turn things around.

■ Practical Idealism

Firms are unlikely to commit themselves to the alternative EMS model if they think it is 'pie in the sky'. Practical experience shows just how hard it is to make relatively small changes in products or production processes. So, how can the vast and seemingly impossible changes needed for ecological sustainability be achieved. And how can the complexity of the task be managed?

■ *Zero Defects? Impossible!*

Goal-directed innovation can be used to achieve win–win solutions in, what appear to be, impossibly demanding situations. A good illustration of this is the Japanese pursuit of 'zero defects' in manufacturing after the Second World War. To compete with the Americans, the Japanese needed to lift product quality while maintaining their cost advantage. To reduce their defect rate, American firms inspected their output. If a higher quality of shipped product was needed the Americans simply increased the proportion of output inspected, enabling them to weed out a higher proportion of defective units. However, in the process, the cost per item despatched was significantly increased. To overtake the Americans, the Japanese decided to adopt a much faster and cheaper improvement process.

The Japanese set 'zero defects' as their goal, even though it was technically and economically impossible to achieve at that time. Innovation programmes were set in motion, inspired by the goal. Rather than taking current best practice as their benchmark, the Japanese used an ideal goal and then 'reverse-engineered' a solution that approached the ideal ever more closely. Instead of controlling quality using inspection, they concentrated on manufacturing improvements to make it highly probable that the product was produced correctly the first time. This quality improvement process *decreased* costs and gave the Japanese a decisive competitive advantage.

■ *The Economics and Management Theory of Inspirational Stretch Goals*

In the *short term*, it is almost always true that the further a goal is pursued, the higher the unit costs will be (see Fig. 4 for a typical static cost curve). But if an effective innovation programme is implemented, new methods or solutions can lower the static cost curve for each time period (see Fig. 5), causing costs to plummet over time (see Fig. 6). Stretch goals are used to inspire both a steady stream of incremental improvements and leapfrog or breakthrough innovations.

☐ *Inspirational 'Green' Stretch Goals*

Du Pont, which has a 190-year history of pursuing a 'zero accidents' policy, now has an accident rate less than one thirtieth of the average for all US industry and one tenth of the average for comparable industries globally. With this experience under its belt, the company is now extending its approach

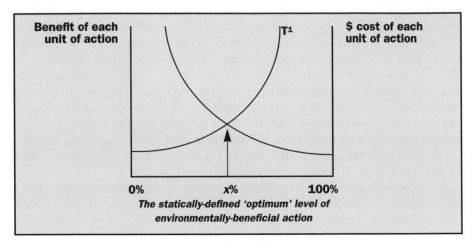

Figure 4: *The Static 'Optimum'*
© Green Innovations, Inc.

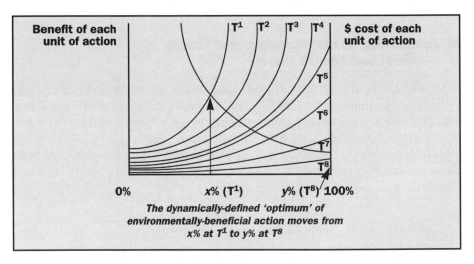

Figure 5: *The Dynamic 'Optimum'*
© Green Innovations, Inc.

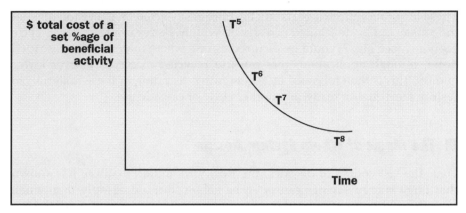

Figure 6: *The Declining Cost of a Given Level of Activity*
© Green Innovations, Inc.

to the environment. In 1994 it established company goals of 'zero waste and emissions'.[18] Many other companies have similar inspirational goals.

So, highly-demanding inspirational goals are already a reality in business. But for a sustainability-seeking firm it is necessary to reformulate the stretch goals. This is because the inspirational goals have to drive the firm's sustainability-empowerment programme and not just its product-and-organisation improvement programme. Consequently, the goals need to focus on what must be done to help *society* to become ecologically sustainable. Appropriate inspirational stretch goals for a sustainability-seeking firm would be to foster the creation of:

- A zero-extinction economy and society
- A zero-climate-damage economy and society
- A zero-soil-degradation economy and society
- A zero-damage-to-natural-ecocycles economy and society
- A closed-cycle economy and society (based on zero pollution and waste and 100% recycling of materials)
- A dematerialised economy and society (based on a stretch goal of reducing resource throughput by 90% in countries with high per-capita consumption[19]
- A renewed-resource economy and society[20]
- A sustainable-population society and economy
- A win–win economy and society that can achieve the above objectives without a major trade-off with income, employment and social justice objectives

These inspiration stretch goals will be more easily acted on if a vision of what they mean can be developed and shared widely. For example, a closed-cycle economy (see Fig. 7) could be pictured as one where materials are cascaded down through applications that can use recycled materials of ever lower quality. This is then followed by regenerative recycling of these materials to restore their quality ready for another cycle of cascade use.[21]

■ *The Magic of Whole-System Design*

Over the last couple of decades, the Rocky Mountain Institute has shown that large energy savings can often be achieved more cheaply than small ones in buildings, motors and many other technical systems through the use of whole-system redesign.[22] Others have also discovered the magic of whole-system design: that is, the potential for win–win solutions. It underpins, for example, the power of TQM techniques, business process re-engineering, Eli Goldratt's Theory of Constraints, cleaner production, eco-redesign and life-cycle assessment.

When societies, economies and technologies become complex, most people focus on incremental improvements in the parts of the system. This leaves a vast reservoir of economic and environmental efficiency gains to be made by those who can handle complexity. If firms are to improve their chances of tapping this reservoir, they must give their EMSs the widest possible scope. For example, it should cover: (a) the development of new technologies, competencies and products; (b) the use of influence to change the rules of the

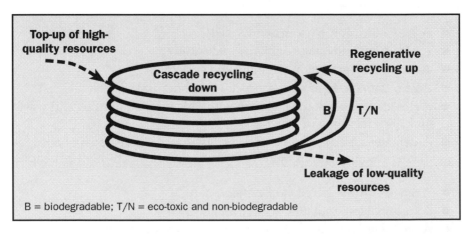

Figure 7: *The Closed-Cycle Economy*
© Green Innovations, Inc.

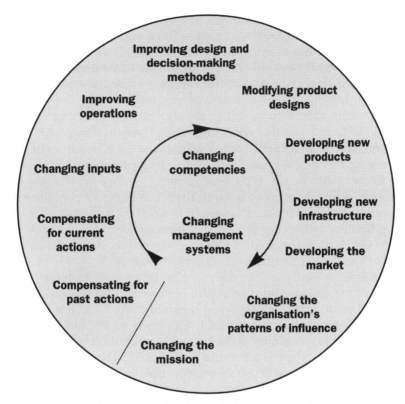

Figure 8: *Areas for Business Action on the Environment*
© Green Innovations, Inc.

game and to mobilise alliances of firms; (c) the lifecycle improvement of products; and (d) the upgrading of production processes (see Fig. 8).

Whole-system design is not just a very good source of productivity gains, it is also the only way that ecological sustainability can be achieved, because, as stated earlier, sustainability is a *system* characteristic and *cannot be achieved by focusing on unco-ordinated changes in parts of the system*. This is why the 'default' application of ISO 14001, which emphasises introspective improvements in the firm's own product and organisational performance, *cannot* result in society becoming ecologically sustainable.

■ *Managing Complexity: Standardised System Frameworks*

Knowing that whole-system design is the key to achieving ecological sustainability with win–win productivity gains is one thing: doing something about it is another. There are two reasons why whole-system design is not undertaken often: one is that most people have not been trained to handle this approach; the other is simply that it is harder to do whole-system design and implement whole-system changes than it is to deal with incremental changes. That is why those looking for fast pay-off action tend to leave the area alone.

One very important way to reduce the difficulty of handling whole systems is to use techniques that reduce the complexity of change. One reason why incremental change is easier than system change is that most of the factors affecting the process are fixed or predetermined. However, once a change to the wider system is proposed, things become much more uncertain because, while there is one present, there can be many possible futures. This uncertainty can be reduced, however, by creating shared visions around which individual actions can coalesce. Commercial-early-mover risk can be reduced even further if shared visions are adopted as standardised system frameworks and are formalised as industry standards.

Standardised system frameworks can also be used to maintain the sustainability of the system over time. A number of standardised system frameworks could be designed which, when combined, are known to be sustainable. Then, as long as each product is designed to fit into one of these frameworks, the sustainability of the system as a whole is likely to be maintained. For example, all products could be required to be recycled through a limited number of standard take-back and reprocessing systems (e.g. kerb-side mixed paper collection, parcel-post-style return system for high-value low-bulk products). A product would only be allowed onto particular regional markets if it could be handled by the available systems or if the product manager was prepared to develop and implement a viable alternative return–recycling system: that is, they take responsibility for creating a new standardised system framework.

An analogous approach is now being used in software development, the construction of very large office towers, sophisticated aircraft, and in the co-ordination of inter-related product offerings from a range of independent companies. In these areas, systems are reaching such high levels of complexity that no one person can comprehend all the detail of the system. To make it possible for many people to work together on these products or projects, it is necessary to treat the components of the products, or indeed whole products, as modules of a larger system. A shared standard is developed, or adopted *de facto*. Then the components or products can be designed and fabricated semi-independently and yet they still work together finally.

It might be appropriate for standardised system frameworks to be developed by joint government–industry–community working groups: national

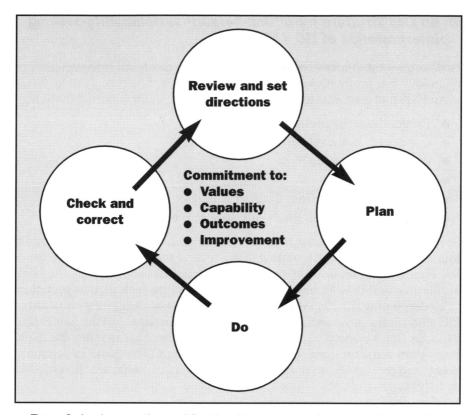

Figure 9: *Implementation and Continual Improvement Cycle for a Sustainability-Seeking Interpretation of ISO 14001/4*

© Green Innovations, Inc.

and international standardisation bodies might play a role, too. Standardised system frameworks should be considered in the EMS process before the EMS objectives and targets are set. The need to fit into existing standardised system frameworks would be determined in the review stage of the EMS. Then, in the EMS strategy process, the need for new frameworks would be considered. Applicable standardised system frameworks would guide the setting of objectives and targets.

■ *An EMS Structure for a 'Non-Default' Sustainability-Seeking Interpretation of ISO 14001*

How can all these ideas be brought together to create an interpretation of ISO 14001 that promotes sustainability-seeking?

An EMS that conforms to ISO 14001 is a mechanism for ensuring that a firm

- Thinks about the environment
- Decides what it wants to do[23]
- Works out how to do it
- Actually does it
- Corrects deviations from the plan
- Reviews its directions for the future so it can do better next time

For a sustainability-seeking firm, it is desirable to group these steps into a four-stage cycle (see Fig. 9), starting with a review and the setting of directions. See Figure 10 for an examination of the differences between a by-the-book application of ISO 14001 and a possible sustainability-seeking interpretation.

Underpinning the EMS cycle for a sustainability-seeking interpretation of ISO 14001 is top-level and organisation-wide commitment to the values that drive the firm's environmental action and commitment to capability-building, real-action outcomes and continual improvement. The issue of 'commitment' and the 'review and set directions' and 'plan' steps are discussed in detail below.

■ *Commitment*

Every step in a sustainability-seeking EMS cycle needs to be underpinned by a commitment to values, capability, outcomes and continual improvement.

☐ *Values*

The key values underpinning a sustainability-seeking approach are:

- Seeing environment as a goal and opportunity rather than a constraint or threat (see Fig. 11)
- A commitment to help society achieve ecological sustainability, globally and locally, for the sake of people and nature now and in the future
- A commitment to a 'no major trade-offs' approach
- The use of the EMS to promote creativity and effectiveness, as well as the 'default' emphasis on control and efficiency

By-the-book application of ISO 14001/4 (BTB)	'Sustainability-seeking' interpretation of ISO 14001/4 (SSI)	Comments
0: Initial environmental review	1: Review and set directions	*Initial* review not required by BTB if EMS already operating
	Values review	BTB equivalent is under 'environmental policy'
Review of views of interested parties	Review of society's environmental needs and society's environmental strategies	Systematic assessment of environmental needs not specifically required by BTB
Identification of environmental aspects	Review of significant negative environmental impacts and aspects for the product-and-organisation improvement programme	If initial review not required, BTB equivalent is under 'planning'.
Legislative and regulatory requirements (14001) and voluntary requirements (14004)	Requirements review (legal, voluntary, etc.)	If initial review not required, BTB equivalent is under 'planning'.
Examination of all existing environmental management practices and procedures; evaluation of feedback from the investigation of previous incidents	Management review	If initial review not required, BTB equivalent is under 'management review'.
1: Environmental policy	Policy development	
	Strategy development	Not required by BTB
	Identification of significant environmental aspects for the sustainability-empowerment programme	Not specifically required by BTB
2: Planning	**2: Plan**	
Environmental aspects		The SSI equivalent is under 'review and set directions'.
Legal and other requirements		The SSI equivalent is under 'review and set directions'.
Objectives and targets	Objectives and targets	
Environmental management programmes	Environmental management programmes	
3: Implementation and operation	**3: Do**	
Structure and responsibility; training, awareness and competence; EMS documentation; document control; operational control; emergency preparedness and response	Same as BTB	
4: Checking and corrective action	**4: Check and correct**	
Monitoring and measurement; non-conformance and corrective and preventative action; records; EMS audits	Same as BTB	
5: Management review		The SSI equivalent is under 'review and set directions'.

NB: The numbering in the highlighted section represents the order of execution.

Figure 10: *Comparing the Five-Stage-Cycle By-the-Book Application of ISO 14001/4 with a Four-Stage-Cycle 'Sustainability-Seeking' Interpretation of ISO 14001/4*

© Green Innovations, Inc.

The adoption of these values represents a growth in environmental maturity along the spectrum from compliance with externally-imposed constraints, to keeping ahead of society's constraints, to pursuing proactively the achievement of society's goals as a commercial strategy.

☐ *Capability*

A sustainability-seeking firm will need skills not required by other firms. They will need skills in

- Adapting the normal commitment to customer service to help the firm focus on society's need for ecological sustainability; for example, by adopting the concept of the '5-in-1 customer' (where products have simultaneously to serve the needs of the direct product user, the local community, people globally, future generations and nature)
- Product conception and design for the promotion of ecological sustainability
- Whole-system design and the creation and use of standardised system frameworks

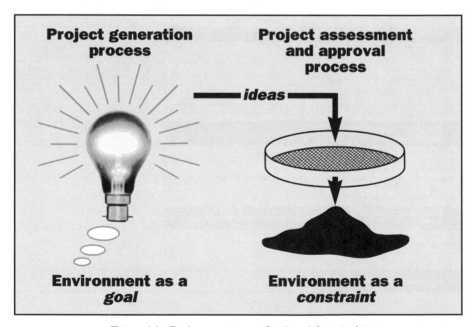

Figure 11: *Environment as a Goal and Constraint*
© Green Innovations, Inc.

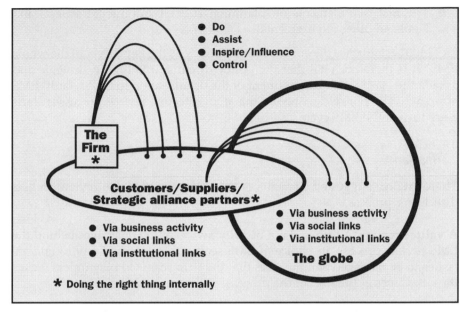

Figure 12: *A Firm's Potential, Positive Knock-On Effects*
© Green Innovations, Inc.

- Problem-solving and opportunity creation for 'no major trade-offs'[24]
- Managing the product lifecycle, often managing via influence rather than control
- Handling non-traditional issues, i.e. biodiversity protection on a life-cycle basis
- The development of strategies to make the pursuit of ecological sustainability commercially and strategically beneficial for the firm
- Exerting influence for positive environmental outcomes on a society-wide basis (see Fig. 12)

■ Review and Set Directions

The 'review and set directions' stage of a sustainability-seeking EMS cycle integrates the ISO 14001–14004 steps of

- 'Management review'
- 'Environmental policy'

- The first two elements of 'planning' (i.e. environmental aspects and legal and other requirements)

ISO 14001 establishes the environmental policy as the key driver of the whole EMS. It is therefore vital that the policy document is highly strategic and avoids the inadvertent commitment of the organisation to the 'default' EMS approach. The policy can only do this if an effective review processes starts *every* turn of the EMS cycle.

☐ The Review

There are five suggested elements of the review part of the 'review and set directions' process.

A values review. A firm will not be fully aware of the objectives behind the EMS review process, its choice of EMS scope, the classes of environmental issues to be managed and the roles that the firm wants to play unless it clarifies its values at the start of the process.

An assessment of environmental needs. A sustainability-seeking firm should not rely wholly on interested parties and stakeholders to draw its attention to society's environmental needs. The firm should take responsibility for proactively identifying these needs as part of its normal customer needs assessment process. But this time, the focus is on the '5-in-1 customer' (see page 232 for definition).

The identification of significant negative environmental impacts and related environmental aspects for the product-and-organisation improvement programme. This corresponds to the 'default' consideration of impacts and environmental aspects. Unlike the by-the-book ISO process, this step should be taken routinely ahead of the policy step so that the results inform policy development. An assessment should also be made of the relative impact of direct versus indirect (lifecycle) impacts.

A review of requirements (legal or voluntary). This is the same as the by-the-book ISO implementation. This review should also routinely feed into policy development.

A management review of the *de facto* or formal EMS. ISO 14001 requires an environmental management system review at the end of each formal EMS cycle and prior to the policy-making step at the start of the next cycle. However, firms implementing an EMS for the first time are not required to assess their *de facto* EMS in the initial review phase that precedes policy-making. Failure to carry out this assessment, however, could lead to weaknesses in the EMS policy and hence in the first formal EMS.

☐ Setting Directions

There are three suggested elements of the direction-setting part of the 'review and set directions' process.

The development of policy. The EMS policy ('environmental policy' in ISO by-the-book jargon) for a sustainability-seeking firm must provide a more detailed framework for the setting of objectives and targets than is the case for a 'default'-style firm. So the policy may need to be longer than the typical inspirational one-pager. If the EMS policy has to be longer or more complex than is appropriate for inspirational and educational purposes, then a separate inspirational summary policy should be produced. The EMS policy can refer to externally-developed sets of guiding principles. A sustainability-seeking firm might wish to refer to the 'guiding principles' on pages 238-39.

Strategy development. Reactive firms do not need strategies; proactive ones do. A sustainability-seeking firm must therefore include a strategy step in their EMS cycle. The strategy should be guided by the EMS policy or at least a good draft of it. Then some of the key insights or directions that emerge from the strategy development process may need to be included in the final EMS policy. The objective of the strategy is to help the firm decide *how* it can best help society achieve ecological sustainability, while at the same time advancing its own competitive or strategic interests. The strategy will need to guide the firm on the best combination of action

- For its sustainability-empowerment and product-and-organisation improvement programmes
- For timely leapfrog change and incremental change
- On the strategic stances that it might adopt (ethical opportunist, catalyst and pioneer—see the section on strategic stances for details [page 237])
- On direct and indirect impacts
- To achieve effectiveness-promoting creativity and quality-assuring control
- On correction and prevention

For each area of action, the firm will need to use the strategy process to decide on the appropriate degree of urgency, magnitude of change and time-frame. Given society's need for leapfrog change and the firm's need for products to be commercially viable at each point in time, it might be appropriate for firms to work on the development of more than one generation of product improvements or new products at any one time.

Overall, sustainability-empowerment programmes (product or influence-based) benefit society more than product-and-organisation improvement programmes, despite the very considerable importance of the latter. Fortunately, in many cases, firms will also gain the greatest long term commercial advantage from pursuing an empowering programme of product development.

The identification of significant environmental aspects for the sustainability-empowerment programme. It has always been possible under ISO 14001 to look at the environmental aspects associated with an empowering programme. However, the structure of the typical or 'default' ISO EMS cycle makes it very difficult to identify them. Environmental aspects for the product-and-organisation improvement programme are relatively easy: for example, when a firm works out what it is doing to cause pollution, it has identified the relevant improvement aspect. To find a sustainability-empowerment aspect, a firm would have to decide which action, out of the infinite possibilities, it should take to help society achieve ecological sustainability. A strategy is needed to decide on the best choice of activity; only then can the firm work out the sustainability-empowerment aspects of its operations that need to be managed well to implement its choice.

■ Plan

The 'plan' stage of a sustainability-seeking firm's EMS cycle needs to be different from the ISO by-the-book cycle. It is desirable to start the 'plan' stage of a sustainability-seeking EMS cycle with the setting of objectives and verifiable targets that cover three programme areas:

- Sustainability-empowerment programme:
  - Via *products*—to provide other firms and the community at large with the goods and services needed to achieve sustainability
  - Via *influence*—to spread sustainability-seeking and to *catalyse* sustainability-directed change to overcome blockages and to reform the business playing field
- Product-and-organisation improvement programme:
  - Lifecycle *product stewardship*—to improve all activities in the supply chain and in the use and disposal of the product or associated materials
  - *Production process improvement*—to improve all of the firm's own activities

- EMS quality assurance programme: to ensure that the commitments in the EMS policy, objectives and targets are followed through and that actions to promote continual improvement are taken. This programme is made up of the 'do' and 'check and correct' elements of the EMS cycle.

Instead of undertaking the identification of 'environmental aspects' for the product-and-organisation improvement programme and the determination of 'legal and other requirements' at the 'planning' stage, as is done in the by-the-book application of ISO 14001, these activities are best carried out for all sustainability-seeking firms in the 'review and set directions' stage. After setting objectives and targets, practical 'how-to' programmes need to be developed for sustainability-empowerment, product-and-organisation improvement and EMS quality assurance. These implementation programmes can contain fast-track actions and actions with longer-term pay-offs.

■ *Strategic Stances*

Sustainability-seeking firms can exhibit a number of different orientations towards environmental issues. The three stances of most interest to a sustainability-seeking firm are:

- **Ethical opportunist:** the organisation generally tries to 'do the right thing' on environmental issues as long as action is clearly feasible and economically viable and does not involve significant risk.
- **Catalyst:** the organisation identifies environmental problems or opportunities that need action and it triggers other organisations to pioneer the solution or to take the necessary action.
- **Pioneer:** the organisation works to make desirable environmental outcomes technically, organisationally and economically feasible. Pioneering can be directed at incremental or leapfrog movements towards inspirational goals.

The **ethical opportunist** firm encourages a general improvement in the environmental performance of all its products and all its processes. Ethical opportunist firms need to have well-developed search skills for finding the best available methods and product ideas and they need to be adept at taking on new methods and technologies. Many firms that eventually adopt an ethical opportunist stance may start out as 'cynical opportunists'.[25] For example, successful experience with 'green' niche marketing or with the application of an EMS may create a positive climate in which the shift can occur.

The **catalyst** stance is reflected in the cultural or policy influence the organisation attempts to achieve in order to promote sustainability. Advertising, customer and employee education programmes, sponsorships and policy advocacy are some of the channels for achieving influence. They are often used at present purely to promote the self-interest of firms rather than the public good as well. However, this need not be the case if firms are actively repositioning themselves to make a living by pursuing the public good. This stance is relevant for all organisations but sometimes it might be the only one that can be pursued actively if it is not possible, at a particular point in time, to make rapid progress developing or improving products or production processes.

The **pioneering** stance is likely to be most successful where complete management units are involved. This allows all the management structures and processes to be aligned with the stance. Such an alignment is necessary if the organisation is to have the sticking power to make the necessary breakthroughs. Pioneering management units can exist within an organisation that is opportunistic overall.

The catalyst stance would almost always be found in combination, i.e. with ethical opportunism or pioneering. Because of the risk and cost associated with pioneering, it would be rare for an organisation to pioneer in everything it did, even where pioneering is a core strategy.

■ Guiding Principles for a Sustainability-Seeking Organisation

The following guiding principles could be referred to in the EMS policy of a sustainability-seeking organisation. Nine key actions that need to be taken by a sustainability-seeking organisation are:

- **Goal:** to take on the timely achievement of global and local sustainability as *one* of its top goals and as a significant area of organisational opportunity

- **Customer service:** to ensure that when products (including services) and production processes are developed or modified to meet the needs of the active users, that they are also designed to serve the needs of the local community, people globally, future generations and nature

- **Management:** to implement an environmental management system that enables the organisation to help society achieve sustainability in a timely fashion, for example, through the use of leapfrog pioneering strategies

- **Market:** to take action to maximise the market for products that contribute to sustainability and as far as possible favour the servicing of other firms and final users that are making a positive contribution to the achievement of sustainability

- **Products:** to change product offerings—to foster sustainability
- **Operations:** to change the way activities are carried out or the way production is undertaken—to foster sustainability
- **Proliferation:** to promote the spread of the sustainability-seeking approach generally: through the private, public and community sectors and up and down the supply chain
- **Co-operation:** to co-operate with other sustainability-seeking organisations
- **Society's rules and structures:** to work to change society's rules-of-the-game and structures, including the standardised system frameworks for products, so that they favour the achievement of sustainability

If sustainability-seeking organisations proliferate, then the chance of achieving sustainability will switch from 'vanishingly small' to 'virtually certain'.

■ Conclusion

For many firms, the last three decades have been a confusing time. No sooner have governments set environmental standards than the community has demanded higher standards. It seems that trade-offs between environmental quality and economic production are not respected and the goalposts keep shifting.

However, once firms adopt a sustainability orientation, this constant change is no longer seen as a lack of faith but as continual improvement leading to a win–win environmental and economic outcome, that is, to sustainable development.

Firms will improve their chances of long-term success if they understand the destination and can position themselves to take advantage of the change. So when firms design their environmental management system, it is essential that they adopt a sustainability-seeking interpretation of ISO 14001. Anything less than this is not best practice.

■ Acknowledgements

I would like to thank Manningham City Council in Melbourne, Australia, for their vision and pioneering spirit in pursuing sustainability-orientated environmental management. No words can express the debt that I owe Kathy Preece for her continuing support.

■ Notes

1. Other bodies include those that provide education services and that engage in research and development.
2. Sustainable development is often wrongly defined as development that takes environment into account in a 'balanced' way, that is, environmental objectives are not pursued too vigorously so as to reduce the perceived risk of a major trade-off with economic, employment and income objectives.
3. J. Houghton, B. Filho, B. Callander, N. Harris, A. Kattenberg and K. Maskell (eds.), *Climate Change 1995: The Science of Climate Change* (New York, NY: Cambridge University Press, 1995).
4. See A. Lovins, M. Brylawski, D. Cramer and T. Moore, *Hypercars: Materials, Manufacturing and Policy Implications* (Snowmass, CO: Rocky Mountain Institute, 1996).
5. Pacific Solar is owned by the largest electricity generator in Australia. See P. Lawley, 'Photovoltaics: A Quantum Jump', presented at *3rd Renewable Energy Technologies and Remote Area Power Supplies* Conference, Cairns, Australia, 1996.
6. See DRI *et al.*, *Potential Benefits of Integration of Environmental and Economic Policies: An Incentive-Based Approach to Policy Integration* (Report prepared for the European Commission; London, UK: Graham & Trotman, 1994); M. Jacobs, *Green Jobs? The Employment Implications of Environmental Policy* (Brussels, Belgium: World Wide Fund for Nature, 1994).
7. For example: European Partners for the Environment / SustainAbility / Wuppertal Institute in Europe; the Natural Step Environmental Institute in Sweden and elsewhere; and Green Innovations, Inc., in Australia are developing sustainability-oriented management tools for business.
8. Given the slowness with which even the leading countries are putting an ecological sustainability-compatible industrial base in place, leading-edge 'green' firms may have to support the establishment of such an industrial base in one or more of the newly developing countries. Historically, major shifts in the industrial paradigm are associated with a geographical shift in the technological frontier or leading edge. (The frontier moved from craft production in the UK to production based on interchangeable parts / mass production in the US, and from there to Japan with the development of lean production.) See D. Wallace, *Sustainable Industrialisation* (London, UK: Earthscan, 1996).
9. See J. Houghton, G. Jenkins and J. Ephramus (eds.), *Climate Change: The IPCC Scientific Assessment* (New York, NY: Cambridge University Press, 1990).
10. The predicament that gives rise to the need for such massive changes is well described in D. Meadows, D. Meadows and J. Randers, *Beyond the Limits: Global Collapse or a Sustainable Future* (London, UK: Earthscan, 1992). This study also makes it clear that the restructuring of the economy needs to begin urgently.
11. If solar electricity rapidly displaces coal generation and is used to produce hydrogen as an oil and gas substitute, then the 5%-per-annum improvement in resource use efficiency could be lowered while still achieving the necessary greenhouse gas reduction targets.
12. A firm can exert influence via advertising, public relations, training, strategic alliances, lobbying, etc.
13. For most heavy industries, issues such as flora and fauna, soil conservation, and

urban planning tend to attract attention as *direct* environmental impacts only occasionally: for example, when capital works are undertaken and after that they are forgotten again. (Mining is an exception, with flora and fauna protection being a continuing issue.) Even where organisations try to manage their *indirect* impacts, most lifecycle assessment techniques do not deal with issues such as flora and fauna, soil conservation and urban planning because they are hard to reduce to a small number of simple, quantified indices.

14. Issues that caused difficulty mainly for society or the natural environment, rather than for the firm, were much less likely to be dealt with.

15. Even in a resource-intensive economy such as Australia's, the sectors of the economy that include firms with high *direct* impact (i.e. mining, agriculture, forestry, fishing, energy production, manufacturing, construction, transport and tourism) only constitute about 40% of the total economy. Since not all firms in these high-impact sectors themselves have high *direct* impact, the 'default' approach to EMSs would not apply well to more than 60% of the Australian economy. (Figures based on data from Australian Bureau of Statistics, *Catalogue No. 52.06: National Income, Expenditure and Product for the June Quarter 1996* [Canberra, Australia: ABS, 1996]).

16. Being driven by society's interests means much more than being 'market-led'. Many firms undertake extensive market research to find out what the public's needs or desires are. However, this research usually only discovers aggregated *private* needs and interests. Even where public opinion-polling deliberately searches for people's views on what is good for the community as a whole, asking for off-the-cuff comments is not good enough because of the complexity of the issues involved. The results of in-depth consultations and research studies would need to be drawn on as well.

17. It is unlikely that certifiers will see the promotion of ecological sustainability as the logical end-point of ISO 14001 continual-improvement programmes.

18. See Du Pont, *Safety, Health and Environment 1994 Progress Report* (Wilmington, DE: Du Pont, 1994).

19. The Wuppertal Institute in Germany has adopted this goal and is promoting it in industry through its Factor 10 Club.

20. Where virtually all the throughput resources (those not recycled in the economy) are derived from renewable sources (e.g. solar energy and biological materials).

21. The *regenerative* recycling of non-toxic biodegradable material is possible now using natural processes (e.g. composting and digestion to decompose biodegradable materials and vegetation growth into breakdown products that can be naturally re-synthesised into useful complex materials). However, a comprehensive system for the regenerative recycling of toxic and non-biodegradable materials is still to be developed.

22. See Lovins *et al.*, *op. cit.*

23. ISO 14001 requires compliance with relevant legislation and regulations. This sets the standard's only environmental performance requirements.

24. Eli Goldratt is actively disseminating methods for achieving 'no major trade-off' outcomes in industry. See The Avraham Y Goldratt Institute, *An Introduction to The Avraham Y Goldratt Institute and the Theory of Constraints* (New Haven, CT: The Avraham Y Goldratt Institute, 1994); and E. Goldratt, *It's Not Luck* (Aldershot, UK: Gower, 1994).

25. A *cynical* opportunist organisation is interested in environmental issues only as a source of commercial opportunities and for no other reason. Cynical opportunist firms add 'green' products to their range as market research shows that there is demand. However, products with poor environmental performance are continued without improvement for as long as demand exists. Cynical opportunist companies are more likely to perform well in the 'green' market if the product development and production staff servicing 'green' niches have personally-held ethical opportunist or pioneer orientations. A team made up entirely of cynics will find it hard to identify 'green' opportunities and is unlikely to have the commitment or insights necessary to make 'green' initiatives work well.

Only the ethical opportunist, catalyst and pioneer stances match the spirit of ISO 14000. However, the ISO 14000 series definition of continual improvement (improvement up to the level set in the EMS policy) and the ability to apply ISO 14001 to only parts of an organisation suggests that a cynical opportunist organisation could successfully meet the technical requirements of the standard and be certified if it wrote its policy carefully.

13 From EMAS to SMAS[1]
Charting the Course from Environmental Management to Sustainability

Andrea Spencer-Cooke

ENVIRONMENTAL AUDITING, management and reporting are vital in driving the transition to sustainable development. As highlighted in the European Union (EU) 5th Environmental Action Programme (5EAP), their importance lies both in addressing 'conventional' issues of environmental protection and in helping to promote more sustainable patterns of production and consumption. Properly applied, these tools can deliver the frameworks and the information necessary for sound decision-making by key actors. They can also help to promote new forms of corporate governance and accountability.

The European Commission's Eco-Management and Audit Scheme (EMAS) is among other national and international attempts, including the British Standard BS 7750 and ISO 14001, to develop appropriate management instruments for achieving improved environmental performance. As the uptake of EMAS and similar initiatives spreads, a number of questions are being raised:

- Why should companies adopt EMAS?
- What can and cannot EMAS deliver?
- What are the optimum conditions for EMAS to be effective?
- How can it be improved?
- How and why should stakeholders be involved?
- What instruments, in addition to EMAS, will we need for the sustainability transition?

These are among just some of the questions being asked in the latest Sustain-Ability report, *From EMAS to SMAS: Charting the Course from Environmental*

Management and Auditing to Sustainability Management.[2] Produced in collaboration with European Partners for the Environment (EPE),[3] this publication explores the implementation of the EMAS Regulation across the European Union, weighs up the strengths and weaknesses of the scheme, reviews its possible application in non-manufacturing organisations and establishes a platform on which to build future instruments for managing the sustainability transition.

The full report *From EMAS to SMAS* is made up of three parts, or 'modules'. Module 1, 'EMAS 1992–1995', maps out the context and content of the EU Eco Management and Audit Scheme, providing a general introduction to environmental management systems and the aims, objectives and structure of the EMAS scheme. Module 2, 'EMAS 1995–1998', looks at the implementation of the EMAS Regulation to date, reviewing progress and uptake across the EU and identifying key issues that need to be tackled in the 1998 review of the Regulation by the European Commission (EC). Module 3, 'EMAS to SMAS 2001–?', goes a step further, placing EMAS in the context of sustainable development, to explore what tools other than EMAS we are likely to need in future.

In this chapter, the main findings of the report are presented. Based on companies' initial experience with implementing the scheme—and stakeholders' initial reactions to the implementation process—a summary is made of the core strengths and weaknesses of EMAS. Eight hot topics for discussion around the Regulation are identified and some key issues outlined for its forthcoming 1998 review by the Commission. This discussion is intended to help pave the way for an improved, expanded EMAS—or 'EMAS-Plus'. The chapter then moves beyond EMAS to look at the wider picture, exploring the range of new indicators required and the role of stakeholder partnerships in managing the sustainability transition.

■ EMAS 1995–1998

The EMAS scheme became officially open for company participation on 13th April 1995. But why should companies adopt EMAS and what can the scheme actually deliver? To answer these key questions, it helps to recall the original purpose of the Regulation and to consider how far EMAS actually goes in fulfilling this. Within the EU 5th Environmental Action Programme (5EAP), environmental auditing is first and foremost intended as an *internal* management tool to monitor performance, and, secondly, as an *external* performance indicator. Ultimately, the success of the scheme depends on the degree to which it fulfils both of these needs.

In the short time since EMAS came into effect, considerable practical experience has been gained by the pioneering companies and organisations that have gained registration to the scheme or otherwise participated in pilot projects testing the application of the Regulation and its extension to other sectors and activities. To capitalise on this experience, the EPE/SustainAbility report included:

- A survey of EMAS uptake across the Union
- An assessment of the strengths and weaknesses of the scheme

☐ *Strengths and Weaknesses*

A key part of *From EMAS to SMAS* involved conducting a 'SWOT' analysis of EMAS—in other words, assessing its strengths and weaknesses, and evaluating the opportunities and threats facing the scheme—to evaluate just how well EMAS fulfils the aims of the 5EAP. EPE companies and stakeholder members, EMAS pilot companies, registered sites, local authorities and accredited verifiers from across the Union all contributed to the SWOT. The SWOT raised a number of critical issues around EMAS as well as providing strong arguments for its adoption. A summary of their views is given in Figure 1.

☐ *Hot Topics and Key Questions*

The SWOT has identified a number of issues that need to be tackled for EMAS to work effectively. In *From EMAS to SMAS*, these are translated into fifteen key questions which provide a useful starting point for reviewing the Regulation.

1. Do the benefits of EMAS outweigh the costs?
2. How can we ensure that a level playing field is achieved in the implementation of EMAS across Member States?
3. Is the bureaucracy and paperwork required under EMAS really necessary or could it be reduced?
4. How can we ensure accessibility, clear definitions and consistent interpretation in the application of EMAS?
5. How can the re-writing of the Regulation in a more readable way be reconciled with the constraints of writing a Regulation?
6. How can we ensure that the objectives for continuous improvement set by sites are not too low and result in real environmental improvement?
7. How do we measure the 'success' of EMAS?

Strengths	Weaknesses
<ul><li>Integration</li><li>Systematic approach</li><li>Formalises good environmental management</li><li>Stakeholder principle</li><li>Transparency and credibility</li><li>Increased staff awareness, morale and team-building</li><li>Harnesses innovation</li><li>Increases company profile</li><li>Improves relations with customers, community, regulators and the parent company</li></ul>	<ul><li>High cost</li><li>No level playing field</li><li>Bureaucracy</li><li>Unclear language</li><li>Lack of environmental performance requirements</li><li>Missing social dimension</li><li>Publicity of the scheme is weak</li><li>Verification seeks failings rather than identifying strengths</li></ul>
Opportunities	**Threats**
<ul><li>Increased awareness</li><li>Better systems</li><li>Lower costs and less regulation</li><li>Improved stakeholder dialogue</li><li>New business opportunities</li><li>Improvement of environmental quality</li><li>Priority-setting</li><li>Attracting investment (e.g. parent company)</li><li>Customer confidence</li><li>Competitive advantage</li></ul>	<ul><li>Ambiguity over roles and definitions</li><li>Non-take-up by companies</li><li>Poor implementation</li><li>Lack of stakeholder credibility</li><li>Exposure of non-compliance</li><li>Being 'knocked off the pedestal'</li><li>Confusion over EMS standards</li><li>Incompatibility of systems</li></ul>

Figure 1: *Summary of SWOT Analysis*

8. Is EMAS the right instrument for introducing social issues and, if so, how does it need to be adapted to fulfil this role?

9. What needs to be done to raise public awareness about EMAS and enhance the profile of the scheme?

10. Can the environmental statement be used to publicise EMAS and, if so, how?

11. How can the constructive side of verification be emphasised?

12. What can be done to promote uptake and ensure a 'critical mass' of companies participating in the scheme?

13. How can stakeholders be assured of the scheme's proper implementation and legitimacy?

14. What can be done to address companies' concerns that bad publicity for non-compliance will outweigh good publicity for participating in the scheme?

15. How can the EMAS/ISO debate be resolved in a constructive spirit?

The challenge in addressing these questions is: how can we build on the scheme's strengths, compensate for its weaknesses, focus on the opportunities it presents and avoid—or insure ourselves against—the threats?

To facilitate constructive discussion, the EPE/SustainAbility report groups the various issues arising from the SWOT into eight EMAS 'hot topics' with key points on which to focus. The framework provided by these eight hot topics does not set out to be comprehensive, but is rather meant as a way of facilitating discussion about how and why the EMAS Regulation can be improved and adapted when it comes up for review by the Commission in 1998. Drawing on the considerable experience gained by implementing organisations and EMAS pilot projects across the Union, it is already clear in many areas that the scheme would benefit from further development, such as, for example, clearer definitions and greater publicity.

The eight hot topics are as follows:

1. Definitions and interpretation (e.g. 'best available technology' [BAT]; 'continual improvement'; guidance on the role of the verifier)

2. Verification and accreditation (e.g. EU guidelines for accreditation and verification; stakeholder legitimacy and credibility of verification)

3. Continual improvement (e.g. guidance on setting, measurement and reporting of targets/environmental quality standards [EQS])

4. Compatibility with other EMS standards (e.g. guidance in annexes and bridging documents; lowest common denominator; trade implications)

5. Stakeholder involvement (e.g. adequacy of existing mechanisms; usability of environmental statement; etc.)

6. The environmental statement (e.g. key audiences; guidance versus no guidance; sectoral benchmarks; relevant information; usability)

7. Costs and benefits (e.g. cost–benefit analysis of EMAS; comparison with other EMSs; value-added verification; etc.)

8. Uptake and critical mass (e.g. publicity of scheme; reducing paperwork and costs; bridging documents; mandatory elements; etc.)

☐ *The Sustainability Laboratory*

It was with this aim of furthering the development of EMAS that these eight hot topics were discussed at the EPE Sustainability Laboratory on *From EMAS to SMAS*, held in Brussels on 30th–31st January 1996. Laboratory participants included representatives from companies, non-governmental organisations (NGOs), public authorities, consultancies and professional organisations from across the EU. Among the outcomes of the workshop were that EPE should focus its efforts primarily on:

- Developing mechanisms for stakeholder involvement in and around EMAS
- Exploring the EMAS–ISO 14000 relationship
- Clarifying the role, scope and format of the environmental statement
- Assessing the costs and benefits of the scheme

It was agreed that EPE would launch an EMAS stakeholder involvement pilot project to explore how stakeholders could become more engaged in the implementation and development of environmental management systems. Dow Europe is leading the EPE pilot project, alongside SustainAbility, to explore how stakeholder engagement lessons learned from Responsible Care might be applied to EMAS (see Fig. 2). One of the project's aims is to pave the way for a more efficient, stakeholder-oriented EMAS—or 'EMAS-Plus'—which goes beyond current EMS requirements.

☐ *From EMAS to 'EMAS-Plus'*

As described above, the EPE SWOT analysis has identified a number of areas in which EMAS can be improved. In 1998, the EMAS Regulation comes up for review by the European Commission, providing a valuable opportunity for stakeholders and companies alike to contribute to the strengthening and extension of the scheme.

Among the core challenges for the long-term success of EMAS is how to ensure that it fulfils its role both as an internal management tool and as an external performance indicator. The *EPE Workbook* presents a number of recommendations for a more effective, new-generation EMAS that builds on strengths and opportunities and takes into account the need for greater stakeholder involvement. The EPE stakeholder involvement pilot project mentioned above is also expected to feed into this process. The aim is a sort of 'EMAS-Plus'.

Put simply, the characteristics of 'EMAS-Plus' describe a scheme that:

EPE Stakeholder Involvement Pilot Project

Dow Europe SA, European Partners for the Environment (EPE), SustainAbility Ltd

Building Trust through EMAS: What Can we Learn from Responsible Care?

Project Overview

As a market-based instrument, EMAS relies on the confidence of stakeholders to operate effectively. This confidence is strongly influenced by the nature of interactions and level of credibility that exists between the EMAS site and its stakeholders. The chemical industry's Responsible Care initiative sees public dialogue as integral to its aims, and signatories already have several years' experience in stakeholder relations. This pilot project offers an opportunity to explore possible synergies between stakeholder relations under Responsible Care and the proposed 'EMAS-Plus' approach to EMAS, emphasising stakeholder involvement.

Project Objectives

The project aims to harness the expertise of the EPE study group to explore the possibilities for extended stakeholder involvement in EMAS or 'EMAS-Plus' as a first step in the development of sustainability management practices. Research will focus on Responsible Care initiatives at Dow Europe's site at Terneuzen in The Netherlands, and will address the following general questions:

1. In which ways does Responsible Care involve stakeholders, and what is the nature of the dialogue?
2. What have been the results of stakeholder dialogue under Responsible Care?
3. What synergies might exist between stakeholder dialogue in Responsible Care and EMAS?
4. Is it possible to identify commonly applicable 'good stakeholder practices'?
5. What are the benefits and limitations to stakeholders of verification?

The pilot project findings are expected to form the basis for a forthcoming EPE handbook on stakeholder involvement and environmental management systems.

Figure 2: *The EPE EMAS Stakeholder Involvement Pilot Project*

- Collates and builds on the experience already gained by EMAS-implementing organisations, to assist EMAS in performing its function as an internal management tool
- Creates and promotes a process of stakeholder engagement to determine how EMAS can best fulfil its role as an external performance indicator

In a similar vein, Philip Sutton of Australia-based Green Innovations highlights in Chapter 12 a number of characteristics of so-called 'leading-edge' EMSs, which go beyond the basic requirements set by ISO 14000. As well as developing the features of 'EMAS-Plus', topmost among the issues that should be discussed are how EMAS can be extended to other non-manufacturing organisations and activities currently excluded from the scheme but identi-fied in the 5EAP as target sectors, namely: energy, transport, agriculture and tourism. Likewise, the contents and format of the environmental statement need to be adapted to meet the information needs of key stakeholders; if environmental auditing is indeed to become a performance indicator 'as important as traditional financial accounts', as the 5EAP claims, it will be necessary to adjust EMAS in a way that reflects the financial community's need for financially-related, corporate-level environmental information.

A strengthened and expanded 'EMAS-Plus' could be an important step in adapting EMAS to meet the demands of the sustainability challenge.

■ From EMAS TO SMAS: 2001-?

☐ Beyond EMAS

With EMAS only just beginning to establish itself in the EU, it may seem pre-mature to be talking about the next generation of tools and systems. Even in its current form, EMAS and its sister standards have a vital role to play in driving greater corporate environmental awareness and improved perfor-mance. Simply ensuring that the majority of businesses have effective envi-ronmental management systems in place will undoubtedly take us well into the next century.

By looking beyond EMAS, the EPE report does not intend to detract from the importance of implementing and developing environmental manage-ment systems; indeed, sound environmental management in business is a necessary step in the transition to sustainability, but it is useful to start think-ing about where we should be heading in charting the course from environ-mental management and auditing to sustainability management.

The ultimate aim of the 5EAP is to create the framework for development that is sustainable: namely, that is economically viable, socially acceptable and environmentally sound. Of all the environmental management system standards on offer, EMAS has the greatest stakeholder support—as con-firmed both by the EPE project and in Chapter 2 of this book, 'Neither Inter-national nor Standard', EMAS is certainly a step in the right direction. But even if every company in the world adopted EMAS tomorrow, we might have a cleaner environment but we still would not be sustainable. Enter the concept of 'SMAS'.

The idea of 'SMAS'—or 'sustainability management and audit systems'—is to tackle aspects of corporate performance that are not addressed by the current generation of EMS standards, but which are a crucial part of sustainable

EMAS	SMAS
● Site-based	● Organisation- and lifecycle-based
● Environmentally-driven	● Sustainability-driven
● Systems-focused	● Close linkage of impact with performance
● Site defines priorities	● Priorities set through stakeholder partnerships
● Assumes need	● Challenges need

Figure 3: *From EMAS to SMAS*

development. Thus existing EMS standards place a strong emphasis on issues such as eco-efficiency and cleaner production, but do not really deal with the question of social impacts, lifecycles or indirect social and environmental effects. Where EMAS is site-based, environmentally-driven and systems-focused, SMAS will need to be lifecycle-based, sustainability-driven and focused on environmental and social impacts rather than systems. Some of these key shifts are summarised in Figure 3.

Exploring the role of EMAS in the context of sustainability, the following central issues arise:

● The effectiveness of EMAS as a market-based instrument.
● The recognition that sustainability goes beyond environmental performance at site level
● The question of what we are actually sustaining

☐ *The Effectiveness of EMAS as a Market Instrument*

Among the central messages contained in the EPE/SustainAbility report is that EMAS is not, in its current form, effectively fulfilling its potential as a market-based instrument, for three key reasons:

● Its voluntary nature
● Its site-based focus
● The lack of guidance on the environmental statement

In order to work effectively as a market instrument, EMAS must meet the information needs of those stakeholders able to exert a positive market-based push towards sustainability, such as customers, consumers and the financial markets. Critical here is not just the type of data, but the *usability* of the information disclosed. This has significant implications for EMAS.

The importance of meeting stakeholder information needs and enhancing the usability of environmental information is also a core message of *Engaging Stakeholders*, recently published by SustainAbility alongside the United Nations Environment Programme (UNEP).[4] Through a collection of stakeholder case studies, *Engaging Stakeholders* demonstrates that the main emerging uses of environmental information are to compare, rank and benchmark company environmental performance. To rank and benchmark, users tend to require information that allows company-to-company comparison, applying a common framework for all companies in a given sector. However, the voluntary character of EMAS, its site-based focus and the lack of guidance on the environmental statement, mean that comparison is greatly restricted: so far at least, only a minority of sites have opted to go for EMAS, the information available is only relevant for particular sites, not for the company as a whole, and there is no common framework for disclosure via the environmental statement.

This experience with EMAS has implications for the design of SMAS. In designing effective market-based sustainability instruments, it is important to consider whether it is better to have a few sites certified to high standards, or a large number of companies providing core information which can be used as the basis for benchmarks and league tables, thereby enabling the market mechanism to function. In considering this question, useful lessons can be drawn with other environmental initiatives such as the US Securities and Exchange Commission (SEC) reporting requirements and Toxic Release Inventory (TRI)—both of which are mandatory and favour a high level of transparency, and both of which have found significant market applications in the screening and benchmarking of companies.

☐ Sustainability Goes beyond Environmental Performance at Site Level

The second central issue, as we consider the role of EMAS in the context of sustainability, is that EMAS is site-based and environmentally-focused. Sustainability, however, is not simply an environmental issue: in addition to ecological and economic concerns, we also have to consider the social aspects of sustainability, such as equity. Moreover, sustainability is a systemic concept. Any one organisation or activity can only be considered 'sustainable' in the light of the behaviour of other players in the system.

While EMAS is arguably a valuable tool for improving environmental performance at the level of organisational units, when it comes to improving the quality of the EU environment as a whole, let alone global sustainability, how adequate is the site-based, environmental focus?

Site-Based Focus. This is a question that surfaces immediately when looking at the application of EMAS to non-manufacturing sectors. Considerable environmental impacts arise through dispersed activities such as transportation, from the indirect effects of policies, and through the use and disposal of products, rather than specific production processes. In order to account for these impacts, it is necessary to take a total lifecycle perspective for products and services, rather than simply focusing on the production phase.

In addition, tackling sustainability is not simply a question of incremental environmental improvements. Sustainable development also comprises social and developmental dimensions. To incorporate lifecycle and non-environmental issues, SMAS will need to move beyond environmental performance at the site level in four different ways:

- Applicable to the organisation as a whole
- Product-based
- Relevant to the service sector
- Closely linked to ecosystemic impacts and ambient standards

From EMAS to SMAS argues that, in addition to the site-based focus of today, it is likely that SMAS will need to emphasise organisational performance as a whole in order to generate meaningful, market-related information. It is the totality of a company's activities that constitutes its footprint, not the performance of one or two 'showcase' sites. A comprehensive organisational framework is therefore a necessary feature of SMAS.

This said, although the organisation may be the most practical entity for tracking progress towards sustainability, in many cases it is exceedingly difficult to determine where the boundaries lie and how responsibilities should be apportioned. Moreover, the most significant effects of organisations often arise through their products and services, rather than their production processes. SMAS should therefore aim to provide a picture not only of production processes, but also of the impacts direct and indirect—of products, such as those arising from distribution, use and disposal.

The importance of moving beyond the production process is particularly relevant in the context of service industries. Within the current EMAS framework, participation is essentially restricted to the manufacturing sector. In the service sector, it is generally the case that direct, site-based impacts are minimal and the most significant impacts are associated with the policies or service 'products'. For example, although a bank has direct environmental impacts through heating, lighting, transportation, office equipment and paper use (so-called 'corporate ecology'), the *indirect* impact of its products—its loans and investment portfolios, or 'product ecology'—is likely to be in orders of greater magnitude.

Applying EMAS as it stands to the service sector is arguably of limited value, since the site-based impacts of a service business are generally minimal. As the experimental applications of EMAS to local authorities and other, non-manufacturing activities have shown, to deal effectively with the service sector, SMAS will need to be based on lifecycle-assessment-type design principles that take into account the *indirect* environmental impacts of policies, products and services. At the root of SMAS is the recognition that the impact of organisations does not stop at the site boundary. To be fully managed, impacts must be viewed holistically in space and time: namely in a global context and from cradle to grave.

Systemic Concept. Lastly, while focusing on the site or corporate level may be important from an organisational perspective, it is secondary from an ecological perspective. The 'ecological bottom line', when it comes to environmental quality and the sustainability of natural resources, is the ambient environmental quality standard (EQS), such as, for example, the pollution concentrations in a river. If several companies emit wastes into a shared environmental 'sink' such as a river, the key consideration from an ecological perspective is the total pollution that results from the combined activities of those companies. Even if one company improves its environmental performance, but that of the other companies deteriorates, the environmental quality—or ecological bottom line—as a whole will still decrease. Although the individual company might be moving in a more sustainable direction, this is of little consequence if the resource is nevertheless destroyed through the combined actions of the companies sharing it: the performance of any one actor can only be labelled as sustainable or unsustainable in the light of behaviour by the rest of the system.

The conclusion that must be drawn is that systems such as EMAS, which focus on individual actors, cannot provide meaningful indicators of sustainability. Instead, SMAS would need to include ecosystem-wide measures of impact and indicators that catalyse more co-operation among companies to protect shared resources.

Social Aspects. Environmental improvement is frequently confused with sustainable development, resulting in a very partial picture of the sustainability challenge. Throughout the United Nations Conference on Environment and Development (UNCED) process, and in the resulting *Rio Declaration* and *Agenda 21* documents, the international community emphasised that sustainable development is not about the environment alone, but also development, and that sustainable development requires the integration of environmental and social components into economic decision-making. The need to move beyond 'eco-efficiency' (i.e. the economic and environmental aspects of sustainability) to include the social dimensions of sustainability (i.e. equity and need) is a core message of the report. Among the necessary shifts required here are a greater emphasis on human capital, ensuring fair trade and promoting new sets of values. Adopting an approach that goes beyond environmental management at site level to embrace the systemic nature and social dimensions of sustainability will require a number of transitions. Some of these are presented in Figure 4.

Systemic thinking tells us that sustainability can only be defined for a complete socio-economic-environmental system and not for its component parts. No person, company or nation can *achieve* sustainable development alone, in isolation from other actors, since others can affect our sustainability just as we can affect theirs. However, it does make sense on a micro-level to speak of individual *progress* towards sustainability. To measure progress, indicators are required and, critically, a shared vision of what is actually being sustained.

☐ *What Are We Actually Sustaining?*

Although there exist widely-accepted definitions of sustainability, such as that of the Brundtland Commission, these do not provide *operational* guidance that can be used in policy-making or as the basis for SMAS. For example, although people might agree that policies should seek to meet the needs of the present generation without compromising the ability of future generations to meet their own needs, they are unlikely to agree in detail about what these needs are, or which are the highest priorities when it comes to the inevitable stage of making trade-offs. This does not mean that sustainability has no meaning, but rather that it has many meanings. Different individuals, groups, institutions and cultures will have different ideas about *what* needs to be sustained as development proceeds, let alone about *how* to sustain it. This plurality of understandings of sustainable development has been broached in an earlier EPE/SustainAbility report, *Towards Shared Responsibility*,[5] which outlined three internally consistent, but quite different, visions of sustainability and how it might be attained.

Thus, for example, sustainability could be interpreted to mean the sustainability of a service, of an industry, or of the biosphere and human community. The question of what is actually sustained, and the strategies and actions

1. Economic vectors

- Quantity to quality
- Eco-infrastructure
- Human capital
- Reduce economic waste
- Integrate environment
- Fair trade
- Global standards
- Getting the prices right

2. Technological vectors

- Dematerialisation and eco-efficiency
- Sustainable technology
- Appropriate technology
- Innovation and information economy

3. Environmental vectors

- Renewable resources
- Preserve biodiversity
- Stay within sink limits

4. Socio-cultural vectors

- Preserve diversity
- Meet basic needs
- Equity
- Access to opportunity
- Employment
- Sustainable consumption

5. Political vectors

- New institutions
- Transparency
- Gender and racial equality
- Visions of sustainability
- New sets of values

Figure 4: *An Overview of Sustainability Transition Vectors*
Source: Environmental Strategies and SustainAbility Ltd, 1996

necessary to achieve this, will vary radically according to which perspective one adopts. This highlights the need for sustainability visions, or scenarios.

☐ Three Sustainability Scenarios

Among the core features of Module 3 of *From EMAS to SMAS* are three sustainability scenarios presenting very different visions of a sustainable world. The aim of this is to help readers visualise what sustainable development means to them.

Scenario 1: No Limits. The first scenario, 'No Limits', emphasises sustainability at the level of individual wants, needs and quality of life, stressing the role of technological innovation and free markets in providing new opportunities for meeting needs more effectively, while at the same time reducing environmental and social impacts. Keywords include: rapid change; technological innovation; adaptation; no limits; cultural diversity; maximising quality of life at the individual level.

The role of SMAS in this scenario would be to:

- Help companies improve profitability by improving environmental performance
- Catalyse the re-examination of processes and spur innovation and new ideas
- Provide transparency and make markets work more effectively

Scenario 2: Orderly Transition. The second scenario, 'Orderly Transition', is less confident about the ability of the market to deliver sustainability and recognises the need for governments to provide guidance to individual voters and decision-makers. Problems must be tackled directly by regulating changes in production and consumption patterns. The emphasis is on identifying key issues through sound science and careful assessment of costs and benefits to identify priority areas for policy intervention. Keywords are: sustainability; stewardship; managerialism; targets; steering; scientific expertise; international negotiation; optimisation; carrying capacity.

The role of SMAS in this scenario would be to:

- Test compliance through audited and comparable data
- Provide feedback to the policy steering and control mechanisms
- Collect information linked to, e.g. economic instruments and ecological tax reform

Scenario 3: Values Shift. The third scenario, 'Values Shift', contains a sense of urgency that is missing from scenarios 1 and 2. What is needed is a significant, immediate shift to sustainable lifestyles with much lower social and environmental impacts. SMAS here goes to the heart of corporate impacts both in terms of companies' use of ecological space, their social and ethical performance and their interactions with the rest of the world, especially developing countries. Products are scrutinised closely as to whether they meet real needs and society plays an active role in defining a vision for sustainability and ensuring that attitudes and behaviour are changed. Keywords are: prevention; urgency; participation; new relationship with nature; decentralisation; equity; community; caring; spirituality.

The role of SMAS in this scenario would be to:

- Provide full transparency to allow monitoring of a company's use of social and ecological space
- Act as a tool to enable trust-building, dialogue and partnership with stakeholders

The purpose of these scenarios is not to predict the future, but rather to prepare stakeholders for a range of possible futures and to help them understand better the objectives and values of partners who hold different visions. Moreover, by encouraging readers to picture what sort of future they identify with, the scenarios can help to define the types of strategies and systems that should be developed under SMAS.

■ *Conclusions*

From EMAS to SMAS set out to help review progress so far achieved in the development and implementation of environmental management systems, most specifically EMAS, and to help chart a course beyond environmental management and auditing to sustainability management. Among the key conclusions of the report is that stakeholders and companies now have a valuable window of opportunity to shape the future development and implementation of the EU Regulation by contributing to the 1998 review of EMAS by the European Commission. A number of areas in which the scheme could be extended and improved have been identified.

Placing EMAS within the wider objective of sustainable development, a number of challenges for the design of sustainability management instruments were spelled out, with the aim of raising questions rather than giving answers.

One key conclusion is that SMAS will require new types of strategic alliances and stakeholder partnerships. Companies cannot achieve sustainability alone; the responsibility must be shared. To chart a course from environmental management and auditing to sustainability management, companies will have to find new ways of engaging stakeholders. Developing SMAS may seem an insurmountable task. Far greater than the challenge of new systems, though, will be the need for all parties to revisit their assumptions, articulate their needs and expectations, find new modes of influence and be prepared to enter into solutions-focused partnerships.

■ Notes

1. 'SMAS' (pronounced 'es-mas'), or sustainability management and audit systems, is a new term, introduced in the EPE/SustainAbility report to describe the broad range of instruments and indicators that will be needed to manage the shift from environmental management to sustainability.
2. EPE/SustainAbility Ltd, *From EMAS to SMAS: Charting the Course from Environmental Management and Auditing to Sustainability Management* (Brussels, Belgium: EPE/SustainAbility Ltd, 1996).
3. EPE is a non-profit organisation, set up in 1994 at the joint initiative of the European Environmental Bureau (EEB), business, public authorities and environmental non-governmental organisations (NGOs). Its mission is to facilitate dialogue and stimulate co-operation between all sectors involved in or affected by the implementation of the EU 5th Environmental Action Programme. EPE serves as a model for constructive dialogue and action in partnership for sustainable development. Further information on EPE can be obtained from: Garsett Larosse, EPE Liaison Office, Antwerpsesteenweg 461, Westmalle 2390, Belgium. Tel: +32 3 312 9383; Fax: +32 3 309 2874; E-mail: epe@ecotopia.be *or* 100013.3231@compuserve.com
4. The UNEP/SustainAbility two-volume *Engaging Stakeholders* report (London, UK: UNEP/SustainAbility Ltd, September 1996) examines progress in company environmental reporting from the angle of the report-user. Drawing on twelve case studies featuring stakeholder groups already making use of company environmental information, the report predicts increased emphasis on comparability, verification and benchmarking as stakeholder demand for company environmental information grows. For further information on the *Engaging Stakeholders* programme, contact SustainAbility Ltd on Tel: +44 (0)171 937 9996; Fax: +44 (0)171 937 7447; E-mail: info@sustainability.co.uk
5. EPE and SustainAbility Ltd, *Towards Shared Responsibility* (Brussels, Belgium: EPE/SustainAbility Ltd, 1996).

Part 3 **Tactical Responses**
Managers at the Greenface

14 Beyond ISO 14001
Ontario Hydro's Environmental Management System

Philip M. Stoesser

WITH THE arrival of the ISO 14001 standard for environmental management systems, its potential impact on the business world is the subject of increased speculation. There is, however, every indication that, in order to compete, companies will need to demonstrate care and concern about the impact they have on the environment. Dirty companies will be held accountable and any company unable to prove that it is protecting the environment will have difficulty competing in the future marketplace.

ISO 14001 is part of the ISO 14000 series of standards on environmental management currently being developed and structured according to the ISO 9000 series on quality management. Similar to ISO 9000, registration under ISO 14001 is expected to become an important business priority, as many companies feel that third-party certification to ISO 14001 will be important to their future business success.[1] Further, any company interested in its corporate image and goodwill will want to be environmentally correct, and ISO 14001 registration will be the vehicle to achieve this goal.[2] Another potential benefit of registration is an opportunity for a company to prove due diligence,[3] which will provide some credibility with financial institutions, insurance companies and customers.[4] But what will it really mean to be ISO 14001 registered? Will a company need to follow the ISO 14001 standard precisely in order to become registered? Does a company have the freedom to change or go beyond the ISO 14001 standard when developing its environmental management system? Since registration requirements are still in the process of full development, these questions currently have hazy answers at best.

The objective of this chapter is to demonstrate that ISO 14001 is just a minimum standard or the 'building blocks' for developing and implementing environmental management systems. However, as a minimum standard, it makes no commitments to areas such as environmental performance, environmental reporting or public disclosure.[5] In addition, there is nothing in the standard that prevents a company from doing more than the minimum. Using Ontario Hydro, a large North American electric utility as a case study, this paper will explain how it is possible to go beyond the specifications of ISO 14001 when developing an environmental management system.

The chapter begins by describing Ontario Hydro's early efforts in developing and implementing an environmental management system and some of the inherent problems associated with this activity. It then compares the ISO 14001 and Ontario Hydro environmental management systems and explains why it was necessary for the company to adopt ISO 14001 as the minimum requirement for environmental management systems. Finally, the chapter describes how ISO 14001 was extended in ten different areas, resulting in a revised corporate Environmental Management System (EMS) that has adopted the ISO 14001 format but has been enhanced to accommodate all of the components of the company's environmental management system.

■ Ontario Hydro: The Company

Ontario Hydro is a publicly-owned electric utility founded in 1906. Currently, it has an installed generating capacity of approximately 34,000 megawatts, which includes 69 hydroelectric stations, five nuclear stations, and six fossil fuel stations. Electricity is transmitted over 29,000km of transmission lines and 109,000km of distribution lines. The utility serves more than 3.7 million customers through a supply system that includes municipal electric utilities, retail customers, and large direct customers. The company's service area includes most of the Province of Ontario (see Fig. 1) with export sales to neighbouring provincial and US utilities.

Ontario Hydro has had a history of enormous growth and expansion in order to accommodate the increasing consumer demand for a cheap, efficient and reliable source of electricity. With regard to the environment, the company has been generally successful in complying with federal and provincial government laws and regulations. However, until recently, the environment did not always play a role in corporate business decision-making, and the promotion of an environmental agenda beyond compliance was the exception rather than the rule.

The 1990s heralded significant change in Ontario Hydro and, like many large corporations, it had to restructure in response to changing market conditions. It is also in the process of adjusting its position from monopoly to

Figure 1: *Ontario Hydro's Service Area*

marketplace, as competitors from other energy service industries have become more aggressive, resulting in customers being presented with alternative energy sources at competitive prices. To meet this challenge, the company has been restructured into 'Business Units', which have been established according to the nature of the company's various business activities, including: generation of electricity; transmission and distribution of electricity; customer services; research and development; and international business opportunities. The company also initiated a movement towards becoming a global leader in sustainable development, which included a recommendation for 'the establishment of environmental management systems, consistent with corporate goals, principles and policies'.[6]

■ Ontario Hydro's First Environmental Management System

In an attempt to implement consistent environmental management systems throughout the company, Ontario Hydro produced a corporate EMS model. This model was developed in 1994 by forming an internal Task Group made up of environmental experts from the respective Business Units. At that time, the concept of environmental management systems was relatively unknown, although there was familiarity with generic management systems, i.e. 'Plan, Execute, Control'. Several quality management system programmes had also been introduced, but they were usually carried out on an ad-hoc basis, and no initiative was taken to become ISO 9000 registered.

Given this situation, a seminar was conducted to educate the Task Group on environmental management systems. Examples of existing national association standards were discussed, such as the draft British Standard BS 7750 and the draft Canadian Standards Association Z-750. The seminar was followed by an environmental management system workshop, which resulted in the identification of sixty potential sub-elements as possible entities for inclusion in an environmental management system. The Task Group narrowed down these sub-elements, resulting in a corporate EMS model that would serve as a guideline for the rest of the company in developing environmental management systems.

As illustrated in Figure 2, the model is comprised of four main elements: Communication, Planning, Implementation, and Assessment, and basically follows a 'Plan–Do–Check–Act' cycle. To indicate continual improvement in the environmental management system, arrows link the three main elements:

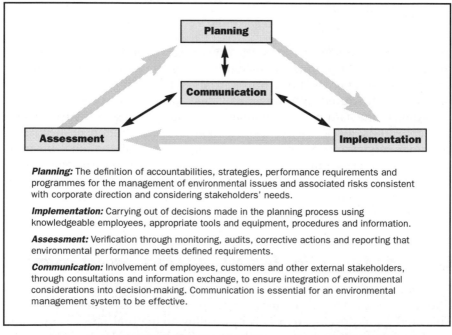

Planning: The definition of accountabilities, strategies, performance requirements and programmes for the management of environmental issues and associated risks consistent with corporate direction and considering stakeholders' needs.

Implementation: Carrying out of decisions made in the planning process using knowledgeable employees, appropriate tools and equipment, procedures and information.

Assessment: Verification through monitoring, audits, corrective actions and reporting that environmental performance meets defined requirements.

Communication: Involvement of employees, customers and other external stakeholders, through consultations and information exchange, to ensure integration of environmental considerations into decision-making. Communication is essential for an environmental management system to be effective.

Figure 2: *Ontario Hydro: Environmental Management System Model*

Planning, Implementation, and Assessment. Communication was placed in the centre of the cycle, with two-way arrows to each of the three other elements to signify the need for effective communications in all areas of the environmental management system. Generic definitions were then developed for each of the elements.

In coming to an agreement on this model, the Task Group felt that an effective environmental management system should emphasise five main areas:

- **Generic Model.** The model had to be generic in order to provide the Business Units with the flexibility needed to develop environmental management systems that would meet their specific environmental requirements.

- **Environmental Policy.** An effective environmental management system depends on having an environmental policy statement in place. When the corporate model was developed, the only official environmental policy statement was outdated and did not apply to the new direction of the company. This shortcoming was identified by the Task Group as a major deficiency that the company had to address if effective environment management systems were to be established.

- **Communications.** Effective environmental management depends on communications being an integral part of the entire environmental management system. Subsequently, the Task Group insisted that 'Communication' be highlighted as a main element in the model.

- **Accountability.** Accountability needed to be included as part of the 'Planning' element, based on the firm belief that before implementation of an effective environmental management system can take place, employees have to know who is accountable for the environment at all levels of the organisational structure.

- **Reporting.** Environmental performance reporting was considered to be an essential part of a environmental management system. Without it, there is no way of determining improvement over time.

■ Implementing Environmental Management Systems

The model was accepted by senior management as the corporate EMS standard and a programme was initiated to develop, implement and document environmental management systems at three levels of the company: corporate, Business Unit, and site/facility. This programme was to be carried out in 1995, but the programme was very slow in developing for the following reasons:

- An overall perception problem in some Business Units regarding what an EMS looks like and whether there are any good examples available from which to learn
- Environmental accountability and organisational structures were well established in some Business Units but less well established in others
- In some cases, there were insufficient resources allocated to carry out the programme
- Environment was still perceived in some areas of the company as an add-on rather than a fundamental re-think of all operations and activities.[7]
- There was no priority for getting the work done, since no actual time line or schedule had been given.
- There was no reward or recognition programme—such as performance contracts for senior management—involving the development, implementation and documentation of an environmental management system.

■ The Emergence of ISO 14001

While the company was struggling to develop its environmental management systems, the ISO 14001 standard emerged and rapidly progressed to final draft status. With the final version now approved, ISO 14001 is expected to be recognised as the official global standard for environmental management systems, which will have a significant impact on the more environmentally-progressive companies who have already developed their environmental management systems. A recent informal survey of eighty worldwide industrial corporations found that a high percentage of the companies have developed their 'own' environmental management systems, which are either similar to, or are about to be adjusted to follow, ISO 14001.[8] For these companies, the impact of ISO 14001 means spending additional resources in order to bring their environmental management systems in line with the standard.

Briefly, the ISO 14001 model (see Fig. 3) identifies five elements: Environmental Policy, Planning, Implementation and Operation, Checking and Corrective Action, and Management Review. To indicate continual improvement, an arrow 'spirals' through the model.

A comparison of Ontario Hydro's EMS model and the ISO 14001 model identifies similarities and differences. Both models resemble flow diagrams and make effective use of arrows to illustrate continual improvement. 'Planning' and 'Implementation' have been identified in both models as main elements, and there are several examples of common terminology such as: programmes, audits and monitoring.

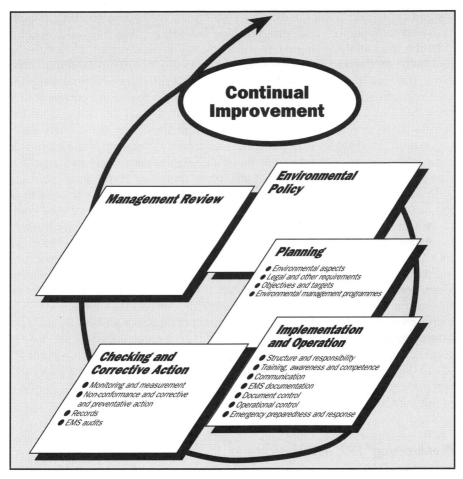

Figure 3: *ISO 14001 Environmental Management System Model*

The comparison also reveals differences between the two models. The ISO model identifies 'Environmental Policy' and 'Management Review' as main elements, yet neither are referred to in the Ontario Hydro EMS model. In addition, a different emphasis has been placed on some components of the models. For example, the ISO 14001 model identifies 'Checking and Corrective Action' as a main element and 'Communication' as a sub-element. Conversely, the Ontario Hydro EMS model identifies 'Communication' as a main element and 'Corrective Action' is included as part of the 'Assessment' definition. Finally, there are numerous terminology differences between the Ontario Hydro EMS model and the ISO 14001 model, such as: 'issues' instead of 'aspects'; 'knowledgeable employees' instead of 'training, awareness and competence'; and 'programmes' instead of 'objectives and targets'.

In the final analysis, the comparison clearly indicates that, although the ISO 14001 model and the Ontario Hydro EMS model are fundamentally the same, they are 'packaged' quite differently. This difference became a significant concern within the company. Questions arose regarding compatibility with ISO 14001 and the potential risks associated with deviating from the standard. In particular, there was apprehension about how third-party auditors would interpret the different emphasis and terminology of the corporate model. However, since there was no official certification body in Canada, it was not possible to get any formal advice on this issue at the time.

To address these concerns, the EMS Task Group was re-convened to assess the future direction of environmental management systems within the company. Consensus was reached that there needed to be a change in direction to be 'more in line' with ISO 14001. Further, the Task Group decided to make ISO 14001, as a minimum, the corporate standard for developing environmental management systems. Despite the consensus, several of the Business Units were uncomfortable with the fact that they had already undertaken significant development of their own environmental management systems using the corporate model, and the prospect of 'starting all over again' was unacceptable. Therefore, the Business Units were given sufficient latitude to continue using the corporate model, provided all elements and sub-elements of the ISO 14001 standard were addressed. Any differences in emphasis and terminology would be cross-referenced in the environmental management system documentation.

■ *'Abusing' ISO 14001*

From Ontario Hydro's perspective, the decision to make ISO 14001 the corporate standard for environmental management systems was very significant, because now the company had a internationally-recognised standard that provided the guidelines for developing environmental management systems. However, it was also recognised that the ISO 14001 standard did not include all of the components that were part of the corporate environmental management system. To address this concern, Ontario Hydro enhanced its corporate environmental management system by following the ISO 14001 standard format while at the same time extending the standard by changing terminology and adding new sub-elements. This 'abuse', as it might be termed, occurred in ten areas of the ISO 14001 standard, as follows.

☐ *Communication*

In the ISO 14001 standard, 'Communication' is included as a sub-element within the 'Implementation and Operation' element. The standard states that there is a need for 'internal communication between the various levels and functions of the organization' and communications are required with 'external interested parties'.

As previously stated, there has always been a strong feeling within the company that 'internal communication' is essential for an effective environmental management system. This need has been identified by making 'Communication' a sub-element within each of the five main elements of the revised corporate EMS.

☐ *Sustainable Energy Development Policy and Principles*

The ISO 14001 standard defines 'Environmental Policy' as one of its main elements. The definition stresses, among other things, the need for environmental commitment, documentation, and communications.

In keeping with its corporate objective of being a leader in sustainable development, the company has developed and endorsed a 'Sustainable Energy Development Policy and Principles' statement which includes seven governing principles and seven strategies that provide direction for the company to become more sustainable and environmentally responsible. To accommodate this new official policy, the revised corporate EMS has replaced the element 'Environmental Policy' with 'Sustainable Energy Development Policy and Principles'.

☐ *Senior Management Commitment*

The ISO 14001 standard states briefly that 'top management shall define the organisation's environmental policy', but it does not provide any further direction to ensure that the environmental commitment from top management is continuous.

Continuous top-down management support for the environment is required for an effective environmental management system. This commitment starts with the Board of Directors and cascades down through line management. Also, this commitment must be continually communicated, both internally to employees and externally to the competitive marketplace, in order to demonstrate that the company is serious about its environmental obligations. To stress the need for continuous top-down commitment to the environment, the revised corporate EMS has identified 'Senior Management

Commitment' as a new sub-element within the 'Sustainable Energy Development Policy and Principles' element.

☐ Environmental Issues

The ISO 14001 standard identifies 'Environmental Aspects' as a sub-element and states that 'the organization shall establish and maintain a procedure to identify the environmental aspects of its activities, products and services that it can control and over which it can be expected to have an influence, in order to determine those which have or can have significant impacts on the environment'.

The meaning of the word 'aspects' is ambiguous in a company such as Ontario Hydro, with a history of using the term 'environmental issues' as a means of identifying the impacts of its activities, products and services on the environment. Consequently, it would be very difficult to introduce 'aspects' into a corporate culture that knows and understands the meaning of 'issues'. To avoid any misunderstanding, the terminology in the revised corporate EMS has been customised by substituting 'Environmental Issues' for 'Environmental Aspects'.

☐ Accountability

The ISO 14001 standard identifies 'Structure and Responsibility' as a sub-element and makes reference to the fact that 'roles, responsibility, and authorities shall be defined, documented, and communicated in order to facilitate effective environmental management'.

As part of its commitment towards leadership in sustainable development, Ontario Hydro is making its employees more accountable for environmental matters. To do this, environmental roles and responsibilities are being defined at all levels of the organisational structure and, in some cases, environmental performance is being measured through the use of performance contracts. To confirm this corporate-wide initiative, the revised corporate EMS has replaced the 'Structure and Responsibility' sub-element with an 'Accountability' sub-element.

☐ Business Plan Decision-Making

The ISO 14001 standard also makes reference in the 'Structure and Responsibility' sub-element to management providing 'resources essential to the implementation and control of the environmental management system'.

At Ontario Hydro, the decisions regarding the allocation of funds for all corporate programmes including environmental ones, are made during the annual business planning process. The budget is finite, and the environment is in competition with other programmes for limited resources. The simple truth is that 'what gets money gets done'. To protect against inadequate funding for environmental programmes, the revised corporate EMS has added 'Business Plan Decision-Making' as a sub-element.

☐ *Training, Education, Awareness and Competence*

The ISO 14001 standard identifies the sub-element 'Training, Awareness and Competence' and states that 'all personnel whose work may create a significant impact upon the environment, have received appropriate training'. But the standard makes no reference to the environmental education of employees.

As part of its commitment to leadership in sustainable development, the company has put a high priority on environmental education of employees, and considers it to be equal in importance to environmental training. To ensure that this emphasis is maintained, the revised corporate EMS has added 'Education' to the sub-element title.

☐ *Environmental Audits*

The ISO 14001 standard declares the need to 'establish and maintain (a) programme(s) and procedures for periodic environmental management system audits to be carried out in order to . . .'. However, there is no reference to other types of environmental audits, such as compliance audits.

Environmental auditing programmes have been conducted within Ontario Hydro for a number of years and currently include three types of audits: environmental compliance, environmental management system, and environmental issue management. In addition, a corporate Environmental Audit Standard has been developed that defines the elements, principles, practices and ethics required for a comprehensive environmental audit programme. Included in this audit standard is a need to identify corrective action plans on audit findings. These plans are considered by the company to be as important as the audit itself. As a means of assuring that the audit programme is not restricted in scope, the revised corporate EMS has generalised the title of this sub-element to 'Environmental Audits' instead of 'Environmental Management System Audits'.

☐ Environmental Performance Reporting

The ISO 14001 standard makes reference, under the 'Structure and Responsibility' sub-element, to 'reporting on the performance of the environmental management systems to top management for review and as a basis for improvement of the environmental management system'. However, it makes no reference to the production of any other environmental performance reports.

Ontario Hydro has a history of producing corporate-level environmental reports, including: the annual environment/sustainable development report; due diligence reports to the Board of Directors and Chief Executive Officer; quarterly environmental reports to the Board of Directors; and the reporting of significant events such as spills to senior management. Further, reports have always been perceived as a good method of communicating, internally and externally, on environmental performance. To make sure that this long-standing corporate activity is maintained, the revised corporate EMS has identified 'Environmental Reporting' as a new sub-element within the 'Checking and Corrective Action' element.

☐ Rewards and Recognition

The ISO 14001 standard makes no reference to recognising and rewarding employees for significant environmental achievements or contributions to environmental improvement.

To make an environmental management system work effectively, employee support is extremely important. One of the methods of encouraging this type of commitment is through the implementation of a rewards and recognition programme. As part of its corporate-level environmental management system, the company has established the 'President's Sustainable Development Awards Program' to recognise formally outstanding employee contributions to environmental and sustainable development initiatives. To acknowledge this programme, the revised corporate EMS has included 'Rewards and Recognition' as a sub-element within the 'Management Review' element.

As a result of these ten areas of 'abuse', the revised corporate-level EMS has taken on the appearance of ISO 14001, but also includes the extensions (see Fig. 4). Currently, the revised corporate EMS is still a draft and subject to change, depending on senior management direction or possible revisions to ISO 14001 following publication, as part of the process of practical implementation. Formal documentation of the corporate-level environmental management system will then follow. At the same time, the Business Units have the freedom to develop their environmental management system in the format

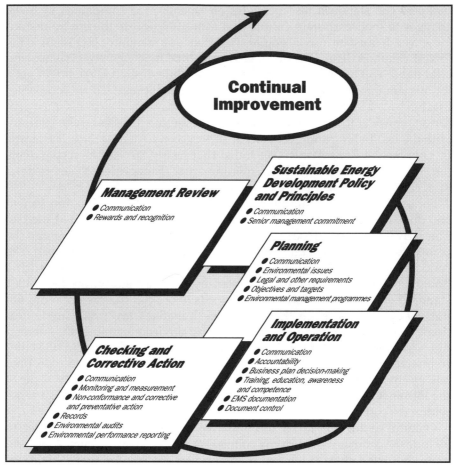

Figure 4: *Ontario Hydro's Revised Environmental Management System*

that meets their specific needs. The only qualifier is the requirement to satisfy, as a minimum, the ISO 14001 standard, an important consideration if future systems are to seek certification against the standard.

■ Conclusion

The ISO 14001 standard provides the world with a recognised specification standard to develop environmental management systems, and the opportunity now exists for the environment to be considered in corporate decision-making. The standard also provides an opening for companies to make changes to their corporate culture and attitude towards the environment.[9] However, it must be also be emphasised that, as a standard, ISO 14001 is the lowest common denominator for environmental management systems, and although it will not get companies to the absolute desired level, it will get a lot of them into the game and at least paying attention.[10] Those companies that have already developed environmental management systems must give serious consideration to making whatever changes are necessary to conform with the ISO 14001 standard. A recent survey, conducted by Arthur D. Little Environmental Consultants, concluded that nearly 86% of the respondents felt it was important that their environmental management systems be at least equivalent to the ISO 14001 standard.[11]

For over two years, Ontario Hydro has attempted to develop, implement and document its environmental management systems with limited success. The arrival of the ISO 14001 standard has brought a new definition into the company of environmental management systems and how they work. The concept of environmental management systems may no longer be such a 'tough sell'. However, as demonstrated by this case study, the ISO 14001 standard did not completely fulfil the needs of the company, since it did not put enough emphasis on the company's initiatives towards communications, senior management commitment and environmental audits; neither did it include business plan decision-making for the environment, the environmental education of employees, environmental performance reporting, and an environmental rewards and recognition programme. Finally, some of the terminology had to be changed in order for the standard to make sense within the corporate culture.

Ontario Hydro has taken the initiative to alter ISO 14001 in order to include the company's entire environmental management system. Any company that has either developed a comprehensive environmental management system, or is about to embark on such development, should consider taking a similar approach. First, they should ensure that the ISO 14001 standard is met as a minimum, and then tailor their environmental management system to the specific needs of the company.

■ *Notes*

1. See Arthur D. Little, 'Businesses Believe EMS Standard will Influence Success', in *Eco.Log Week*, 1 December 1995, p. 3.
2. See J.H. McCreary, *ISO 14000: A Framework for Co-ordinating Existing Environmental Management Responsibilities* (Rochester, NY: Nixon, Hargrove, Dewars & Doyle, 1995).

3. A Canadian court decision (*R. v The City of Sault Ste Marie* 85 DLR, 3d, p. 185) described 'Due Diligence' as follows: 'The Defendant must establish on the balance of probabilities that they were **duly diligent**, that is, they must establish that they exercised **all reasonable care** by establishing a *proper system* to prevent the commission of the offence and by taking **reasonable steps** to ensure the *effective operation of the system*. The availability of the Defence to a Corporation will depend on whether such **due diligence** was taken by those who are the directing mind and will of the Corporation, whose acts are therefore in law, the acts of the Corporation itself.'

4. See D. Bisson, 'ISO 14000: What is it and Why was it Developed?' (Conference Proceedings, Canadian Environmental Auditing Association, Toronto, Canada, 26–27 October 1996), p. 22.

5. See Benchmark Environmental Consulting, *ISO 14000: Uncommon Perspective: Five Questions for Proponents of the ISO 14000 Series* (Brussels, Belgium: The European Environmental Bureau, October 1995), p. 4, and Chapter 2 of this book.

6. Ontario Hydro, *A Strategy for Sustainable Energy Development and Use for Ontario Hydro* (unpublished internal report; Toronto, Canada: Ontario Hydro, 1993), p. 72.

7. See W. Wehrmeyer and S. Rees, 'Who's Afraid of Local Agenda 21?', in *Greener Management International*, No. 10 (April 1995), p. 81.

8. See Asea Brown Boveri, *Survey on Implementation of Environmental Management* (unpublished survey; Växjo, Sweden: Asea Brown Boveri, 1995).

9. See McCreary, *op. cit.*, p. 6.

10. See 'ISO 14000: Boon or Bust?', in *Business and the Environment*, Vol. 7 No. 1 (January 1996), p. 3.

11. Arthur D. Little, *op. cit.*, p. 3.

15 EMAS Implementation at Hipp in Germany

Matthias Gelber, Bernhard Hanf and Sven Hüther

I N THIS CHAPTER, the authors look at Hipp, which was the first industrial food producer in Europe to receive recognition under the Eco-Management and Audit Scheme (EMAS). Hipp is the world's largest user of organically-grown raw material and can claim to be led by ethically-motivated company owners, making the company a rich case study, especially with its long traditions in environmental management. The chapter describes the company's experiences in the course of implementing an environmental management system (EMS) in line with their already existing environmental activities. Some lessons of practical relevance are highlighted, and the framework of EMAS, as applied in Germany, is set out.

■ EMAS in Germany

EMAS has been seen by many companies in Germany as a means to express and market their high environmental profile. A real struggle has been taking place to be the first in a particular sector of industry to attain EMAS, resulting in 169 German companies registered to EMAS by May 1996. The market forces—which the European Commission hoped would arise in order to motivate companies to participate in the voluntary scheme—have been highly visible, resulting in considerable uptake of EMAS by German industry. Beside anticipating increasing pressure from their retailers to have a site registered for EMAS, Hipp saw in EMAS the opportunity to give their long history of environmental management a publicly-recognised framework.

Hipp's main competitor in the German market, Alete, originally made a statement that it would not consider the implementation of EMAS, but, shortly after Hipp was verified, they announced their own verification to EMAS. This development may be seen as surprising in the light of the early resistance of German industry to the scheme and the fact that voluntary EMSs represent quite a different approach from the classical German legislation-driven environmental technology focus. However, the degree to which companies implement an EMS and hence derive an environmental programme varies considerably. Hipp had been practising environmental management long before EMAS was on the horizon, as evidenced by awards in 1989 and 1992 for proactive environmental management from an industrial working group. Criteria for these awards included: compliance with environmental legislation; documented purchase guidelines for products and raw materials with lower environmental impacts; and other issues describing elements of an EMS. However, the additional potential for improved environmental performance, realised through the structured EMAS approach, has been considerable. It has resulted in a comprehensive environmental programme.

In contrast to the UK, Germany has not developed an environmental management standard, nor adopted one. This has meant that the EMS Hipp implemented was guided by the text of the EMAS Regulation alone.[1] Because the Regulation itself is not very well structured—and, in some areas, open to interpretation—many questions had to be clarified. This was done by consulting the individual verifier, the EMAS accreditation body in Germany (DAU), the IHK (Chamber of Commerce) and some of the numerous published EMAS guides derived from various funded pilot projects. Additional support came from collaboration via a national sector-specific association of which Hipp is a founder member.

■ Hipp

Hipp is located in northern Bavaria at Pfaffenhofen, employing over 800 people. In 1994 the turnover was DM300 million (£135 million). The headquarters of the company are in Pfaffenhofen, and associate companies are Hipp Austria and Bio-Familia in Switzerland. Products range from baby milk to bottled and powdered baby food, as well as tea drinks—in the UK, some of the company's products are retailed by Boot's the Chemist. The foundation for Hipp's business success is its corporate philosophy, which aims to achieve consistently superior quality with minimal environmental impacts. Mothers' concerns about the health and well-being of their children is given utmost priority, with around 10% of employees working on quality assurance alone. The organic raw materials are purchased from renowned

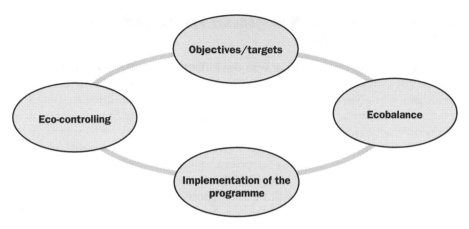

Figure 1: *The Eco-controlling Loop*

organic producer associations. Fruits, crops and vegetables have been culti-
vated on unadulterated soil—without the use of pesticides—and processed
at Bio-Familia in Switzerland since 1956 in collaboration with Dr Hans
Müller (who is the founder of organic biological cultivation).

In 1959 the parents of the current partners Claus, Georg and Paulus Hipp
used the experiences gained in Switzerland to direct their own farming
methods towards organic agriculture. The first farmers to be contracted have
been producing raw material for Hipp according to organic cultivation
methods since 1963.

In 1992 a full-time environmental manager was employed, and one year
later a comprehensive waste programme was developed with targets of minimisation, separation, re-use and disposal. Since 1993 an ecobalance[2] has
measured all the company's materials and energy streams, which was the
starting point for a comprehensive eco-controlling[3] concept (see Fig. 1). The
results have been encouraging. Water consumption of 22.6 litres per kilo-
gramme of product produced in 1989 was reduced to 14.5 litres by 1994;
around 96% of waste from production is recycled as fertiliser or as feedstock
for animals; and altogether the amount of disposable waste has been reduced
by 68% in the last five years through minimisation and re-use. In total, these
measures have resulted in estimated savings per year of DM300,000 on water
and effluent, DM200,000 on energy and DM100,000 on waste. Labour savings have also been achieved. However, the investment required was significant. For example, DM30 million was spent on developing a sterilisation
process with water re-circulation.

■ The Environmental Management System and Environmental Policy

The environmental programme was developed using the results and information provided by the ecobalance, the initial environmental review, and a weak point analysis; it contains measures and responsibilities that lead to continuous improvement in environmental performance. Systematic control of enforcement and effectiveness of the environmental measures will be achieved through an environmental audit,[4] undertaken in accordance with the EMAS Regulation, to assess the extent to which environmental objectives and targets have been met. Hipp plans to undertake the audit in early 1997. The next EMAS verification will be in Autumn 1998.

The environmental policy contains a commitment to protect the natural environment, reflecting the strong ethical motivation of the company owners.[5] This standpoint has always been the key factor behind the priority environmental protection is afforded at Hipp, although it is also seen as a strategic business opportunity. It was seen as crucial to involve the different departments in developing the policy and the programme in order to gain their support during the implementation stage. About thirty people from middle management level have been involved in drawing up the policy, and this assisted in securing wide approval throughout the company. Several drafts were circulated, and amended with the help of comments received. When objectives require additional investment and a significant change in attitudes and behaviour, resistance will be encountered, but Hipp have found that early and thorough involvement of employees is an important factor in the success of such an ambitious programme.

■ Ecobalance

Hipp aims to evaluate its environmental impacts with the help of an ecobalance. The underlying concept behind this method (as with a product life-cycle analysis) is that there is a cycle of processes all resulting in environmental impacts: extraction of raw material; production use; and disposal. All materials and energy entering the company will sooner or later leave again in a different form: the amount of input equals the amount of output, with the inputs being transformed during the production process into outputs.

The history of ecobalancing at Hipp started even before EMAS was on the company agenda, when Hipp realised, in 1992, that its data on environmental aspects was neither comprehensive, structured, summarised nor interpreted. It was known, for instance, how much water was used by the whole company, but it was not possible to trace the points of consumption exactly. Weaknesses were not properly analysed and, therefore, the control of environmental performance was very difficult. In 1993, Hipp decided to collect and summarise the data with the help of an environmental software programme.

Employee involvement while conducting an ecobalance was of central importance, making individuals in the organisation aware of their own behaviour and their contribution to the overall environmental impact of the firm and thus strengthening motivation.[6] During the development and implementation of the ecobalance, a project team of two external advisors and three Hipp employees was formed. The ecobalance is viewed from four different levels: the state of the environment; the Hipp site; process groups; and individual processes.

The corporate balance captures all the in- and outgoing material and energy streams of the Hipp site in Pfaffenhofen, compiled on a yearly basis since 1993 with the help of a balance sheet format. The corporate balance does not show where energy is used and where raw materials on the site have been consumed, nor who are the producers of pollution on the site. With the help of process balances, these major points of consumption and pollution can be allocated. The whole company is separated into five areas: production, infrastructure, storage, transport and administration. Additionally, materials moved between the different areas, but which did not leave the site, were registered as intermediate steps. Special attention was given to the processes in the production area, because here the most significant use of auxiliary materials, water and energy was found, as well as the highest sources of environmental impact.

A major advantage of the ecobalance as a core tool during the implementation of EMAS has been the relative ease with which the data demands of the public environmental statement could be put together. The requirement of the Regulation (Article 5) to produce summarised data on waste, emissions, energy usage, raw material usage, water usage and noise was also easy to fulfil by making use of the already-existing data contained within the ecobalance. Thus computerised data collection provides a comprehensive integrated environmental information system underpinning continual public environmental reporting. Additionally, this system will provide a comprehensive database recording the development of overall environmental performance from when the system was introduced. Consistency is obtained via a standardised mode of data collection, which makes third-party assessment of the public environmental statement and the relevant data much easier and more efficient.

The next step is to allocate energy consumption to the relevant cost centre and, later, to the cost carrier. Until now, electrical energy consumption has been one of the few areas where Hipp has not seen any significant performance improvement over the years. Only with proper allocation is efficient consumption control—and activities aimed at achieving a reduction of consumption—possible. This is aided by increasing the potential of employees to be motivated by making the data continuously available. The current practice of not measuring environmental costs at source does not motivate the various departments to reduce consumption. The real product costs are not allocated because they are determined by traditional accounting methods. First results of the analysis at Hipp have shown that, for the electrical energy needed for production of individual products (cost centres), a difference of a factor of between 3 and 4 exists. These are Hipp's first experiences with environmental cost management, which, in connection with the information on the materials and energy streams in the company, can support other controlling instruments. This means that, for an investment appraisal, environmental costs—which are far from being insignificant—can be taken into consideration.

■ *Environmental Performance Indicators*

The absolute figures derived from the ecobalance provide only limited scope for comparison of continuous development of environmental performance and, also, comparison to other companies is difficult. However, by linking the absolute figures to a given quantifiable unit, such as a tonne of product produced, comparison and assessment of developments becomes possible. Hipp succeeded in expressing the following indicators for the last three years per mass unit of product produced: water, energy, effluent, carbon dioxide, sulphur dioxide, nitrogen oxide and disposal costs.

■ *Ecobalance and ABC Analysis*

An additional relative impact assessment of the ecobalance is performed through the ABC Bewertung (ABC assessment), which was developed by the IÖW (Institute for Ecological Research in Economics) in Berlin and Prof. Dr Volker Stahlmann from the Nürnberg Polytechnic.[7] This method takes into account the fact that environmental problems are generally very complex and difficult to judge, and that environmental threshold values often derive from political compromises. Generally, there are no clear-cut answers on the significance of environmental aspects. However, the management of a company has to set priorities concerning their environmental activities. The ABC assessment uses the following six ecological criteria for deciding the priorities according to which environmental issues related to the business will be tackled:

1. Legislative and political requirements related to the environment
2. Acceptance in society
3. Accident and risk potential
4. Internalised environmental costs
5. Negative external effects
6. Disposal and recycling

A qualitative ranking of 'A' for a specific issue indicates an urgent need for corrective action; 'B' is less acute and can be tackled during a longer time frame; 'C' means that the issue is harmless according to current knowledge. By comparing all the scorings for the six ecological criteria relating to the different environmental aspects of the business, management has an environmental decision-making tool, which provides the following advantages:

● The results can be easily presented in table format, which provides a helpful overview before priorities are set (colours can be used for the presentation instead of 'ABC', with red, A, being used for the aspects requiring urgent action, yellow for B, and green for C).

● The qualitative rankings for different products and processes can be related and compared to one another.

● Internal and external changes of circumstances—detected through stakeholder analysis—provide no difficulties for this relative approach and can be easily integrated into the methodology.

● Auxiliary materials and cleaning liquids have been included in the analysis. Some chemicals, such as Neomoscan RD, which is a cleansing liquid containing chloride, were proven to be in urgent need of attention.

■ Employees and Environmental Management

The human factor plays an important part in environmental management and in the responsible use of resources. Hipp is very aware of this, and tries to involve its employees significantly, having traditionally aimed for environmentally-aware employees capable of skilled labour at a professional level. Co-operation by every employee in environmental management at Hipp is expected, which is made explicit by the fact that every employee has to sign an environmental declaration containing a commitment to prevent action or behaviour that damages the environment. The environmental policy is mailed to each employee's home along with their salary.

Even before EMAS was implemented, various environmental working groups were founded within the Hipp company: a transportation committee with the aim of improving the company's transportation issues relating to employees and products; a hazardous material group with the aim of ensuring the safe handling of such substances; and an eco-group, established to allow environmentally-proactive employees to meet four times a year. However, the initial environmental review revealed that Hipp still had some weaknesses in this area and that overall comprehensive training of staff had not been achieved. There is now a strategy in place to deliver environmental training through the following means:

● Vocational training
● Information via the Hipp newspaper
● An environmental report presented to each employee
● A hotline for environmental protection
● Continually-updated environmental information on the various notice-boards
● Continually-updated environmental training of the purchasing department
● Participation of the environmental department in production and management meetings
● Presentations by the environmental department to external audiences

■ Organic Cultivation and the Collaboration with Suppliers

The platform for collaboration with Hipp's suppliers has been the company's specifications on organic cultivation in accordance with the guidelines of the relevant associations. Meanwhile, these have been supplemented by the EU Regulation 2092/91 dating from 1991 concerning organic agriculture. Originally, the relationships between Hipp and its suppliers were seen as pioneering, but now such activities are more common, having developed into well-organised supply systems with good market structures. The following criteria are applied to Hipp's suppliers:

● A chemical analysis of the soil decides on whether arable land is suitable. Additionally, fields must be far from harmful emission sources and heavy traffic. Even wind direction is taken into consideration.
● In the organic cultivation, only untreated seeds are permitted. This is double-checked in Hipp's own laboratory.
● From sowing-time until harvest, the plantation is continually controlled by Hipp's own cultivation advisors.
● The use of chemical or synthetic sprays is absolutely prohibited. This means that farmers have to remove weeds by mechanical means or by

hand. This is more costly, but protects the environment and conserves raw materials.

● The pest control methods used are strictly natural ones which are defined and documented, such as the use of stinging nettle liquid manure.

● By rotating crops, natural soil fertility and ecological balance is achieved.

● By avoiding the use of inorganic fertiliser and cultivating crops organically, the health of both the produce and the soil is maintained.

Overall, the proportion of organic raw material used has grown from 18.4% in 1990 to 69.1% in 1995. It is hoped to increase the proportion of organic raw material even further, but this is difficult to achieve, as it would make some fruits and spices prohibitively expensive. Hipp's products are already 15%–20% more expensive than their direct competitors, and there is only a limited margin for increased prices in the name of superior quality. However, so far, the market in Germany *has* shown that it is willing to pay a higher price for organic produce, with Hipp holding over 50% of the country's market share. The company was bombarded with hundreds of additional telephone enquiries during the current BSE crisis. Hipp has used beef and veal from free-range registered organic farms in Germany for many years.

Hipp has always had a very strong and influential relationship with its suppliers. The use of organic raw material has already displayed the extent of commitment towards achieving ecologically-sound business practices throughout the supply chain. By addressing environmental issues on the supply side, negative impacts can be minimised throughout the entire product chain without having to rely on end-of-pipe measures. Hipp's environmental policy 'expects suppliers and the service industry to fulfil environmental standards comparable to Hipp themselves'. One of the objectives of Hipp's environmental programme is to purchase increasingly from suppliers with an EMS in place and to verify their co-producers to EMAS—environmental checklists have been developed in order to assess their suppliers. An additional objective—which is required by EMAS—is to provide suppliers with information on Hipp's environmental management.

■ *Association of Ecological Food Producers*

Three medium-sized companies—Ludwig Stocker Hofpfisterei (organic bread), Neumarkter Lammsbräu (organic beer) and Hipp (baby food)—formed the Association of Ecological Food Producers in 1994. The association has also been joined by Meyermühle (flour) and Dairy Scheitz (milk products). These family businesses are pioneers in organic food production and have been concerned for many years with the implementation of ecologically-sound business policies. Altogether the members process raw materials from approximately 1,000 farms covering collectively around 10,000 hectares.

The following objectives are pursued by the association:

- Support for organic agriculture
- Development of a market for organic food
- Advocating on behalf of organic food producers
- Contact with interested members of the public

The members meet twice a year to exchange current experiences and plan joint activities, with workshops covering topics such as environmental management, quality management, logistics, laboratory results and joint business activities. This close collaboration and open exchange of environmental expertise has facilitated environmental benchmarking. During 1995/96, members of the association are collaborating with one another over the implementation of EMAS, with the objective of verifying all members to EMAS, an activity that has proved very helpful in interpreting the Regulation and taking sector-specific issues into consideration.

■ Validation of the Public Environmental Statement

The accredited individual EMAS verifier who performed the validation of the public environmental statement at Hipp was Prof. Dr Helmut Wirner.[8] The validation was a two-step process beginning with a one-day preliminary audit focusing on gathering documented evidence relating to the central criteria of the Regulation that have to be covered. Hipp's management held an in-depth discussion of the relevant points with Prof. Wirner, which clarified the issues surrounding the requirements of the Regulation. The register of environmental legislation had to be slightly supplemented with information on existing licences, and operating procedures had to incorporate topics such as training needs, scenarios for accidents, subcontractors on the site, synergies with quality management, and proactive maintenance. The documentation was then analysed off-site. Before the final visit of the external verifier, the management board at Hipp agreed on the proposed environmental objectives and the environmental programme. The set objectives go far beyond legislative requirements and try to eliminate weak spots in order to achieve further improvements in the standard of environmental performance at Hipp.

For the principal audit, the management received a detailed two-day programme from the verifier prior to his arrival. On 15th November 1995, thirty-minute interviews were conducted with representatives of the following functions: board of directors, site management, quality management, research and development, packaging engineer, purchasing department, biological raw material co-ordination, energy and supply, logistics, sales, personnel, production and environmental management. The content of the interviews was related to function-specific environmental aspects— such as product lifecycle and objective-setting through the management appropriate to the leader of research and development—and corporate aspects such as the environmental policy.

On 16th November, a site tour of two hours' duration, with an emphasis on production and the storage of hazardous material, was conducted. The environmental programme, environmental risk management and the public environmental statement were tackled during a meeting with the environmental department. Two minor non-conformances of the public environmental statement were discussed and subsequently altered: the organisational chart did not clearly indicate the connection between the environmental manager and the board; and some formalities, such as the date for the next public environmental statement and the name of the external verifier were missing. A final report for the management was submitted by the verifier and a debriefing conducted with all those involved. The validated public environmental statement was passed on to the Chamber of Commerce for Munich and upper Bavaria, which is one of the competent bodies in Germany.[9] A statement from the local environmental enforcement agency, concerning the site's compliance with legislation, was ordered before Hipp was entered into the register as a site on 6th December. A press conference on 19th December 1995 announced Hipp's EMAS validation publicly to the press.

The public environmental statement has been distributed to employees, and the different business functions have passed the statement on to suppliers, major customers, business partners and local institutions. Other external requests for the statement have received a copy free of charge. The company was recently awarded second place in the category of medium-sized enterprises by the IÖW, which ranked the environmental reports published in 1995.

■ Costs for EMAS

Apart from the environmental statement, published with the help of an agency, and support from students in the installation of the ecobalance, Hipp did not involve any external consultants; only two additional placement students were present on site for a limited time period. This had the advantage of keeping the knowledge and the information relating to EMAS in-house and the costs relatively low. The system is up and running mainly due to the involvement of the company's employees, who have taken ownership of the system. Altogether the implementation of EMAS took about one year in man-hours, costing about DM100,000 of which a large portion was needed for establishing the ecobalance. Included in this is DM13,900 for the external verifier and DM800 for the competent body.

■ Synergies between Quality Management and Environmental Management

The company is currently in the process of implementing ISO 9000. In areas where it is relevant, Hipp is attempting to use synergies between the EMS and QMS, environmental aspects having been integrated in many quality guidelines in order to achieve environmental protection at the beginning of the product chain. Therefore, in areas such as new product developments, evaluation of suppliers and packaging guidelines, information and training of employees and planning and testing of new production processes, quality documents exist with integrated environmental aspects. On the other hand, separate environmental guidelines exist in areas such as waste segregation, storage of hazardous material and treatment and control measurements of the water supply.

■ Conclusion

Through verification to the European Eco-Management and Audit Scheme, Hipp has experienced additional benefits by giving their already existing environmental management a more systematic organisational framework. This has achieved a minimisation of risks and an overall internal and external increase in the transparency of their business. The identification of employees with the company has significantly increased during this process. It has been important to make the EMS fit into the already-existing culture and environmental management tradition of the business in order to exploit the innovative and strategic potential of EMAS relating to the whole product chain and the different business functions. It is important to set objectives and targets that go beyond a short-term financial payback period in order to receive public credibility, which is vitally important today. In conclusion, the fact that 1,500 requests for the environmental statement were received in the first three months of its publication clearly demonstrates that the public interest in EMAS in Germany is not to be underestimated.

■ *Notes*

1. Even though translated versions of BS 7750 and Committee Draft ISO 14001 have been available, most German companies have focused directly on the text and the approach given by the Regulation.

2. This refers to input and output balances, widely used by companies in Germany in order to identify their significant environmental aspects and the related environmental impact through an inventory of mass and energy flows inside the boundaries of an organisation. For further cases and a description of the approach, see C. Jasch, 'Environmental Information Systems in Austria', in *Social and Environmental Accounting*, Vol. 13 No. 2 (Autumn 1993), pp. 7-9; and M. Gelber, 'Eco-balance: An Environmental Management Tool Used in Germany', in *Social and Environmental Accounting*, Vol. 15 No. 2 (September 1995), pp. 7-9. In recent German environmental management literature, there have been debates on whether EMAS requires such a comprehensive data inventory or whether a more qualitative approach with some selected measurements might be sufficient.

3. The concept of eco-controlling is a control loop, making use of a monitoring base within the ecobalance. A so-called 'weak point analysis' provides the foundation for the implementation of measures aimed at improving environmental performance. Checking and corrective action are part of the eco-controlling loop, which is derived from the concept of financial controlling. Continually trying to steer performance with the help of environmental performance indicators deriving from the ecobalance is the method by which eco-efficiency is attempted. A publication by the Umweltbundesamt (German Department of the Environment), *Handbuch Umweltcontrolling* (München, Germany: Verlag Franz Vahlen, 1995), drew together all the leading German researchers in the field in an attempt to define the concept. This book has now become the standard text on eco-controlling in Germany.

4. Article 3 point (d) in the EMAS Regulation clarifies that, in order to be registered in the scheme, a company must 'carry out, or cause to be carried out, in accordance with Article 4, environmental audits at the sites concerned'. This means that it is not necessary for organisations to perform the environmental audit prior to their first EMAS registration. This has been practised in Germany, but in the UK, due to the fact that the general practice has been to go for certified EMS first (BS 7750), people seem to perceive that one cannot register for EMAS without having first performed the environmental audit. This opinion was recently confirmed to one of the authors during a conversation with an accredited UK EMAS verifier, who indicated that he would require the environmental audit for the first EMAS verification. The NACCB (now UKAS) document, *The Accreditation of Environmental Verifiers for EMAS* (London, UK: NACCB, July 1995), does not mention this point and only gives guidance that the environmental verifier shall check 'whether the audit is carried out in accordance to Annex II of the Regulation', which can lead to the assumption that it is needed for every verification. On the other hand, the *SCEEMAS Participants Guide* of the UK DoE (London, UK: HMSO, November 1995) confirms on page 20 that the environmental audit does not need to be performed prior to the first registration of a site for EMAS. There seems to be a need for clarification on this point in the UK's EMAS verification and implementation

practice. Clarification could make it more attractive for organisations to achieve EMAS on a shorter timescale and reduce the perception that EMAS is so 'much more hard work' due to the public environmental statement involved. However, one should be careful to not interpret this as a way of getting a quick and easy EMAS registration, because the mandatory initial environmental review requirements for EMAS according to the Regulation are very comprehensive.

5. A strong ethical motivation from the company owner can be found as a characteristic profile of the most famous German case studies in environmental management. Most of these companies are SMEs with a long tradition of family ownership, such as the recent ACCA environmental reporting award-winner Neumarkter Lammsbräu.

6. This approach is called in the literature a 'sociological' implementation of an eco-balance, as opposed to a 'technocratic' implementation, where an external consultant comes in and collects and interprets the data without the organisation itself being significantly involved (see J. Böning and S. Brückl, 'Regional-orientierte Ökobilanzierung', in *UmweltWirtschaftsForum*, Vol. 3 No. 2 [June 1995], pp. 12-18).

7. A more comprehensive description of the methodology can be found in K. Christiansen *et al.*, *The* PREMISE *Manual: A Manual for Preventative Environmental Management in Small Enterprises* (unpublished).

8. In contrast to schemes in other countries, such as the UK and Denmark, where only organisations have been accredited so far, there are individuals accredited in Germany as EMAS verifiers. This issue led to a fierce debate at a meeting of the European Organisation for Testing and Certification in January 1996. EMAS verifiers in Germany have to show expertise in environmental policy, EMS and sector-specific environmental issues through an oral exam, which about 46% of the applicants have failed so far. An academic qualification and three years of on-the-job experience related to environmental responsibilities are required by the German Accreditation Council for Environmental Verifiers (DAU).

9. The regional Chambers of Commerce are the competent bodies in Germany, and they number in total a surprising 136. This was the result of trying to give EMAS a notion of being part of a regional network between the regulators, chambers of commerce, industry and the public. However, this system is being increasingly criticised as too bureaucratic and expensive.

16 Training

Preparations for Maintaining Effective Environmental Management Systems[1]

Gabriele Crognale

T O ESTABLISH a common understanding of what environmental management is, let us take a closer look at the main components that comprise a generic environmental management programme: the **structure**, **responsibilities**, **focus** and **ultimate goal**.[2] As an insight into this chapter—which provides examples of various types and levels of training for individuals engaged in some form of environmental responsibility—it is worth examining closely the main components of a generic environmental management system.

☐ The Structure

Environmental management may first come into being as a corporate document containing the mission statement and/or environmental policy providing the blueprint for the organisation to follow. From there, the documents are set in motion by the group of individuals charged with the execution of these corporate goals and maintaining a constant vigilance towards adhering to the regulations governing their industrial and/or manufacturing processes. These environmental management employees may also find themselves interacting with individuals within other business units of the organisation with whom there may be an interdependency. For example, while the environmental staff may be responsible for managing and properly disposing of chemical wastes at a manufacturing facility, if the purchasing or inventory department does not maintain an open flow of communications with the environmental department, employee health and safety and/or environmental controls may be compromised at some point. Proper training in various areas of concern can go a long way towards increasing the effectiveness and conscientiousness of these workers.

The Responsibilities

Taking a micro view of the individuals within environmental management, one may see individuals in various functions responsible for various facets of manufacturing-process-generated waste streams, such as hazardous wastes being handled, aqueous wastes being monitored and discharged, air emissions being monitored, appurtenant mechanisms being inspected, checked, calibrated, and the like, in order to address an array of various regulatory and organisational controls. It is understood by various forward-thinking environmental and other top managers within major manufacturing and chemical-related industries that proper and regular training of such individuals has now become a *de facto* way of doing business. In other words, these people understand that their investment of time, effort and resources in training does eventually provide tangible and measurable results arising from their cumulative efforts. These efforts can enhance their particular operations, and/or other operations within the companies in which they work. There is some link between training efforts and notable differences in employee activities. The degree of difference may depend on the effectiveness of the training provided and the receptiveness of the audience being trained. (See also the section 'Summing It Up: Lessons to Be Learned' on page 305.)

In addition, such training is seen as a prerequisite to providing greater assurance to management that incidents will occur less frequently, that employees will have a greater understanding of their responsibilities, and, in general, fewer excursions with some degree of environmental significance will occur in the work area. For some tangible proof to substantiate this premise regarding training, one need look no further than one's own environmental management programme.[3] In addition, since these workers are not necessarily isolated from workers in other business areas, their responsibilities could sometimes overlap with each other and provide an additional layer of shared responsibilities. For example, workers in inventory receiving could be bringing speciality chemicals into a facility that may require particular attention, handling or paperwork requirements, that could become regulatory issues for the environmental staff if not managed or co-ordinated properly. This could be especially true in situations where chemicals are imported from another country that may require some toxic chemical notification.[4]

☐ The Focus

Within any regulated organisation, whether it be in the US or elsewhere, one of the primary focuses of environmental management is to maintain a programme that meets or possibly exceeds regulatory compliance. Besides the benefit of not being subject to repeat regulatory inspections at any governmental level and possibly being levied more stringent enforcement actions and fines, maintaining a compliant programme helps to ensure worker health and safety, can decrease the risks of releases that could be harmful to the environment, decrease future clean-up costs, could save monies in unnecessary waste of raw materials and excessive waste generation, and could actually become good PR[5] for an organisation. In short, maintaining regulatory compliance can positively impact the organisation's bottom line, even if, at first, it may be perceived as taking away from the (corporate) bottom line.

☐ The Ultimate Goal

While there may be variations of this general theme throughout various organisations, maintaining an environmental management system and programme that functions properly in all its areas of concern while providing for continuous revisions to improve upon existing conditions pretty much sums up the ultimate goal of many regulated organisations. Such goals may be embodied in the Chairman's message from a large publicly-held company's environmental report and in such a company's environmental policy and mission statement. Of course, each organisation's financial conditions may dictate the extent to which this ultimate goal is achieved. Smaller organisations (such as those referred to in the US as 'mom-and-pop' operations) may not have a large cash flow, and may need to budget their resources to achieve their stated goals.

■ How Training Lends Itself to a Proper EMS

While some pundits may argue as to whether the EMS standards can provide additional benefit to regulated industries, or that companies trying to decide whether to seek certification under ISO 14001 should first determine the standard's real value (and, as a result, some organisations may have adopted a 'wait-and-see' attitude to the standards and certification), it is important to stress that the standards were conceptualised and developed to provide a consistent and global environmental management 'guidance document' for use by any organisation.

Let us also take one step back to revisit the standard's development process. Since the standards were developed by environmental professionals from industry, governmental agencies and consulting and law firms as members of the various Technical Advisory Groups of ISO's Technical Committee 207, the embodiment of the standards consists of the collective experiences of these individuals from their day-to-day environmental compliance-related activities. The standards are not meant to replace or replicate existing environmental management systems at an organisation since many of these predate the ISO standards; rather, they are intended to provide a plan for those organisations that desire to develop or revise their EMSs to conform to ISO 14001. In addition, another advantage of maintaining an EMS in conformance with ISO 14001 would be to allow such organisations to sustain their EMSs in a more comprehensive and effective manner.

With that as a backdrop, let us now take a closer look at how good, effective and well-thought-out environmental training can make a difference over time to enhance the effectiveness and sustainability of a proper EMS.

☐ A Few Points to Consider

In the US, environmental training is a prerequisite required by the Environmental Protection Agency (EPA) under several federal regulations, such as: the Resource Conservation and Recovery Act (RCRA) and the Occupational Safety and Health Act (OSHA), to name but two. Specifically, in RCRA, a whole section is devoted to personnel training, which states in part that:

> Facility personnel must successfully complete a program of classroom instruction or on-the-job training that teaches them to perform their duties in a way that ensures the facility's compliance with the requirements of this part.[6]

Similarly, in ISO 14001, there is a reference to training allowing employees to be made aware of their environmental responsibilities within an operating facility. Within Section 4.4.2; 'Training, awareness and competence', there is a reference (see pages 186-87) to training that reinforces the importance of an organisation's responsibility to identify training needs and to ensure that all employees whose work may have a significant impact upon the environment have been trained. (Also important is the competency of the individuals performing their respective duties: see related examples in the section 'Summing It Up: Lessons to Be Learned' on page 305.)

In addition, environmental compliance audits, or audits of the EMS, can point to deficiencies related to inadequate or improper training. Similarly, continual improvement of the EMS can also hinge upon proper, focused and effective employee training, although the main driver is usually top management commitment and the diligence of environmental managers in following through with previous audit recommendations to ensure continual improvement. As a further link, continual improvement is reinforced in the embodiment of the future environmental performance evaluation standard (ISO 14031),[8] which includes a reference to matters for consideration as environmental performance indicators (EPIs) within the management area, such as: training, legal requirements, resource allocation, documentation and corrective action—which have, or can have, an influence on an organisation's performance.[9]

☐ Training: A Key Component within an EMS: The US Perspective

The reference to 40 CFR 265.16 in the US regulations was highlighted to point out the importance that the EPA places upon the proper training of personnel engaged in some aspect of hazardous waste management. Similarly, EPA-regulated industries also recognise and acknowledge the importance of specialised training, and spend considerable sums of money to provide such training to their affected employees. In addition, at the professional and some technical levels, environmental management programmes at various EPA-regulated organisations budget funds each year to allow staff to attend professional enrichment training, such as are provided at various training courses, conferences and symposia across the country—budget-dependent, of course—to sustain as well as increase their proficiency in the environmental area. These are commonly referred in the UK as 'continuous professional development'.[10]

Industries take all relevant, timely and appropriate training and technical sessions seriously as a means to enrich further the knowledge base of those professionals whose responsibilities also include imparting their skills to younger and less experienced staff and technical assistants at their respective facilities. Further discussion of both formal and informal training and various techniques is contained in the next section of this chapter.

In general, US industries regulated by the EPA view training as an integral part of doing business. This training may be an informal in-house training session dealing with: proper handling of hazardous waste; proper sampling techniques in taking water samples; properly calibrating monitoring equipment in sampling air emissions; or any other such beneficial and/or required training. Alternatively, it may take the form of more formalised environmental regulations courses such as: environmental audit training; ISO 14000 training; or attending a technical paper session at some industry symposium. In whatever form, if such training can impart beneficial knowledge to the affected employees in the execution of their work responsibilities, the chances are, monies may be budgeted for such training opportunities.

Referring back to the nuts-and-bolts training required by the federal regulations, we can take a moment to focus on the specifics relating to hazardous waste management that also impact upon employee health and safety considerations and effects upon the environment. It is important that line or shop employees working with hazardous materials and/or chemicals; technical employees responsible for managing any such wastes generated; and professional environmental employees all understand, to the best of their abilities, their duties, responsibilities and related regulatory requirements. We can probably all guess what might happen should any of these personnel deviate from their duties to some degree. At the extreme end, various trade publications, environmental newsletters, courses and books have been published that highlight pertinent case studies, such as a worker possibly overlooking a valve or improperly grounding a drum of flammable material, or handling a drum of mislabelled hazardous waste, culminating in dire consequences.

☐ Environmental Management System (EMS) Audits

Within the framework of maintaining a proper EMS, training, or, more specifically, focused training with clear objectives and expectations from management, can go a long way towards achieving such maintenance. One of the 'litmus' tests for gauging this hypothesis is an audit of the EMS. As the Introduction to ISO 14010 states:

> Environmental auditing has established itself as a valuable instrument to verify and help improve environmental performance . . .

Also, as stated in the Introduction to ISO 14011:

> Organizations of all kinds may have a need to demonstrate environmental responsibility. The concept of environmental management systems (EMS) and the associated practice of environmental auditing has been advanced as one way to satisfy this need. These systems are intended to help an organization establish and continue to meet its environmental policies, objectives, standards and other requirements . . .

Thus, in utilising auditing tools, top management has the ability to gauge whether some deficiencies in the EMS can be attributed to inadequate or improper training, or some other subjective factor. For example, the audit could be structured based on one or several clearly defined management objectives, which may include determining:

- Whether the auditee's EMS is being properly sustained

- Whether findings may relate to opportunities for potential improvement(s)
- Whether the management review process can effectively sustain a proper EMS
- To highlight the focus of this chapter, whether scheduled and ongoing training of various organisation personnel can help in some or all of these objectives

In this context, let us once again refer back to the USEPA training requirement. As a former regulator and as a consultant having performed on numerous environmental audits, the author recalls instances that could have alluded to and been associated with one of several deficiencies. In those instances, the root causes of the noted findings could have been related to improper or insufficient training of the workers; a casual worker attitude; or, perhaps, the workers may have not completely understood some or all of the training received due to some other barrier.

The intent of the US regulations was clearly meant to allow such employees to work in a prudent and safe fashion while striving to minimise any adverse impacts to themselves, to others and to the environment. Maintaining a paper trail would be one way to track such training and help to ensure that their facilities would be maintained and operated to minimise the possibility of a fire, explosion, or any unplanned sudden or gradual release of hazardous waste or hazardous waste constituents to the environment that could threaten human health or the environment.[11]

During an EMS audit, then, there could be instances where a seasoned and experienced auditor[12] could note deficiencies in the EMS that could be interpreted as findings pointing to training deficiencies. For example, employees may receive training in the proper disposal of solvents or other materials at their workstations, but, for one reason or another, may choose to dispose of such materials in inappropriate locations which may have dire consequences for the organisation's top management. Without citing specific examples, readers may be aware of such instances from their own experiences or by having read of enforcement actions by regulatory agencies in which various organisations were found to have allowed the inappropriate disposal of such wastes, either by accident or intentionally.

As another example, let us say that if one of management's objectives at an organisation was to evaluate training for its effectiveness, the EMS auditor's findings could point to such deficiencies as action areas for improvement. In similar fashion, if one of management's objectives was to identify continual improvement opportunities within the EMS, again, the auditor's ability to delve deeply into problems attributable to training deficiencies could be instrumental for the auditor in producing clear and focused findings. These, in turn, could be instrumental in developing EMS audit recommendations for improvements to the EMS.

☐ Continual EMS Improvement Linked to Environmental Performance Evaluation (EPE)

Finally, in the context of continual improvement of the EMS and how proper training plays an integral part, the forthcoming ISO 14031, a standard on environmental performance evaluation, is also linked to the underpinnings of the EMS audit and the EMS itself. As stated in the introduction of this standard's working draft:

> EPE and environmental audits help the management of an organization to identify areas for improvement. EPE is an ongoing process of collection and assessment of data and information to provide a current evaluation, as well as trends over time, about the environmental aspects of an organization's activities. In contrast, environmental audits are conducted periodically to verify conformance to defined requirements. Examples of other tools that management could use to provide information for EPE include life cycle assessment (LCA) and environmental reviews.[13]

Within the context of evaluating an organisation's environmental performance, one of the areas to be evaluated by EPE is the management of an organisation that includes the people, practices and procedures at all levels of the organisation. In evaluating the management of an organisation, the environmental performance indicators (EPIs) in the management area play an important role. These should provide information during the EPE process regarding the organisation's capability and efforts in managing management matters such as *training*, legal requirements, resource allocation, documentation, and any corrective action that may have an overall influence on the organisation's EP. One item in particular that bears noting is that effective EPIs '. . . may help to identify root causes where actual performance exceeds or does not meet relevant environmental performance criteria',[14] which was previously highlighted as one of the key items focusing on potential training deficiencies that may have been identified in an EMS audit report.

Let us look at another example regarding the linkage between EPE, auditing of the EMS and effective training. The relationship between management and operations can overlap with respect to training of employees. Among the indicators within management are included: the conformance indicators, which can relate to environmental audits; and the systems indicators, which can relate to improvement suggestions, or the number of training programmes. Similarly, among the indicators within operations are included such factors as: employee training being utilised to measure any decreases in frequency of incidents involving employee injury or material or waste releases into the environment; stressing proper employee maintenance of process equipment during specific maintenance training being utilised to measure emission reductions—such as Toxic Release Inventory emissions required yearly by the USEPA—or other similar measurement opportunities that can be directly be linked to training, EMS auditing and EPE.

■ *Proper Training: The Heart of Maintaining an Effective EMS*

The previous sections within this chapter provided a general overview with respect to some of the elements associated with training, which, for the sake of clarity, are further broken down by the author into his interpretation of the two main training types.

- **Formal training**, such as: training to allow an employee to understand the organisation's focus, goals and objectives; and to understand the responsibilities associated with one's work product(s) and potential hazards, and related generated waste(s)

- **Informal training**, such as: training to allow an employee to do the job correctly, safely and effectively; managing waste(s) properly (as previously noted in the USEPA training requirement); learning from supervisors'/peers' experiences; interactions with third parties such as auditors, consultants, regulators and others; research into other conduits (books, magazines, newsletters, etc.); interaction with other business units within the organisation; and learning from examples such as news stories, headlines, benchmarking other organisations, where actual case studies can provide invaluable information from which one can gain additional insight.

The ultimate end result of such training exercises can be that better informed individuals can help maintain a more effective EMS.

☐ *Peering into Training One Layer at a Time: The Onion Concept*

First and foremost, part of the foundation of an effective training programme during the development stage of training is to step back, evaluate and understand the organisation's focus, goals and objectives, and examine how a training programme, or a revised training programme, will help to achieve these management milestones. An integral part of this review is an understanding of the costs to the organisation for developing and executing such training in relation to the organisation's stated goals, etc. The development of such programmes also needs to be tempered with the reality of fiscal constraints that may hinder some organisations. Assuming that such obstacles can be met and overcome in order to provide training that can realise a favourable return on investment for the organisation, training programme developers and organisers can then move forward to the next phase of the formal training portion: developing and conducting training programmes that focus on the root of the subject matter; and the job specifics of the targeted trainees and their related regulatory requirements and organisational directives.

Within this **formal training** stage, employees are provided with various organisational and regulatory required training for their specific duties, often relating to the correct operation of machines. In the US, OSHA requires various types of operational training such as

- The proper wearing of personal protection equipment (PPE)
- The understanding of Lockout/Tagout or machine guarding in the use of mechanised equipment[15]
- The correct way of grounding drums of volatile compounds
- The correct identification of drums of hazardous waste prior to handling
- The correct labelling of canisters of combustible gasses as empty before they are shipped by any carrier

The above are but a few examples. There have been instances where improper labelling of gas canisters or improper labelling of hazardous waste drums have caused accidents with catastrophic results.

The underpinnings of an effective training programme are highlighted within the formal training stage, although the effectiveness of the many hours devoted to classroom and off-site training can only be gauged after the fact. This may occur as early as the completion of course evaluation forms by course participants in the classroom or in another off-site setting; or an internal inspection of a hazardous waste area; a manager's noting of an employee's disregard for the appropriate usage of PPE at his or her workstation; or after an unfortunate incident dealing with a forklift operator, to give but a few examples. Conversely, this may occur as late as an auditor's finding during an environmental audit, or as a potential regulatory deficiency noted during a regulatory inspection, or after an incident.

If we take a step back to view the 'bigger picture' with respect to training and its overall effectiveness on workers, it can be that during the after-the-fact or follow-up time frame of actual training, or the **informal training** stage that the 'onion concept'[16] can come into play in helping to gauge the effectiveness of any particular training programme. Referring back to some of the examples portrayed previously, the main concern of managers should be to impart to their employees the importance of performing their duties in a safe, proper and effective manner, and, if there are any problems, misunderstandings, or other items requiring clarification, these need to be discussed as soon as possible. For example, the discussions could address such issues as

- Whether problems have arisen as a result of a misunderstanding of the regulations
- How certain tasks could be executed in an improved fashion
- Whether some problems may be related to PPEs that are uncomfortable to wear and possibly lead to excursions

All are items that may need to be reviewed further. Such items could be a concern to management should they be the cause of problems with some employees.

Furthermore, follow-up, or post-training, evaluations of workers being trained could also help management determine whether various regulatory responsibilities have been effectively imparted onto the workers. As a few illustrative examples, routine spot checks could verify whether wastes are properly being managed, or are being correctly segregated to avoid increasing waste streams being handled, and whether the appropriate regulatory requirements are being followed by the various workers handling the wastes for ultimate disposal, among others.

Another key area in formal training includes the opportunities for workers to learn from their supervisors' and peers' experiences. This learning can arise from sharing different methodologies that they may have come across in previous work environments, or possibly in utilising various software tools that help to streamline their tasks, or by the experience gained with their having been with the organisation over several years.

Similarly, the experience gained in any interaction with third parties may not necessarily be something that elicits unfavourable memories, such as may be the case in an unsettling environmental audit and/or regulatory inspection—the 'us-versus-them' concept in some audit exercises. Rather, turning any such circumstances into learning opportunities can actually become additional extensions of informal training. The benefits arising from this may include:

- A better understanding of what auditors may be looking for
- Developing a more focused scope that better addresses management's objectives
- Replacing certain auditors who may not possess all of the attributes specific to auditors,[17] including some personality traits

These could be among the factors during an EMS audit—whether conducted by corporate staff or external third-party auditors—where invaluable lessons could be learned from any such experiences.

Equally as important may be any experiences gained in dealing with a regulator during an inspection, which may involve exercising any one of a number of negotiating traits or techniques during the inspection. Such exercises may be instrumental in diffusing any potentially adversarial positions during the inspection, and possibly lead to a favourable inspection, depending upon any regulatory issues or interpretations to be resolved during the process that can be successfully argued by facility personnel.[18]

Additional interactions with consultants, whether in the execution of an audit, other client-related work, or in a classroom or seminar engagement setting can provide valuable learning experiences, as can any interaction with community groups, dependent upon the subject matter. This too can be tracked and gauged as part of a continual training evaluation process tool by management.

☐ *Additional Training Opportunities*

As part of any training programme, whether formal or informal, or both, there is some degree of research that the trainee needs to undertake to help fill in the gaps, so to speak, in the training 'curriculum'. The information that can be gained in conducting this research (through, for example, acquiring and reading timely books on the subject, trade magazines, or pertinent newsletters; or surfing pertinent home pages on the Internet, or other such conduits) can help answer or address a particular question or issue, or provide an insight into relevant happenings on the environmental management horizon. The Internet can be an especially helpful tool to the seasoned professional and the novice alike—with one pre-condition: one needs first to know how to navigate it and know which Home Page to seek to obtain the necessary information.

Also, part of the informal training programme involves the ability to interact effectively and communicate with some of the other business units within the organisation. Personnel in purchasing, for example, may not be aware of certain regulatory requirements in the purchasing of domestic or foreign chemicals; likewise, personnel in inventory or shipping and receiving may not be aware of certain safety precautions that may be required of certain materials, or that every chemical in use within the organisation requires some form of Chemical Material Safety Data Sheet (MSDS) to be kept at various locations where these materials are used, and in the environmental department; or other such related situations.

The training process also requires that trainees are made aware of additional learning opportunities by keeping abreast of timely news stories in national and local newspapers, and hometown weeklies. To a trained eye, these papers can provide a wealth of information that can be passed on to other staffers as a general item of interest, or as a circulated memo.

The final topic within the focus of this chapter is the opportunity for staffers to learn from networking with their peers in other organisations, such as in an industry networking group, or other such round-table discussion groups.[19] Benchmarking the exemplary actions of forward-thinking organisations can also prove to be invaluable as a long-term training tool for employees, provided the components of the benchmarking study of the subject organisation are in tune with the objectives of the organisation wishing to learn from the study. One important point to carry through in training programmes of the various individuals within the given subject area is that the more informed the individuals are, the better they can be equipped to maintain and sustain a more effective environmental management system.

■ *Summing It Up: Lessons to Be Learned*

The focus of this chapter was to evaluate the training of individuals engaged in one or a series of responsibilities dealing with environmental management within a regulated organisation and how this training can be an effective and integral component of an effective environmental management system.

If an organisation is truly dedicated to its environmental management programme and is focused in its objectives to help achieve, maintain and sustain its environmental management goals in the light of the ISO 14000 standards, true diligence must be exercised on a regular basis; there is no real shortcut to sustaining an effective EMS.

To help capture the essence of those bases that should be covered as part of an organisation's training programme, training objectives, and in maintaining an effective EMS, the main points of this chapter can be summarised as follows:

- Effective employee training takes time, and continuous effort: it is not a one-shot deal. The importance of clear communication between workers, their trainers and supervisors is important to the success of the training programme's objectives, and in the subsequent review or follow-up.
- Make a concerted effort to delve into the effectiveness of the training programme as a means of ensuring that proper training is being provided and that trainees understand the subject matter and are putting it into practice—the 'onion concept' analogy. Use any information gained from any follow-up exercises to gauge this effectiveness.
- Employees need some form of motivation, or other subjective factors to maintain the momentum of training objective(s), such as role-playing or other types of training games that can be played.

- Real-life examples, or role-playing, helps to bring the points raised in training programmes home. Creativity is a key.

- The organisation must stand behind the environmental management programme governed by the EMS to help keep the momentum going. This is not something to be perceived by workers as mere 'window dressing' on the part of the organisation. Organisations truly committed to sustaining effective environmental management programmes must do as they say, or 'walk the talk'.

- Do not rest on any laurels; keep going. This is ingrained within the text of ISO 14001 and within ISO 14031 as part of the continual improvement aspect. Without such a concerted and diligent effort, true improvement cannot be achieved.

- In summation, environmental management programmes are not for the faint of heart. Good luck with your efforts.

■ Notes

1. This chapter is dedicated to the memory of Piero Trotta, Environmental Audits Program Manager at EniChem, SpA, until his untimely death on 2nd July 1996. He believed the EMS was the heart of an effective environmental management programme and that it required much diligence to make it successful. I couldn't agree with Piero more. He was a dedicated environmental professional and a true friend.
2. While the focus of this chapter, and the book in general, deals with environmental issues, it is understood that the theme of this discussion could also apply to health and safety issues, since there is some overlap in the roles and responsibilities of the individuals in each of these disciplines. In like fashion, the similarity in the architecture and strategies between ISO 9000 and 14000 can allow for an integrated, rather than fragmented, linkage between the quality and environmental standards.
3. In many of the inspections or environmental audits conducted by the author, many of his findings could be directly attributed to either improper or inadequate training of individuals responsible for some part of the environmental programme, or perhaps in the make-up of the environmental management system itself, which might be attributable to some disconnected system. Further discussion of this aspect is provided in the section on Environmental Performance Evaluation.
4. The Environmental Protection Agency, under the Toxic Substances Control Act (TSCA), regulates such chemicals imported into the US.
5. In the US, since about 1990, there has been a increasing trend for publicly-held companies to issue initial and follow-on voluntary environmental reports for their shareholders and the general public. While not all developed in a standard format, these organisations should be applauded for their willingness to share this information with the public. Some of these organisations have gone one step

further in their reporting by committing to the PERI Guidelines—those of the Public Environmental Reporting Initiative.

6. *Code of Federal Regulations (CFR)*, 40: *Protection of Environment*, Parts 260 to 299, Paragraph 265.16: 'Personnel Training', p. 315.
7. Cf. *Ibid.*, Subpart C: 'Preparedness and Prevention', Paragraphs 265.30–265.37, pp. 317-18.
8. At the time of going to press, as a Working Draft of ISO 14031 (EPE) dated 5th December 1996.
9. *Ibid.*, p. 10.
10. The author can attest to the experiences that can be gained at some of these off-site offerings, having attended and spoken at various conferences, training courses and industry-sponsored symposia over the past several years.
11. See *Code Of Federal Regulations*, Subpart C: 'Preparedness and Prevention', Paragraph 265.31, p. 317.
12. Among the topics that have generated attention in the qualifications criteria for EMS auditors include whether such auditors will be aptly qualified to perform such audits. As a means to 'raise the bar', the Registration Accreditation Board (RAB) developed a certification criterion for EMS auditors from which auditors could be certified. At the time of going to press, this criterion was still in draft form.
13. Working Draft of ISO 14031 (EPE), 'Introduction', p. 3.
14. *Ibid.*, p. 9
15. In one tragic instance, the author recalls an incident where a worker was instantly killed after a piece of his clothing was caught on the roller of a piece of equipment with which he was working, and did not have any emergency power shut-off, or other machine-guarding procedure. In another instance, two workers were killed when a seemingly empty tank they were welding exploded; they had not taken any precautionary Lockout/Tagout steps in accordance with company procedures prior to commencing work.
16. The author penned this the 'onion concept' in his environmental management training courses to provide an analogy of how training should be evaluated as a measure of employee understanding and the effectiveness of the trainer and the training materials. Without such in-depth scrutiny, the trainer's efforts and the trainee's time could be ineffectively utilised.
17. See ISO, 14012.2, 'Personal Attributes and Skills', p. 2.
18. See S. Millstone, M. White, G. Crognale *et al.*, *The Greening of American Business* (Washington, DC: Government Industries Inc., 1992), pp. 59-65; G. Crognale, 'Agency Inspections and Preparing for Administrative Proceedings', in *Proceedings of the Air and Waste Management Association's 85th Annual Meeting Exhibition, Legal and Regulatory Issues*, 21–26 June 1992, PE 92-124.01.
19. See Millstone *et al.*, *op. cit.*, pp. 183-85.

17 Eco-Management and Audit Scheme for UK Local Authorities

Three Years On

Nigel Riglar

A T THE Earth Summit in Rio de Janeiro in June 1992, world leaders signed a global action plan called Agenda 21. Over two-thirds of Agenda 21 cannot be delivered without the commitment and co-operation of local government, and the key role of local authorities is set out in Chapter 28.[1] Each local authority has been encouraged to adopt its own individual Local Agenda 21 by 1996—this is to be its own sustainable development strategy at the local level.

The first key element in any Local Agenda 21 plan is a commitment by the local authority to 'putting its own house in order' through a systematic examination of its own environmental practices and service delivery. Leading by example is essential if local authorities are to fulfil the second element, which is to work in partnership with the community to develop a process of awareness-raising, consultation, measuring and reporting on the move towards, or away from, sustainability.

It is the former element with which this chapter is concerned: in particular, the question of how local authorities go about improving the environmental performance of their operations, and integrating sustainability into the delivery of plans, policies and programmes. The experience of local government in the UK has important messages for other local and provincial authorities around the world. The context may change; the problems will not. Local authority management of the environment is not a new phenomenon. Indeed, after social services and education, it accounts for the third biggest area of spending. Local authorities in the UK have traditionally had statutory responsibility for regulating and managing many aspects of the environment, including: land use planning; development control; environmental health (e.g.

local air pollution control; noise and nuisance regulation); household waste collection; recycling; disposal and regulation;[2] and nature conservation (local nature reserve designation and establishment of country parks). In addition, local authorities are large owners of property, operating schools, for example, as well as social service establishments, sports halls, swimming pools, museums, libraries and depots. Consequently, they experience the environmental impacts that accompany the management of such facilities, e.g. energy use, discharges to air, land and water, etc.

It was not until the late 1980s and early 1990s, however, that systematic environmental management became a recognised discipline within UK local government. Early attempts to become 'green' involved not just specific environmental initiatives (such as nature conservation or recycling), but also attempts to build environmental awareness into all activities of the authority. 'Green charters', or policies, were produced and 'environmental audits' conducted. Emphasis was placed on the corporate nature of environmental effects, with particular weight being placed on internal issues such as energy management and purchasing.

Much of the work methodology was invented by local authorities for themselves, often through trial and error. This was a creative, if sometimes frustrating process, particularly for authorities with a lot of commitment but limited resources. Many felt that, while every authority may be different, there was a need for a broad framework to help change the complex environmental agenda into an understandable and—more importantly in a time of constrained resources—manageable process. It was argued that such a framework was needed to ensure that environmental management became fully integrated into the core business of local authorities.

At the same time that local authorities were searching for an effective framework, EMS standards such as BS 7750 and the European Union's Eco-Management and Audit Scheme (EMAS) Regulation were being developed. It was soon realised that these standards provided the missing link. EMAS was deemed the most appropriate, and thus began the development of the Eco-Management and Audit Scheme for UK Local Government (LA-EMAS), which is discussed in detail in this chapter. In essence, it was the public reporting element of EMAS—the Statement—and the subsequent validation of this document that was seen as a major means by which a local authority could demonstrate its commitment to Local Agenda 21, and communicate its environmental credentials to all sectors of the community.

Written with practitioners in mind, the aim of this chapter is to summarise the collective experience of UK local government over the last three years, while attempting to provoke the reader into thinking about the implications of pursuing LA-EMAS, or a similar formal recognition, and showing how registration can be successfully achieved. To these ends the following topics will be examined:

- A review of LA-EMAS history
- Differences with the industrial scheme
- Current performance and trends
- Implementation strategies
- The main drivers for change
- The future challenges

■ LA-EMAS: History

As stated above, it was during the early period of UK local government 'greening' that the European Union was developing its Eco-Management and Audit Scheme (1836/93). It was felt by many that this Regulation would provide the management framework that was needed by local authorities. One problem existed, which was that the Regulation applied only to the manufacturing and other assorted sectors, not to local government. However, Article 14 of the Regulation allowed for the inclusion of other sectors on a pilot or experimental basis. The main differences between the industrial and local authority schemes are discussed below. This clause was invoked, and late in 1993 a project, jointly funded by the UK Department of Environment (DoE), the Local Government Management Board (LGMB) and the Scottish Office, was initiated. The aim was to adapt the Regulation for use in local government and to provide practical guidance on how to implement the scheme. The adapted regulation was piloted in seven local authorities, with a further seventeen providing a reference group to examine and comment on draft guidance. This latter reference group also included the District Audit Service, which has responsibility for ensuring that local authority services are delivered in an efficient, effective and economic way, and Directorate General XI of the EU, which, among other issues, has responsibility for European environmental policy, including the development of EMAS.

The outcome of this project was the 250-page publication, *A Guide to the Eco-Management and Audit Scheme for UK Local Government.*[3] This contained the adapted regulation for UK local government and a great deal of 'how-to-do-it' advice, including worksheets and checklists, developed during the project. Although this publication continues to be the main reference work for local authorities wishing to implement LA-EMAS, its importance has waned over the last eighteen months, as practical experience of implementing the scheme has increased.

The initial project steering group developed into the now formally established Local Government Eco-Management and Audit steering group (LoGEMA). A subgroup of the Local Government Environment Forum (consisting of DoE Ministers and senior members of the local authority Associations), LoGEMA is made up of leading environmental practitioners from local government and representatives from the UK Competent Body for EMAS (the DoE). This group has overseen the establishment of the LA-EMAS Help-Desk (discussed below); the publication in February 1995 of *DoE Circular 2/95*[4] (Welsh Office 9/95) which set out formally the requirements of LA-EMAS; and the formal launch of the scheme in April 1995. It is perhaps important to point out that EMAS was introduced through an EU Regulation, which is the highest level of European legislation, to ensure the standardisation of implementation across all member states. Thus it is primary legislation. The Circular referred to above is described by the DoE as being for guidance only, and, as such, external verifiers assess the performance of a local authority against the main Regulation, and not the Circular.

■ LA-EMAS Help-Desk

The Help-Desk is jointly funded by the DoE and LGMB. Established in January 1995 and employing two full-time staff it is charged with:

- Promoting the use of environmental management systems (EMSs) which can be subject to third-party validation, and, more specifically, LA-EMAS
- Providing the practical advice and assistance that local authorities will need to reach these standards

The Help-Desk operates out of the Environment Unit of the LGMB which is responsible for the UK's Local Agenda 21 initiative.

There has been a great deal of international interest in the UK's LA-EMAS work, particularly from member states of the EU, but also from further afield including Japan, Norway, New Zealand and Australia. Although the powers and responsibilities of local authorities vary around the world, it is the opinion of the author that the principles of LA-EMAS are applicable to any local authority that wishes systematically to examine, evaluate, manage, monitor and report on its environmental performance. At the very least, LA-EMAS provides a management tool that allows an authority to demonstrate that it is maximising environmental improvement from the limited resources it has available.

The Help-Desk is aware of three LA-EMAS or 'eco-auditing' studies currently under way in Europe. In Norway, research is in progress into 'Municipal

Environmental Auditing'. The Finnish Local Authority Association is undertaking a two-year pilot project to examine the feasability of introducing LA-EMAS. Finally, a number of Danish local authorities have produced 'green accounts' in line with the relevant national legislation on this subject.

■ *Differences Between the Local Government and Industrial Schemes*

The local authority scheme is closely modelled on the industrial version, which is more commonly known throughout the EU. There are, however, three principal differences. First, the industrial scheme applies to a company's 'sites'. For local authorities, a more convenient and appropriate unit of management is the department, division or service function, e.g. planning, environmental health or social services. Thus the Circular uses the term 'operational unit' to describe these.

Secondly, the industrial scheme allows a single site within a company to seek registration. The only thing that is required at the level of the whole company is a satisfactory environmental policy. LA-EMAS similarly allows a single operational unit to register. However, there are more obligations at a corporate level. An authority must not only have a corporate policy, it must establish a 'Corporate Overview and Co-ordination System'. This is a set of management responsibilities, structures and procedures for environmental management at the corporate level. In addition, the scheme requires that registration of individual departments or divisions is an intermediate step, as the whole authority must commit itself eventually to seek corporate registration within a defined and agreed timespan.

Finally, the industrial scheme is principally concerned with the control of production activities: polluting emissions, production of solid waste, use of energy and resources, and so on. Like all productive organisations, local authorities have these 'direct effects' on the environment. However, unlike industry, they also have major indirect environmental effects associated with their service delivery and it is the management of these that is the additional focus of LA-EMAS. Figure 1 explains the differences between direct and indirect effects, and Figure 2 is used to identify the various operational units of a local authority that would be involved in managing the various direct and indirect environmental effects associated with energy use. As the reader will note, the majority of a local authority's impacts are through service delivery, and are therefore indirect. Thus, for example, the use of energy in a council's HQ buildings, although very important, pales into insignificance if the structure plan (strategic land-use plan) is unsustainable and is actually encouraging

Issue	Direct effects	Indirect effects
Examples	Use of vehicles in own activities	Effect of transport/land-use policies on the commmunity's use of cars
Range	A small number, largely the same for all processes and activities (e.g. energy use, water use, purchasing, transport, waste and pollution production)	A large number, different for each area of policy or service (e.g. planning, education, social services, environmental health)
Direction of effect	Negative	Can be negative, more often positive, neutral or potential
Quantification?	Yes—at least in principal: measurement may present practical problems	Often not, and dependent on judgement
Nature of action required	Reduction, mitigation, less damaging methods	Prevention, persuasion, facilitation
Examples of action	Recycling own office waste paper	Providing recycling facilities; encouraging the public to use them
Degree of control/influence	Success is under authority's direct control	Success depends on response or collaboration of others (e.g. business community, individual citizens)
Nature of targets/indicators	Action taken, organisation's own performance level, quantity of organisation's own impact	Action taken, amount of behavioural change caused, performance levels of other agents, total impact of all agents in the area

Figure 1: *Direct and Indirect Effects*

energy use in the wider community (e.g. the need for longer car journeys to access day-to-day services).

■ *LA-EMAS: Current Performance and Trends*

Local authorities have been working with LA-EMAS since late 1993. The pace of work has quickened over the last twelve to eighteen months since the

Activity	Main operational units with responsibility	Dominant type of environmental effect
Energy use in authority buildings	Energy management unit, leisure services, social services, education	Direct
Energy use in council housing stock	Housing/tenants	Direct/indirect
Council house repairs and maintenance programme, e.g. installing new boilers	Housing	Direct
New housing provision, e.g. building quality and energy efficiency standards specified	Housing	Direct
Private housing—renovation grants	Environmental health	Indirect
Enforcement of building regulations, including energy specifications	Building control	Indirect
Authority capital programme; the design and construction of energy-efficient schools, leisure facilities, etc.	Design and construction services	Direct
Environmental appraisal of land use and transport plans to ensure transport use minimised and services easily accessible without car	Forward/strategic planning	Indirect
Fuel use by transport (fleet), or staff using private cars for authority business	Fleet management unit, personnel (paying travel allowances)	Direct

Figure 2: *An Examination of the Various Units and Types of Environmental Effect Associated with the Management of Energy within a Local Authority*

Environmental Review	Now	April 1997
There is limited awareness of environmental effect and legislative requirements. No review work has been done.	**(1)**	**(1)**
Limited review work has been carried out. Collectively, the authority is aware of some of the direct environmental effects that relate to its operations; however, these have not been quantified or qualified in any detail. As yet, resources may not have been made available to undertake the work, and there may be a lack of senior management/political commitment. There is currently no corporate mechanism for undertaking a review, and there may be a preceived lack of in-house expertise. It is generally unclear whether managers are aware of all of the significant indirect environmental effects and legislative requirements associated with their service delivery.	**(2)**	**(2)**

Figure 3: *Extract from the LA-EMAS Survey: First Two Paragraph Choices for Performance against the Environmental Review*

publication of the Circular and the launch of the scheme that opened the way for verification and registration. In order to monitor the uptake of the scheme, the LA-EMAS Help-Desk undertakes an annual survey. The results of the 1995 and 1996 surveys will be discussed below. However, this will only be a partial discussion of the 1996 results as, at the time of writing, the survey was still under way.

In addition to the LA-EMAS work, the LGMB undertakes an annual Local Agenda 21 (LA21) survey which covers wider sustainable development issues, but also includes questions on environmental management systems. Comparisons between the two data sets will be made as and when appropriate. The return rate for this annual survey has been static in recent years, with the last two achieving 50%, equivalent to just under 250 local authorities.

The LA-EMAS survey is based on local authorities carrying out a self-assessment of their current performance, with regard to their implementation of the scheme. They are required to select from a choice of four, the paragraph that best describes the current status of each particular component of LA-EMAS (e.g. policy, statement). In essence, a score of '1' represents no environmental management and '4' compliance with LA-EMAS (see Fig. 3 for first two choices). The first survey was undertaken in May 1995 and a 48% return rate was achieved. The 1996 survey used the same format, but in addition to assessing current performance it required a prediction of performance twelve months hence. At the time of writing, a return rate of 28% had been achieved and it is likely that the response will eventually approach the 1995 figure.

The matrix displayed in Figure 4 is included in the survey and provides local authorities with a simple summary sheet on which to report and record progress. Further questions were asked regarding implementational approach and registration plans, and it is the answers to these that will be examined first.

■ *Approaches to Implementation*

The survey results reveal three general approaches to EMS implementation:

1. Formal commitment to external validation (i.e. a recommendation to achieve LA-EMAS verification or ISO 14001 certification formally has been endorsed by the full council)

2. Councils implementing the principles of LA-EMAS but waiting to see if third-party validation will add any real value (i.e. have yet to make a decision on external validation)

3. Councils in the early stage of EMS implementation and unable to identify future third-party validation strategies

The results, as a total number and/or as a percentage of respondents, for these three categories are shown in Figure 5. The results reveal an increase in the proportion of authorities in the first two categories comprising those actually implementing EMSs, and a decline in the third category. This would suggest that there has been an increase in EMS activity, with rising numbers of councils committing either part or all of their organisations to LA-EMAS verification or EMS certification.

Local government in the UK is a mix of single- and two-tier administrations. Single-tier administrations, often referred to as unitary authorities, exist in the major cities and Scotland and Wales, and are responsible for all local authority services. In the remainder of the country, a two-tier system exists with the larger county councils delivering strategic functions such as planning, social services and education, and the smaller district councils delivering local services, such as environmental health or leisure. Typically, in any one county, there will be between four and ten smaller districts. In recent years, there has been a drive by central government towards unitary authorities. In the last two years, the result has been a drop in the total number of local authorities from 542 in 1995 to 478 in 1996.

Level	Environmental policy	Review	Action programme
4	Formal corporate environmental policy and objectives adopted. Regular corporate policy review mechanism. Environmental policy integrated and consistent with all policies. Clear budgetary provisions.	Complete and regularly updated corporate register of all significant direct and service effects (software and hard copy). Formal register of statutory environmental responsibilities completed and available in each department.	Corporate implementation of environmental action programme. Commitment to, and effective demonstration of, continual environmental improvement. Mitigation or proposed action for all effects identified in the register of environmental effects. Performance monitoring mechanisms implemented.
3	Formal policy adopted. Irregular corporate policy review mechanism. Environmental policy poorly integrated with existing policies. Resource commitments met from existing budgetary provisions.	Examination of all significant service and direct environmental effects. Corporate recording (quantified and costed) of energy, water, purchasing and waste effects. Some analysis (software/ hardware). Draft register of departmental statutory responsibilities.	Corporate agreement to environmental action programme and performance monitoring. Objectives, targets, responsibilities and performance indicators determined.
2	Informal or draft environmental policy. Corporate and departmental environmental objectives inconsistent and limited. No formal reporting, recording or review mechanism.	Informal records of environmental effects and statutory responsibilities kept. No corporate mechanisms for assessment or analysis of environmental effects.	Statutory responsibilities identified and implemented. Uncoordinated, limited environmental activities undertaken. Corporate objectives, targets and performance indicators to be determined.
1	No environmental policy approved.	Limited awareness of the environmental effects and statutory responsibilities of direct or service activities' effects.	No environmental action programme.

Management system (EMS)	Audit	Public statement
EMS records, officer/member responsibilities, actions and targets. Environmental responsibilities integrated into job descriptions. Environmental Information System present, including regular, concise and openly-available progress reports. Environmental assessments of budgets. Corporate environmental training programme.	Assessment of environmental policy compliance. Verification that objectives, targets, environmental programmes and EMS adequate and completed within defined timescales. Identification of incomplete progress; identification of further objectives and targets. Review and update environmental policy and programme. All statutory compliances met.	Formal consultation, revision and completion of public statement (PS) on environmental performance. Annual progress summary. Made freely available and widely accessible. Commitment and timescale for completion of next PS. Submission of PS to accredited verifier.
Informal record of EMS responsibilities. Members/officers of the environmental committee, environmental co-ordinator present. Regular reporting of progress against environmental objectives/targets. Limited officer/member environmental awareness training.	Completion of corporate appraisal of action programme. Evaluation of contribution towards targets. Informal identification of further actions.	Consultation document outlining significant environmental effects, policy, actions, performance review, statutory and management responsibilities. Informal internal/external consultation procedure on document contents.
Management responsibilities for implementation of action programme identified but poorly supported. Informal officer working group/lead officer for environmental issues present. Informal progress reports. Ad hoc environmental training.	Limited, unco-ordinated internal progress recording of individual actions. No performance evaluation or analysis of contribution towards target reduction figures.	Environment committee/working group reports available. Statutory public registers updated and available (Schedule B premises under the EPA 1990).
No understanding of environmental responsibilities, liabilities or resources.	No system to audit.	No public environmental information available.

Figure 4: *LA-EMAS Summary Matrix*

	Formal	**Implementing**	**Early**
LA-EMAS 1995	17 (7%)	127 (51%)	105 (42%)
LA-EMAS 1996	14 (11%)	72 (57%)	39 (32%)
LA21 survey 1995	19	84	93
LA21 survey 1996	32*	82	86
* includes five authorities with units seeking BS 7750/ISO 14001 certification			

Figure 5: *Results of LA-EMAS Survey*

If a figure of 478 is used for the total number of authorities, post local government re-organisation, then it is possible to approximate the proportion of local authorities in the UK that are addressing their internal EMS requirements (all three categories). The surveys suggest that the figure is between 26% (LA-EMAS 1996 survey) and 42% (LA21 1996 survey). The former figure will almost certainly rise as more survey returns are received and is likely to approach the 46% figure (assuming 542 authorities—pre-reorganisation) revealed by the 1995 LA-EMAS survey.

The 1996 results have also been analysed by authority type for the three categories shown in Figure 6.

Unitary authorities and county councils appear to be taking a more 'customised' approach to implementing an EMS: in other words, instituting the principles of LA-EMAS but not necessarily seeking registration (Category 2: 'Implementing'). In district councils there appears to be more commitment to achieving registration (Category 1: 'Formal'). This is probably a feature of the combined factors of the size of these authorities, the complexity of their environmental impacts, and, subsequently, the cost of verification. Thus the prospect of verification for a 500-employee district is, at least on first inspection, more palatable than for a county council employing 20,000. However,

	Formal	**Implementing**	**Early**
County councils	8%	62%	30%
District councils	16%	52%	32%
Unitary councils	3%	70%	27%

Figure 6: *1996 LA21 Survey*

	1996	**1997**	**1998**	**1999**	**2000**
1995 Survey	5	7	4		
1996 Survey	4	5	5	–	–
LA21 Survey 1996	7	9	5		

Figure 7: *Timescales of Category 1 Verification/Certification*

the London Borough of Sutton is proof that a large, complex authority can achieve verification without massive disruption and at a reasonable cost.

■ First Operational Unit Registration

In terms of verification, councils in the first category identified that they would have their first operational units verified/certified in the timescales shown in Figure 7.

Authorities in the second category in Figure 6 ('Implementing') predicted that they would have their first operational units verified in the timescales shown in Figure 8 (assuming that these auhtorities will eventully seek verification or certification).

In 1995, 51% of respondents predicted that they would have units ready for verification by 1997. The 1996 survey reveals that only a handful of authorities will be in a position to achieve this target. It is likely that in May 1995 there was a certain amount of 'bravado and naïveté' applied in answering the questionnaire, borne out of an ignorance of the work that seeking registration to LA-EMAS entails. The 1996 results represent a more realistic analysis by authorities.

Adding together the 1996 figures for Figure 6's Categories 1 and 2 ('Formal' and 'Implementing') gives a relatively clear indication of the numbers of authorities that will have operational units registered under LA-EMAS by 2000. The data suggest 100 authorities or about 20% of the total: this should

	1996	**1997**	**1998**	**1999**	**2000**
1995 Survey	20	51	24	21	–
1996 Survey	–	9	22	21	20

Figure 8: *Timescales of Category 2 Verification/Certification*

be taken as a minimum figure. It is likely that, as other external pressures increase, as a result of the work of the Audit Commission (the UK national body responsible for monitoring the effectiveness, efficiency and economy of service delivery in the public sector), or of changing funding criteria (both discussed later), that the number registering under LA-EMAS will increase.

■ Time to Register the Whole Authority

Authorities were also asked to estimate the amount of time that would be involved in taking the whole council through to registration. The responses were as follows:

County council	4–10 years (89%)
District council	1–6 years (64%)
Unitary authority	4–10 years (79%)

Not surprisingly, the larger the authority the longer the time that respondents predicted they would need to achieve registration. So a county council registering its first unit in 1996 would not achieve recognition for the whole authority until 2000 at the earliest, and 2006 at the latest. LA-EMAS therefore has to be seen as part of a longer-term process aimed at changing the culture and behaviour of an organisation. It is perhaps prudent to question whether such timescales are beyond the 'attention spans' of most elected politicians and/or chief officers (the most senior tiers of management). In this case, implementation may need to be broken down into digestible parcels of work. However, it can be done: the London Borough of Sutton has a programme in place that will ensure that the whole authority is registered in under three years.

In district councils the timescales are even shorter among those authorities committed to verification (Fig. 6 Category 1: 'Formal'). Over 90% of these councils predict that they will have achieved registration for the whole authority in under three years.

■ Operational Unit Size

The data for the first two categories has been analysed to identify the operational unit size that is most commonly being worked with. The results were:

County council	Department (56%)
District council	Authority (43%)
Unitary	Department (62%)

Clearly, the smaller the authority the more likely it is that a corporate approach to EMS implementation will be taken. This is slightly misleading, in that even in large authorities, taking a unit-by-unit approach, there is a recognition of the need for a corporate overview and co-ordination system. Thus authorities taking a department-by-department approach are forced into putting their corporate house in order by establishing corporate management responsibilities—both at officer and elected member level—corporate information systems and the environmental review of budgets.

■ EMS Performance

Many authorities are still involved in the rhetoric of environmental management. This is clearly revealed by the LA-EMAS survey. As outlined earlier, the survey asked councils to score performance in relation to the six stages of LA-EMAS, assigning scores of 1–4 , where:

1 represented no environmental activity
2 represented an unco-ordinated approach but with some progress
3 represented good environmental performance with further progress being made
4 represented compliance with LA-EMAS

Respondents were asked to identify current performance and predict performance in 1997. Using this data for the 75 authorities that had completed the survey in both 1995 and 1996, it is possible to generate an 'average' profile of performance (see Fig. 9).

It is clear there has been little change in local authority performance between 1995 and 1996. A number of reasons can be cited. First, a more informed realism about the requirements of LA-EMAS has probably resulted in a 'rounding down' of scores. Secondly, pressure has been created by a major reorganisation of local government; and thirdly, Compulsory Competitive Tendering (CCT) has resulted in a number of local authority functions, including housing, highway maintenance and waste collection, being legally market-tested. Finally, many councils now focus on the community aspects of LA21. The 1995 and 1996 profiles hover around the '2' line: in other words, an unco-ordinated approach to environmental management. The predictions for 1997 show a significant increase in performance and a shift towards more co-ordinated and effective environmental management systems and performance: in other words, a score of '3'. It will be interesting in 1997 to reflect on whether authorities have lived up to their expectations. The predicted change suggests that there has been a realisation in a large number of authorities that an EMS is a necessary, if not very 'sexy', component of the

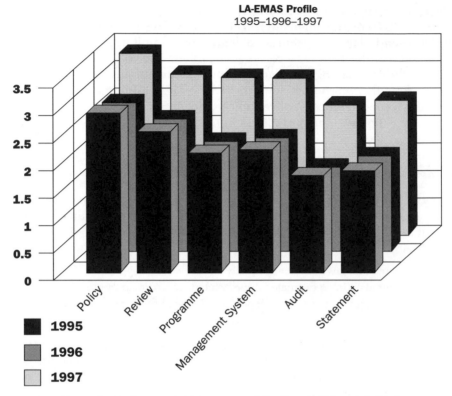

Figure 9: *Environmental Management Profile of UK Local Authority*

LA21 process, and essential if councils are going to maintain their credibility through responding to the issues being identified in the LA21 process.

The weakest link in the EMS chain continues to be auditing. Councils are not able to demonstrate whether they are delivering their policies, meeting their targets or delivering value-for-money environmental management. It is likely that the Audit Commission's 'Environmental Stewardship' work will force an improvement in this situation (examined in more detail later in this chapter).

■ *The Link to Sustainability*

Question 25 of the LA21 survey examined those local authority sustainability activities that help and support others. Respondents were asked to identify a range of 'usefulness' criteria which were subsequently scored: a score of '4' representing 'very useful' and '2' equivalent to 'useful'. The results revealed that EMSs helped and supported the following:

i   Corporate/member commitment (3.65)
ii Cross-departmental officer co-ordination (3.67)
iii Internal training and awareness-raising (3.56)
iv Integration of sustainability into departmental strategies, policies and activities (3.49)
v   State of the environment reporting (2.87)
vi Indicators (3.02)

Elements **i–iv** above indicate that an EMS is essential if a council is to respond fully to sustainability. A systems approach is necessary if the issues to be managed are to be accurately identified, responsibility and resources assigned, monitoring of performance undertaken and environment/sustainability integrated into all areas of activity. However, as we have seen, local authority performance in terms of implementing management systems and auditing arrangements is poor. This fact is probably not surprising given the patchy penetration of ISO 9000. Elements **v** and **vi** above focus more on the external side of sustainability. Authorities will need to find ways to link their internal EMS's performance indicators to wider sustainability monitoring and reporting activities. This would allow councils to reflect the true environmental outcome of their service delivery. For example, measuring improvements in air quality rather than simply the amount of resources that are being invested in such activities.

It is the integration of environment that is the key issue if local authorities are going to change corporate culture and behaviour. Question 13 of the LA21 survey examined the part that sustainable development policies play in an authority's strategies, policies and service delivery. A score of '2' was used to represent a 'significant influence'. Only the land-use planning, waste management and explicitly environmental services categories broke this significant influence threshold.

Clearly, local authorities are not integrating environment into all areas of service delivery. Most, however, seem to recognise that an EMS is essential if the environment is to be truly managed as a mainstream business issue. Yet many seem reluctant to implement an EMS and are even more wary of allowing external verification of their performance. This situation is understandable to a certain extent, given that the environment is still a relatively new discipline, and that experience with ISO 9000 is patchy, but it is still, however, a disappointingly short-sighted attitiude.

■ Implementation Strategies

As we have seen, increasing numbers of local authorities are implementing LA-EMAS. There is no 'right' or 'wrong' way. Rather, there are a host of practical issues that need to be considered. Thus every authority's strategy is different. In practice, local authorities tend to take one of five approaches to implementing LA-EMAS, summarised below.

☐ No Internal EMS Work or LA21 Focus

Over 50% of councils fall into this category. However, it does not mean that the authority is not indulging in any environmental activity. Obviously, it will be fulfilling its statutory obligations in terms of environmental health, waste management and planning. In addition, an authority in this category may have made a conscious decision to concentrate on the community development aspects of Local Agenda 21. However, the avoidance of internal EMS work cannot go on indefinitely. Communication and consultation with the community will raise public awareness and expectation with regards to local authority services. If councils are to respond fully, they will need to examine carefully their management practice and organisation. A failure to do so could result in a loss of credibility in the eyes of the community.

☐ The Unit-by-Unit Approach

This is the implementation of an EMS on a unit-by-unit basis. However, this runs the risk of producing an EMS that focuses very heavily on service delivery and misses wider corporate issues such as energy management or purchasing. These effects are not necessarily significant for individual units, but probably are for the whole authority and therefore need to be managed at the corporate level, which, as previously discussed, brings with it a range of organisational and management problems.

☐ The Corporate 'Issues' Approach

This guarantees a strong corporate response but runs the risk of focusing too heavily on direct effects, and misses significant areas of indirect service delivery effects. If this approach is to be successfully adopted, it must look at both the direct and service effects of the issue, and by definition will need to involve a wide range of operational units. The corporate issues are likely to include transport, pollution, open space and nature conservation, energy, purchasing and waste. For example, a district council wishing to improve its environmental performance with regard to transport would need to include the following types of unit: forward planning (environmental appraisal of the local plan); personnel (travel allowances); design and construction services (local roads); direct services organisation (transport fleet); environmental health (local air pollution); and development control (requesting of environmental assessments for new roads).

☐ *Project-Oriented Approach*

There may be many individual environmental projects within this sort of authority. Often these projects fulfil a short-term political need: 'to be seen to be doing something'. Although very laudable, this approach can create problems. Projects may not be focused on truly significant environmental effects. Although they may improve the environment, they do not necessarily change the culture, behaviour or organisation of a local authority over time, something an EMS inevitably leads to. Thus it is possible to have a series of high-profile, environment-improving projects being delivered by an authority that is not able to demonstrate the environmental sensitiveness of the bulk of its day-to-day service delivery.

☐ *Combination Strategy*

This draws on the strengths of the 'unit-by-unit' and 'corporate issues' approaches. It involves a unit-by-unit approach to service effects and a corporate response to the main direct effects. In reality, it is this approach that is most appropriate for an authority that will eventually aim to seek registration, as achieving verification will involve demonstration of the fact that a corporate overview and co-ordination system exists. It may in practice be impossible for a verifier to assess an issue, even if it does pick up service and direct effects. Thus it is unlikely that a council would be verified for its environmental purchasing or transport management (unless all of the effects involved are managed by one unit—and this rarely happens in reality). The difficulties with regard to the number of units involved in verifying an authority's energy management is highlighted in Figure 2.

■ *LA-EMAS: The Main Drivers for Change*

The benefits of the scheme are numerous and well documented[5] and involve cost savings, compliance with policy and legislation, improved service delivery and raised staff awareness. This section examines a number of currently contentious issues in more detail.

☐ Securing Funding

Environmental performance is becoming an increasingly important issue in the assignment of funding within the UK and Europe. Currently, a number of funding agencies, including the DoE and EU, are asking for environmental performance information. It could be argued that the funding agencies are missing a trick, in that, instead of asking for specific environmental information, which takes time and expertise to assess fully, they would be better off examining the environmental management credentials of the applicants: for if these had third-party validated EMSs, the funding agency could be assured that the environment had been fully taken into account when the application was drawn up. Thus in the near future it is highly likely that questions such as: 'Is your authority registered under LA-EMAS?' will appear on grant application forms.

☐ Operational Efficiency

The increasingly complex environmental management agenda is being faced by a wide range of local government officers, many of whom have probably never received any formal environmental training. Thus local authorities tend to respond reactively to environmental issues as and when they arise. It is not possible to say how much staff time has been wasted as a result of local authorities being prosecuted under environmental legislation or having to redesign major capital projects because an environmental assessment was not done at the feasibility stage and problems subsequently arose. It is, however, probably safe to conclude that the total is substantial and rising. LA-EMAS makes authorities ask the right people the right environmental question at the right time. One of the major benefits of the scheme is the reduction in the amount of staff time wasted on managing environmental issues that could and should have been foreseen. The London Borough of Sutton estimates that the pay-back period for staff time used in implementation is less than one year, realised through improved management control and operational efficiency.[6] In this instance, the authority employed a full-time EMAS co-ordinator to advise and support the process of implementation.

☐ Audit Commission Initiatives

As stated earlier, the Audit Commission is the national body responsible for monitoring the efficiency, effectiveness and economy of public-sector service delivery. The Local Government Study Group is currently undertaking a project entitled 'Environmental Stewardship'. It has been signalled that one of the possible elements of local authority value for money auditing in 1997/98 could be the cost-effectiveness of environmental management, reflecting the ethos that 'good environmental management is simply good management'. In particular, the focus is on how authorities have organised themselves to respond to the environmental challenge and performance in terms of energy management, waste management, planning and transport and aspects of environmental health. The LA-EMAS approach has been highlighted by the Commission for the way in which it helps local authorities improve the effectiveness and efficiency of environmental management and protection. It has already been suggested in some quarters that a future Audit Commission performance indicator could reflect those authorities that have units registered, and those that do not. The spectre of league tables aside, it is clear that those authorities with EMSs will find it much easier to respond to requests for information regardless of whether these are from the Audit Commission or other stakeholders, such as individual citizens, pressure groups, central government or other regulators.

■ The Future of LA-EMAS

This final section represents a look into the LA-EMAS crystal ball and attempts to identify future issues, developments and directions.

☐ Extension of Scope

The 1998 review of the EMAS regulation 1836/93 will almost certainly see an extension of scope, to include local government and the service sector, across the EU. It is likely that LA-EMAS will be used as the basis for service-sector schemes, as most of the effects are service delivery related and therefore indirect. For example, a bank's impact through its use of electricity is minimal when compared to the potential it has to influence the environmental performance of others through its investment or lending decisions.

☐ Should EMAS be Mandatory?

People, either as individuals working within organisations or interacting with one, are very carefully protected by health and safety legislation. In contrast, no organisation in the world is adequately made to account for the way in which it uses common resources and degrades the common environment. EMAS has to be the first, albeit voluntary, step in this direction. Health and safety was, and continues to be, legislatively led, and without this command-and-control philosophy, it is questionable whether organisations would have come so far. In this light, it is sensible to question whether, if the uptake of a voluntary environmental initiative such as EMAS is poor, it will be necessary for the EU to make it mandatory.

Making EMAS mandatory would probably not be appropriate for the 90% of companies across the EU that have under 250 employees. One solution to this problem—in the opinion of the author—would be to make EMAS mandatory on all organisations over, say, 500 employees. Arguing that organisations of this size have the necessary expertise and resources at their disposal, a duty should then be placed on these organisations to work closely with their supply chains and drive improved environmental performance and third-party validation through this mechanism. Obviously, through purchasing power and the delivery of economic development services, local government will have a major role to play in working in partnership with business to promote improved environmental performance, particularly in small companies.

Currently, there are two major EU-funded projects examining the role of local authorities in promoting EMAS to small and medium-sized enterprises (fewer than 250 employees). Interestingly enough, the UK and Denmark are partners in both projects, with Ireland and Greece being the other member states involved. Both projects have revealed that local authorities have a major role to play in raising awareness about environmental issues and in assisting companies to undertake the LA-EMAS process. One of the major ways in which a local authority can effectively undertake the latter is to have gone through the LA-EMAS process itself, and therefore be in a position not only to lead by example, but also to be able to provide technical advice and guidance on setting up management systems that can be subjected to third-party validation.

☐ LA-EMAS and Sustainability

The use of EMSs can only help organisations improve their environmental performance. Often, these operations act from a defensive standpoint: compliance with legislation, securing funding, avoiding liability and maintaining a reputation. Achieving LA-EMAS registration means that a local authority is better placed to manage its environmental effects, it does not mean that it is necessarily a sustainable organisation. However, to survive in the future, sustainability has to be the long-term goal Thus research[7] is already in progress into the feasibility of SMASs—Sustainability Management and Audit Systems. These would be systems aimed at integrating the management of economic, social and environmental issues, e.g. ethical purchasing. To a certain extent, LA-EMAS is already part way to being an SMAS. For example, an environmental health department exists primarily to protect human health, which is a sustainability issue. It achieves this aim through the management of environmental effects such as local air pollution control, noise and nuisance regulation. A social services department examining its use of transport on environmental grounds may find that any reductions in use will affect people's ability to access essential quality-of-life services. Indeed, such a department may wish to increase transport use, which would potentially be in non-compliance with an authority's environmental policy to improve sustainability performance in terms of community access to essential quality-of-life services. Thus, discussion in three-to-five years' time will be focusing on how an EMS can be used to manage the change to a sustainable organisation.

■ *Conclusion*

Just under half of UK local authorities are currently working with LA-EMAS. By the end of the century, at least 100 councils will have units registered. This figure will undoubtedly rise as other organisations, such as national and European funding agencies, insurance and financial institutions and the Audit Commission, begin to exercise their influence in terms of requiring their customers (the local authorities) to show that they have demonstrable environmental credentials. Registration to LA-EMAS would show this clearly. Within this same timescale, the implementation of LA-EMAS may even become mandatory. In this climate, it is clear that local government will continue to use the scheme. A period of consolidation is needed to ensure that local authorities in the UK have, as far as possible, minimised their negative impacts on the environment and, more importantly, maximised their positive effects. In turn, only once this has been done will it possible for local authorities to consider how to use their existing EMSs to become truly sustainable organisations. Only when this is achieved will local authorities have fulfilled their Agenda 21 obligations, and this will prove to be an important example for local government worldwide.

■ Notes

1. Local Government Management Board, *A Step-by-Step Guide to Local Agenda 21* (Luton, UK: LGMB, 1994).
2. Since 1 April 1996, this regulatory function rests with the UK's Environment Agency.
3. Central and Local Government Environmental Forum, *A Guide to the Eco-Management and Audit Scheme for UK Local Government* (London, UK: HMSO, 1993).
4. Department of the Environment, *The Voluntary Eco-Management and Audit Scheme (EMAS) for Local Government (2/95)* (London, UK: HMSO, 1995).
5. Local Government Management Board, *LA-EMAS Case Studies* (Luton, UK: LGMB, 1996).
6. Personal communication.
7. See Chapter 13 of this book, 'From EMAS to SMAS: Charting the Course from Environmental Management to Sustainability'.

18 Environmental Management Standards
What do SMEs Think?

Ruth Hillary

D URING a twelve-month period in 1994–95, a pilot demonstration project was undertaken to investigate the experience of seventeen small and medium-sized enterprises (SMEs) implementing the Regulation (EEC No 1836/93):[1] the Eco-Management and Audit Scheme (EMAS).[2] The purpose of the pilot study was to identify the barriers and opportunities faced by SMEs when they implemented formalised environmental management systems (EMSs) such as EMAS, ISO 14001 and other similar initiatives. The pilot study also provided recommendations on the type of assistance necessary to aid SMEs' achievement and registration to EMS standards.

Participating SMEs were divided into three working groups: sole enterprises (see Fig. 1), cross-sectorial enterprises located in Blackburn, UK (see Fig. 2) and textile enterprises (see Fig. 3), with the aim of demonstrating the potential support options available to SMEs. Each participating enterprise was given a code to ensure confidentiality.

All enterprises worked through the stages of EMAS shown in Figure 4 at their own pace; some used national EMS standards such as BS 7750.[3] Both Group 2 (cross-sectorial enterprises) and Group 3 (textile enterprises) met monthly. Group 2 was assisted by the Blackburn Business and Environment Association (BEA) and Group 3 was supported by the Textile Finishers Association (TFA) and the Confederation of British Wool Textiles (CBWT).[4] Group 1 enterprises were unassisted.

This chapter details the final results and evaluation of the pilot study's Stage 2—on-site implementation. It provides general information on all seventeen enterprises and their nominated sites and shows the practical experiences of SMEs implementing EMSs. The chapter summarises SMEs' views

SME	Sector and NACE code*	Employees		Turnover (ECU)	Ownership
		Company	Site		
S1	Manufacture of furniture NACE 36.14	34	34	1	Limited: sole owner
S2	Manufacture of insulation NACE 26.14	154	144	8	Limited: family-owned
S3	Manufacture of metal joints NACE 28.12	16	16	2	Limited: sole owner

* Commission of the European Communities, *Council Regulation (EEC) No 3037/90 of 9 October 1990 on the Statistical Classification of Economic Activities in the European Community* (OJ L293, Vol. 33; Brussels, Belgium: CEC, 24 October 1990).

Figure 1: *Group 1—Sole Enterprises*

SME	Sector and NACE code[1]	Employees		Turnover (ECU)	Ownership
		Company	Site		
A1	Manufacture of wallpaper NACE 21.24	476[2]	200	53[3]	Limited: family-owned
A2	Press felt NACE 17.54	450[2]	450	30	Division of PLC[4]
A3	Geogrid NACE 25.23	350[2]	85	40	Limited: management-owned
B1	Confectionery NACE 15.84	180	180	15	Limited: management-owned
B2	Copper and aluminium powder NACE 27.45	385[2]	260	26	Division of PLC[4]
B3	Rubber moulding NACE 25.13	50	50	3	Limited: sole owner

1. See Fig. 1.
2. Employee numbers above the SME employee threshold of 250.
3. Turnover is above the EU SME turnover threshold of 40 million ECU.
4. PLC indicates that more than 25% capital is owned by a larger enterprise.

Figure 2: *Group 2—Cross-Sectorial Enterprises*

SME	Sector and NACE code[1]	Employees		Turnover (ECU)	Ownership
		Company	Site		
KIF1	Dyeing and finishing NACE 17.30	190	190	15	Limited: management-owned
TFA2	Dyeing and finishing NACE 17.30	300[2]	15	33	Division of PLC[3]
TFA3	Dyeing and finishing NACE 17.30	35	35	1	Limited: family-owned
TFA4	Dyeing and bleaching NACE 17.30	86	86	46[4]	Limited: family-owned
TFA5	Dyeing and finishing NACE 17.30	187	187	9	Limited: family-owned
CBWT1	Wool spinning and weaving NACE 17.12 and 17.22	150	150	7–13	Limited: family-owned
CBWT2	Manufacture of wool products NACE 17.73	279[2]	94	24–26	Division of PLC[3]
CBWT3	Wool spinning and weaving NACE 17.12 and 17.22	236	236	17	Limited: family-owned

1. See Fig. 1.
2. Employee numbers above the SME employee threshold of 250.
3. PLC indicates that more than 25% capital is owned by a larger enterprise. KIF refers to Knitting Industry Federation; TFA refers to Textile Finishers Association; and CBWT refers to Confederation of British Wool Textiles.
4. Turnover above the EU SME turnover threshold of 40 million ECU.

Figure 3: *Group 3—Sectorial Enterprises*

on EMSs and details the barriers and opportunities such standards hold for them. The chapter also summarises key recommendations drawn from the SMEs' implementation of EMSs which are designed to increase small-firm participation in EMAS and other EMSs.

■ General Enterprises Information

Seventeen sites were selected from 33 positive responses. Figure 5 summarises turnover, employee numbers and ownership of the selected enterprises.

For the purpose of the pilot study, an 'SME' is defined as an enterprise employing fewer than 250 employees, with a turnover of not more than 20 million ECU per year and with no more than 25% capital owned by a larger enterprise. Small enterprises are further defined as those with fewer than 50 employees and with a turnover of not more than 5 million ECU.

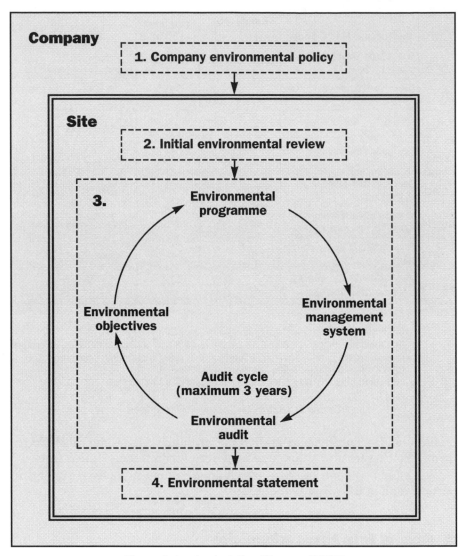

Figure 4: *Implementation Stages of EMAS*

On 7th February 1996, the Commission adopted a recommendation concerning the definition of SMEs.[5] The Commission's definition is broadly consistent with the one above, but differs in that its turnover threshold is set at 40 million ECU for fewer than 250 employees and 7 million ECU for fewer than 50 employees.

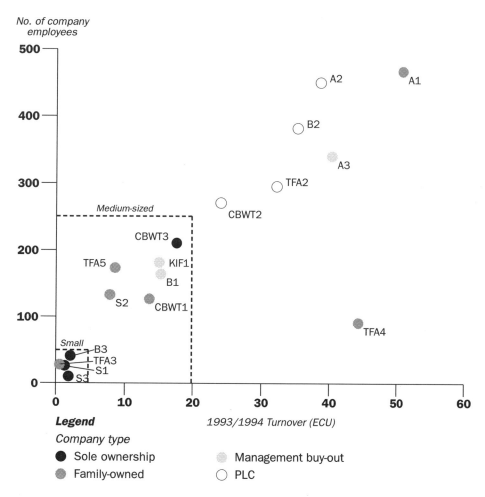

Figure 5: *General Enterprise Information*
(Turnover, Employee Numbers and Ownership)

Four of the enterprises are classified as small and six are medium-sized enterprises. The remaining seven enterprises employ fewer than 500 employees.

■ *Implementation of EMAS*

By completion of the pilot project, all seventeen enterprises had made progress in implementing EMAS or another similar initiative comparable to ISO 14001 (see Fig. 6).

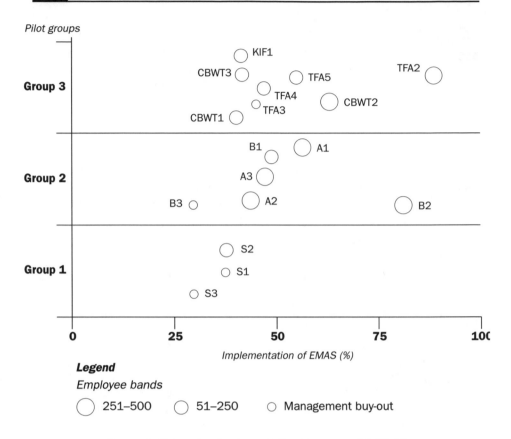

Figure 6: *Enterprise Implementation Progress with EMAS*

The larger (251–500-employee) enterprises with a greater turnover (21–50 million ECU) have made better implementation than the smallest enterprises. There is not a clear correlation, however, between size/turnover and EMS implementation progress. Enterprises around the 50% implementation mark cover the complete ranges of both size and turnover. Both enterprises TFA2 and B2, the most advanced enterprises, had relatively well-developed EMSs before entering the pilot.

Factors that assisted implementation were as follows:

- The existence of structured management systems that address EMSs
- The ability to draw on external consultancy services and external assistance

- Genuine and long-term top-management commitment to implement environmental management
- The use of a team approach to effect implementation
- The rapid involvement of staff at all levels in the environmental programme

Factors that inhibited implementation were as follows:

- Instability in management structure and organisation
- The number of functions for which an implementor had prior responsibility in addition to assuming the role of EMS implementor
- Isolation of the implementor from key decisions and lack of authority
- Implementors lacking training and the tools and techniques to implement EMSs
- Marginalisation of environmental issues from core business activities
- Involvement in other major management projects, e.g. the implementation of ISO 9000

■ *Implementation Routes to Achieving EMAS*

Enterprises may adopt one of four implementation routes to achieving EMAS (see Fig. 7). The choice of route adopted by enterprises influences the promulgation strategy adopted by the Commission and the EU Member States.

Route 1, the direct route, is preferred by SMEs that have no knowledge or a negative view of ISO 9000, and so, consequently, do not understand or mistrust the standards route to achieving EMAS. Route 1 was more typically chosen in Group 3 (sectorial enterprises), which had the lowest occurrence of ISO 9000 certifications. Three textile enterprises (TFA3, CBWT1 and CBWT2) pursued Route 1.

Route 2, utilising a national EMS standard, is of interest to enterprises that want to take a staged approach to achieving EMAS or are uncertain about seeking direct registration to EMAS. A number of representatives expressed concern about the reporting requirements under EMAS. One enterprise (TFA2) pursued Route 2.

Route 3, utilising ISO 9000 solely, is the least likely route to be used by enterprises in EU countries where an EMS standard already exists, i.e. Ireland, Spain and the UK. It may be more likely in Member States that do not have national EMS standards or where the standard has not been widely promoted. Only two enterprises, S1 in Ireland, and TFA4, a textile enterprise, followed this route.

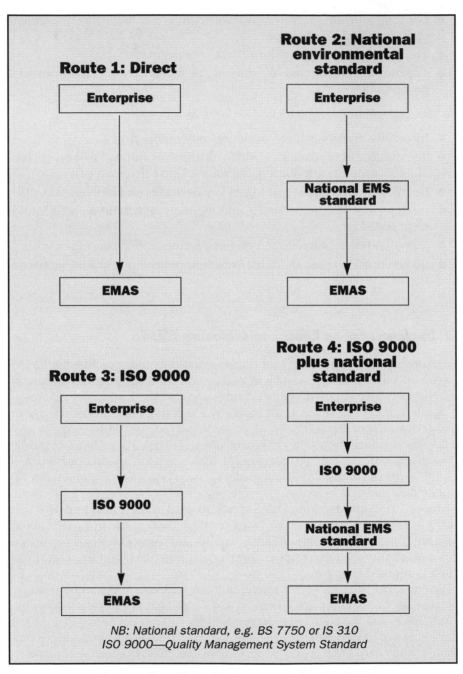

Figure 7: *Four Alternative Routes to Achieving EMAS*

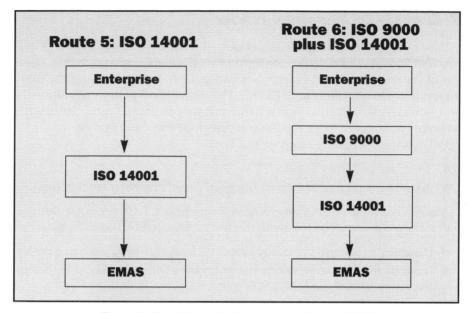

Figure 8: *Two Alternative Routes to Achieving EMAS*

Route 4, ISO 9000 plus a national standard, is the most common route for enterprises certified to ISO 9000, where a national EMS standard has been adopted and promoted, for example, in the UK and Ireland. Eleven enterprises have adopted this route, i.e. all six SMEs in Group 2 (cross-sectorial enterprises), two from Group 1 (sole enterprises) and three textile enterprises from Group 3.

Two alternative routes are available to enterprises once the European and/or international EMS standard ISO 14001 has been formally accepted by the European Commission (see Fig. 8). Route 5, utilising the ISO 14001 standard, and Route 6, ISO 9000 plus ISO 14001, are similar to Routes 2 and 4 respectively. Both Routes 5 and 6 open up the opportunity for enterprises to take a staged approach to EMAS in EU Member States that do not have an adopted EMS standard. Route 6 is particularly attractive to enterprises certified to ISO 9000.

A proportion of enterprises will not seek registration to EMAS, but will be satisfied to achieve certification to ISO 14001.

■ Company Environmental Policies

Company environmental policies developed without reference to EMAS or a national EMS standard do not conform to such schemes' requirements. Nine out of the seventeen pilot enterprises had policies before entering the pilot project that did not conform to EMAS. The twin EMAS policy commitments of continual environmental performance improvements and compliance with relevant environmental legislation were neglected in these policies.

Policies were originated for two reasons:

- To satisfy customer demand
- A top manager had identified the need for an environmental policy

Managers found writing an environmental policy for their organisations particularly difficult. The following reasons were cited by the authors:

- Difficulty in identifying the appropriate language to be used in a policy
- Inability to take a comprehensive overview of environmental aspects of the business because of a lack of knowledge
- A tendency to focus only on those environmental issues that were regulated, therefore causing gaps in the policy to appear
- Concern that the policy would not satisfy a verifier/certifier

Policy authors often neglected the following areas:

- Product planning: design; packaging; transportation; use and disposal
- Visual impact
- Environmental performance of contractors and suppliers
- Water management and savings

The reasons behind their neglect of the above issues included:

- Management saw their products as having little environmental impact or had not identified the opportunities of product redesign as an avenue to improve site environmental performance.
- Industrial location was predetermined, often at long-established sites which were therefore perceived to be unchangeable; even though positive employee and local community benefits could be identified arising from visual improvement at a site.
- Contractors were perceived to be outside management's control.
- Economic factors meant water was a relatively free resource for many sites because they operated their own bore holes.

■ *Continual Improvement*

A commitment to continual improvement is one of the central principles of EMAS and ISO 14001; however, all representatives raised the issue of how continual environmental improvements were to be defined. Most enterprises focused on achieving improvements in their emission outputs. Opportunities for setting objectives and targets in management areas and inputs to their process were neglected. The strong emphasis on controlled emissions reflects the existing management system which had been directed towards regulatory requirements.

Key questions raised by representatives relating to continual improvement are:

- How quickly does improvement need to occur?
- Does improvement need to occur in all areas?
- What rate of improvement needs to be demonstrated?
- What standards of improvement is a site working towards?
- Who sets the standards for improvement?
- Is a site required to go beyond its legislative standards?
- Can an environmental verifier or certification body question the improvement targets set by site management?

□ *EVABAT*

Article 3.a of the Regulation states that environmental performance improvements should aim to reduce 'environmental impacts to levels not exceeding those exceeding the economically viable application of best available technology' (EVABAT). No enterprise had defined EVABAT within the context of the selected sites' activities.

Five enterprises made reference to environmental technology in their environmental policies (see Fig. 9). Few SMEs were adequately aware of the range of new technologies applicable to their processes and fewer had the resources both human and financial to source technologies EU-wide.

The cases of enterprises B3 and TFA2 illustrates the difficulties SMEs face when trying to assess EVABAT. B3 produces vacuum-pressed rubber mouldings which are cleaned by the abrasive action of agitating the rubber moulding in water and fine granite gravel. Currently, the enterprise uses around fifteen old cement mixers, with a unit price of around £20 (55 ECU). These are open and need to be constantly topped up with water. They take ten hours to clean the rubber mouldings completely. An alternative type of technology

Enterprise	Direct reference to technology	Quote from relevant policy in which reference to technology is made or implied
Group 1: Sole Enterprises		
S1	No	'developing clean production processes' (1st and 2nd)
S2	No	
S3	No	'new processes will be selected with a view to minimising environmental impact' (2nd); 'reduce use of hazardous material in design and manufacturing of products' (3rd)
Group 2: Cross-Sectorial Enterprises		
A1	Yes	The policy will be reviewed to 'ensure it reflects adequately the latest developments in technology' (2nd)
A2	No	'Will strive to continually improve performance in line with improvements in technical and practical capability' (2nd)
A3	No	
B1	No	'We will try to eliminate pollution by checking our processes and considering alternative methods' (2nd)
B2	No	
B3	No	
Group 3: Sectorial Enterprises		
KIF1	Yes	'Consider and introduce (where applicable) technology or process for the optimum minimisation of water and air pollutants' (1st)
TFA2	Yes	'makes use of best available technology' (2nd)
TFA3	Yes	'implementing actions on the principal of utilising the best available technology not exceeding excessive cost' (1st)
TFA4	No	
TFA5	No	'Improve boiler efficiency to reduce dust' (1st)
CBWT1	No	'Within operations we will aim to reduce energy consumption, minimise waste and maximise reprocessing' (2nd)
CBWT2	Yes	'Purchase and use . . . technology and equipment in a way that helps to preserve natural resources' (2nd)
CBWT3	No	'Set and publish environmental objectives and targets which are technically achievable' (3rd)

NB: 1st, 2nd and 3rd refer to the different version of the SME's environmental policy.

Figure 9: *Company Policies that Make Reference to Technology*

is available from Germany at a cost of £10,000 (12,600 ECU). It is a closed centrifugal unit that completes its cycle in fifteen minutes using significantly less water and energy and creating no fugitive spills. Application of this new piece of technology would be economically non-viable for B3.

TFA2 illustrates another aspect of EVABAT. The enterprise was established in 1989, and dyeing technology applied in the process was specifically designed to use low liquor dosing. Management at the site believe the technology to be unique. Searching throughout the EU for other comparable technologies would be impractical and fruitless, as the managers do not anticipate changing technology specifically designed for their manufacturing process.

■ Environmental Reviews

An effective environmental review should generate sufficient information to enable a site to develop its registers of legislation and significant environmental effects. Prior to entering the EMAS pilot, ten of the seventeen sites had undertaken initial environmental reviews. None of these reviews completely fulfilled the requirements of EMAS.

Environmental reviews produced prior to the pilot fell into three categories:

- Review undertaken by external consultants (six SMEs). These were broad in scope and were of limited use in establishing registers of legislation or environmental effects.
- Internally-generated reviews (three SMEs). These were limited in scope and generally focused on known consented emissions and contained quantitative data.
- Corporate environmental reviews (one SME). These used extensive questionnaires to check the site against the corporate environmental policy.

All sites were advised to revisit or undertake an initial environmental review based on the environmental issues listed in Annex 1.c of the Regulation. In general, site implementors lacked the necessary tools to undertake the initial environmental review. They were directed to use questionnaires and interviews with personnel and encouraged to use input–output diagrams. As with the approach taken to writing environmental policies, implementors generally focused on legislatively-controlled outputs and neglected inputs. Implementors were encouraged to direct as much effort to inputs as outputs and to investigate their sites using input–output diagrams (see Fig. 10).

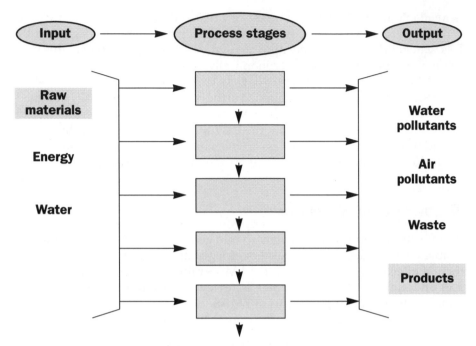

Figure 10: *A Simple Input–Output Model*

■ *Register of Regulations*

At the end of the pilot study, eight enterprises had compiled a register of regulations and other policy requirements. SMEs in Group 2 (cross-sectorial enterprises) were the most successful in compiling their register because they had access to a model register and were linked into a quarterly updating service of the register run by the Blackburn Business Environmental Association (BEA). Compilation of the register of legislation was perceived as a daunting task by all participating enterprises because:

- The vast quantity of environmental legislation meant that the SMEs were uncertain about their ability to cover comprehensively all legislation relevant to their businesses.

- Uncertainty concerning how the SME would keep the register up to date. The SMEs particularly doubted their ability to track EU legislative developments.

Group 2 (cross-sectorial enterprises) were the most proactive in identifying and developing assistance for SMEs compiling their register of legislation.

By the end of the pilot study, Group 2 enterprises had reviewed all relevant legislation relating to their selected site, and had developed, or were in the process of, developing a register of legislation. The successful approach used by Group 2 is summarised as follows:

- A general register of legislation was developed and provided in hard or disk copy form to implementors.
- Each company implementor worked with a BEA representative to determine which legislation from the general register was applicable to his/her site activities.
- BEA developed a twofold updating procedure:
  a. Three-monthly updates with clearly-marked changes to the general register will be sent to each representative on disk. Representatives are then responsible for determining relevance and updating their enterprises' own registers.
  b. If any operational changes occur in an SME, this will trigger the representative to contact the BEA to determine if any additional legislation applies and therefore needs to be included in the SME's register.

Group 3 (sectorial enterprises) had access to a sector application guide (SAG) that listed key pieces of legislation. Representatives were advised to use the list of legislation as a starting point to develop their own registers of legislation; however, some simply accepted the list as their registers. Representatives were not aware of their obligation to understand how the legislation applied to their sites and what measures they needed to introduce to ensure compliance. In recognition of this limitation, all textiles enterprises were provided with the Group 2 generic register of legislation.

The three CBWT members received individual assistance in reviewing the relevance of the legislation. The general register of legislation was further adapted to make it applicable to the textile sector by producing a separate, shorter textile register of legislation.

■ Register of Significant Environmental Effects

At the end of the pilot project, five SMEs had completely compiled their registers of significant environmental effects and eight SMEs were working on their registers of significant environmental effects. All implementors felt limited in tackling the environmental effects registers because of the lack of guidance in the Regulation. All implementors expressed concerns about:

- The scope for identifying indirect environmental effects

- The criteria to determine significance
- The ability of accredited environmental verifiers to challenge the content of the register of environmental effects and to question the criteria used to determine significance

Determining the criteria to establish and attach significance to environmental effects was problematic for all SMEs. None approached the task in a systematic way. Approaches used by implementors to determine significance included:

- A filter system by which each environmental effect is tested against a set of descriptive criteria
- A risk system that ranked environmental effects against scored risk parameters
- The personal opinion of a top manager
- The combined judgement of a selected team of management

All SMEs wanted further guidance on a recognised approach to establishing their register of significant environmental effects because:

- Implementors found the task difficult.
- Implementors feared that the accredited environmental verifiers could inappropriately challenge the contents of a site's register of environmental effects without fully understanding the site activities.
- Management could define apparently logical criteria which could effectively exclude important environmental effects of the site's activities from the site's register of environmental effects.

Additional guidance, in particular in SAGs, can be identified.[6] Although useful, none has the authoritative approval to give implementors the confidence to develop their registers of environmental effects or their procedures for identifying and evaluating their environmental aspects in ISO 14001. As with the register of legislation, SMEs believe their only recourse is to employ consultants, thus adding to the cost of implementing EMAS, ISO 14001 and other similar initiatives.

■ *Environmental Statement*

During the pilot project, all implementors expressed concern about the environmental statement requirements of EMAS. The two issues that stimulated most discussion among implementors were the statement's content and audience. Content and format were perceived as requiring more guidance because implementors did not know the scope of information and data. All implementors had not identified who the audience for their statements would be or how widely the statement should be disseminated. One enterprise (CBWT2) suggested a paragraph in the company's annual report would suffice.

Few implementors recognised that the environmental statement would form part of their communication strategy and was the culmination of efforts to demonstrate to the public the site's achievement of EMAS and environmental performance improvements.

By completion of the pilot project, no enterprise had produced an environmental statement; however, B1 and TFA2 wanted to produce their first environmental statement. Both have now produced environmental statements and successfully registered to EMAS.

■ *EMAS and National Standards*

By completion of the pilot project, all enterprises had raised concerns about the relationship between EMAS, national standards and the new international standard ISO 14001. Concerns centred on the following issues:

- Recognition by the Commission of a national standard as meeting the equivalent requirements of EMAS
- Relationship between the activities of certification bodies and accredited environmental verifiers, especially with regard to factors of cost, duplication of effort and drain on management time
- The impact of the ISO 14001 standard, its stringency and its potential to make both BS 7750 and EMAS obsolete
- Questioning the usefulness of making the additional effort to become registered to EMAS, particularly with regard to factors of additional cost, uncertainty of EMAS's added value in the marketplace, and EMAS disclosure requirements

Twelve of the seventeen enterprises intended to implement a national EMS standard (BS 7750) prior to deciding to seek registration to EMAS (see Fig. 11). All six enterprises in Group 2 (cross-sectorial enterprises) intended to use BS 7750 as their EMSs. B1 is currently undergoing assessment to be certified to BS 7750. Only four textile enterprises in Group 3 (sectorial enterprises), considered using BS 7750. Only one (TFA2) of the twelve enterprises implementing a national standard did not have a certified ISO 9000 quality system in place; the remaining eleven either had an accredited certified ISO 9000 system or are in the process of implementing ISO 9000 and seeking certification.

The preference for using an EMS standard route to achieving EMAS in the UK is strongly related to whether or not the enterprise was already certified to the international quality management systems standard ISO 9000.

Enterprise	Certified to ISO 9000	Implementing EMS standard
Group 1: Sole Enterprises		
S1	In progress	No
S2	Yes	Yes
S3	Yes	Yes
Group 2: Cross-Sectorial Enterprises		
A1	Yes	Yes
A2	Yes	Yes
A3	Yes	Yes
B1	Yes	Yes
B2	Yes	Yes
B3	Yes	Yes
Group 3: Sectorial Enterprises		
KIF1	Yes	Yes
TFA2	No	Yes
TFA3	No	No
TFA4	In progress	No
TFA5	Yes	Yes
CBWT1	No	No
CBWT2	No	No
CBWT3	Yes	Yes

Figure 11: *Companies Implementing ISO 9000 or EMS Standard*

■ Integration of Systems and Links with ISO 9000

EMAS was intended to integrate with existing management systems and it was the aim of all seventeen enterprises to integrate their EMSs with existing health and safety, quality or general procedure systems (see Fig. 12). Ten enterprises intended to integrate their EMS with ISO 9000; six with health and safety systems; and one with general procedures. The integration preference depended on the prevailing corporate culture.

Enterprise	Health and safety (H&S)	Quality (Q)	Stand-alone
Group 1: Sole Enterprises			
S1			✓ (GP)
S2		✓	
S3		✓	
Group 2: Cross-Sectorial Enterprises			
A1			✓ (Q)
A2			✓ (Q)
A3			✓ (Q)
B1		✓	
B2			✓ (Q)
B3			✓ (Q)
Group 3: Sectorial Enterprises			
KIF1			✓ (Q)
TFA2			✓ (H&S)
TFA3	✓		
TFA4	✓		
TFA5	✓		
CBWT1			✓ (H&S)
CBWT2	✓		
CBWT3			✓ (Q and H&S)
TOTAL	4	3	10

NB: GP refers to general procedures

Figure 12: *EMS and Links with Other Systems*

Only half of the textile enterprises had ISO 9000, and a number of textile enterprises had strong misgivings about ISO 9000. Two enterprises (TFA3 and CBWT1) had undertaken to implement ISO 9000, but later rejected it because they perceived it as being document-driven and having little to do with improving quality. Associating EMAS with the quality management systems standard was not encouraging for these enterprises.

Ten implementors had been initially developing and implementing their EMSs as stand-alone systems rather than integrating them with existing systems. Six enterprises had the aim to link their stand-alone EMSs with a quality management systems standard; two intended to link their environmental systems with health and safety; one with general procedures; and one intended to integrate all three systems.

The reasons given for initially developing a stand-alone system include:

- Until the new system matures, integration may disrupt existing systems.
- Developing a stand-alone system initially raises the profile of the environment as an issue within the enterprise.
- Integration would be attempted once assessment criteria for certification to both quality and EMSs became clear.

■ Costs and Resources

The human and financial resources devoted to implementing EMAS varied considerably between enterprises (see Fig. 13). There appeared to be little correlation between turnover and financial resources devoted to EMAS implementation. A stronger relationship existed between the amount of human resources devoted to implementation and the number of staff employed by an enterprise.

EMAS implementation in SMEs is dependent on human resources more than financial resources.

Implementation in SMEs is often a staged, stop / start approach. Successful implementation to registration will typically take between 12 and 24 months. At the pilot outset, implementors wanted to know how much EMAS implementation would cost at their sites; however, once they had started the implementation process, they found it difficult themselves to put a price on implementation. A number expressed the view that, once top management had committed the enterprise to implementing an EMS, they were prepared to pay for its implementation. The important aspect for many SMEs was to focus on the benefits they anticipated reaping once successful registration / certification was achieved.

A number of enterprises received external assistance in the form of:

- Grant-aided environmental reviews, e.g. City Challenge funding up to 50% of costs to undertake environmental reviews and the Industrial Research and Technology Unit in Northern Ireland, which supports up to two-thirds of the cost of an Environmental Audit

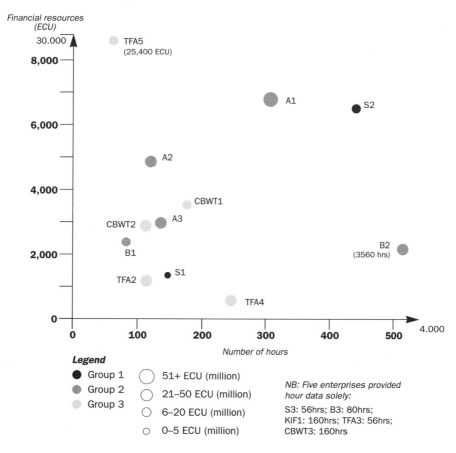

Figure 13: *Human and Financial Resources*

- Counselling from the Environmental Help-Line (four-hour site visit plus a site report)
- Information via Business Environment Association or Trade Association

The quality of the assistance available varied considerably. For example, some environmental reviews did not cover EMAS requirements (although they made no claim to), but also did not stimulate action in SME management because they were too general.

■ *Recommendations*

The pilot study results provided concrete examples of enterprise experience in the implementation of EMSs and EMAS. The pilot showed the barriers and opportunities facing smaller companies when they seek to formalise the environmental aspects of their processes and activities into structured EMSs such as EMAS, ISO 14001 or other similar initiatives. Recommendations were made on the key findings of the pilot and are shown in the Appendix. The recommendations are directed at institutions and governments assisting SMEs. The recommendations are directed at elevating the internal and external barriers SMEs face when attempting to implement formalised EMSs and EMAS.

■ *Conclusions*

The pilot study identified internal company factors that assist and inhibit the implementation of EMSs such as EMAS. The major internal assisting factors included: genuine and long-term top-management commitment; utilisation of a team approach; rapid involvement of staff at all levels; and recourse to external assistance. Significant internal inhibiting factors included: instability in management structures; involvement of the company in other major management projects, e.g. ISO 9000; marginalisation of environmental issues from core business activities; and the nature of the implementor, i.e. if he/she is multifunctional and/or isolated from key decisions and lacks authority.

All companies found the initial stages of EMAS the most taxing: i.e. the initial environmental review and the establishment of the registers of significant environmental effects and legislation and policy requirements were found to be difficult. A model register of legislation and an updating service was developed to aid SMEs in this process. The pilot study identified SMEs' most limited resource as human rather than financial; hence those support services that most effectively boosted human capacity are the most beneficial in aiding EMS implementation. Assistance strategies should support self-help groups hosted by the appropriate sector and local business groups. Such host organisations could be aided in their efforts by placing with them roving environmental managers specifically designed to assist EMS implementors.

Enterprises may take four routes to achieving EMAS registration. When an enterprise is certified to ISO 9000, or where a strong national EMS standard exists, companies are more likely to take a standards route to achieving EMAS registration. A proportion of SMEs will stop once certified to a standard such as ISO 14001 and not seek EMAS registration. This proportion will be influenced by future developments with the international EMS standard ISO 14001. Awareness of EMAS among SMEs is low, if non-existent. Confusion exists between the relationship between standards such as ISO 9000, BS 7750 and ISO 14001 and EMAS. Communication and promulgation strategies need initially to raise awareness by providing positive messages on the benefits of EMAS and EMS standards, but these messages need to be tailored for selected audiences, as well as clarifying the relationship between standards and EMAS.

The pilot study concluded that EMAS or comparable EMS standards are not unattainable for SMEs, but that SME implementation will be characterised by a staged stop/start approach. The staged approach is mainly a result of the limited human resources available for implementation. Group support services are effective at aiding implementation by enabling the pooling of experience, the transfer of knowledge between companies and the development of appropriate services. Sector and business organisations would be appropriate bodies to facilitate self-help groups with support from larger companies, technical specialists and the authorities.

■ Notes

1. Commission of the European Communities, *Council Regulation (EEC) No. 1836/93 of 29 June 1993 Allowing the Voluntary Participation by Companies in the Industrial Sector in a Community Eco-Management and Audit Scheme* (OJ L168, Vol. 36; Brussels, Belgium: CEC, 10 July 1993).
2. The pilot project was funded by the Commission of the European Communities and the UK Department of the Environment.
3. See R. Hillary, *The Eco-Management and Audit Scheme: A Practical Guide* (Cheltenham, UK: Stanley Thornes, 1994).
4. Special thanks to these organisations and to their representatives who assisted the pilot scheme.
5. Commission of the European Communities, *Draft Commission Recommendation Concerning the Definition of SMEs* (Brussels, Belgium: CEC, 8 February 1996).
6. See R. Hillary, 'Sector Application Guides' (Special Report), in *Environmental Policy and Procedures*, No. 14 (July 1996).

Appendix:
Key Findings and Recommendations from the Pilot Study

Environmental Policies: Key Findings and Recommendations

Key finding	Recommendation
1. Environmental policies developed without reference to EMAS rarely satisfied its requirements.	● Provide tools to enable enterprises to assess their own policies against EMAS. ● Ensure grant-aided assistance facilitates the review of environmental policies.
2. Top management had environmental 'blind spots' in their policies.	● Ensure grant-aided assistance effectively brings to the attention of management all relevant environmental issues. ● Support sector application guides that encourage effective review of environmental issues.
3. Policies were generated in response to customer enquiries.	● Support supply-chain enquiries that are reasonable and responsive to SME needs.
4. Not all environmental policies were systematically communicated to employees.	● Develop material that suggests ways of effectively communicating environmental policies.

Continuous Improvement: Key Findings and Recommendations

Key finding	Recommendation
1. The definition of continuous improvement concerned all implementors; most believed the external certifiers or verifiers would define continuous improvement.	● Provide information on the role of the verifiers and certifiers to enterprises, consultants and the verifiers and certifiers themselves. ● Address key questions raised by implementors in promotional material.
2. No enterprise defined its site's environmental performance improvements with reference to the economical viable application of best available technology (EVABAT).	● Encourage a broad definition of EVABAT. ● Clarify how continual improvement is achieved, i.e. by using a variety of techniques, one of which is technology.
3. Implementors had some difficulty in defining their own continual improvement standards.	● Support trade associations' efforts to assist their member via sector application guides.
4. Opportunities to set improvement targets for inputs were often neglected.	● Promote material that supports a holistic analysis of a site. ● Encourage regulators to communicate the benefits of input minimisation.

EVABAT: Key Findings and Recommendations

Key finding	Recommendation
1. Enterprises did not define continual improvement with reference to EVABAT.	● Review applicability of EVABAT. ● Provide support to establish EVABAT by industrial sector.
2. Focus on technology directs the enterprise towards emissions control rather than taking a holistic approach to its environmental impacts.	● Provide case studies that illustrate the different routes to achieving environmental performance improvements.
3. SMEs do not have the resources to source EVABAT solutions EU- wide.	● Provide a clearing house of EVABAT solutions. ● Support sector definitions of EVABAT.
4. EVABAT constrains SME implementation of EMAS and does not necessarily deliver the challenging targets it intended.	● Review value of EVABAT at the 1998 Regulation Review.

Environmental Review: Key Findings and Recommendations

Key finding	Recommendation
1. Enterprises with existing environmental reviews are unlikely to conform to EMAS requirements.	● Advise enterprises to pattern match their environmental reviews against EMAS requirements.
2. Existing grant-aided environmental reviews generally do not conform to EMAS.	● Provide advice to support organisations on what an EMAS environmental review should contain.
3. SMEs require a range of tools to undertake environmental reviews.	● Develop environmental review workbooks and promulgate existing ones. ● Support associations' development of review tools.
4. The initial review is a resource-intensive activity for an SME.	● Structure grant-aided regimes and assistance to provide more support for initial stages of EMAS.

Register of Legislation: Key Findings and Recommendations

Key finding	Recommendation
1. Producing the register of regulations was a major task for all SMEs.	● Provide a tool to facilitate the development of register.
2. SMEs do not have the internal resources to keep up to date with environmental legislation.	● Encourage external updating services upon which SMEs can draw. ● Support trade and business associations' efforts to provide assistance on legislation to their members.
3. Sector application guides (SAGs) can provide lists of appropriate legislation, but are not exhaustive.	● Support the development of SAGs.

Register of Significant Environmental Effects: Key Findings and Recommendations

Key finding	Recommendation
1. All implementors had difficulty in establishing their site's register of significant environmental effect, in particular scoping boundaries for environmental effects and establishing criteria for significance.	● Provide guidance on effective methods of establishing a register. ● Grant-aid reviews that generate full lists of environmental effects. ● Support sector efforts to develop list of sector environmental effects and methods for establishing significance.
2. Few enterprises approached attaching significance to environmental effects in a systematic way.	● Provide case studies of best practice and worst practice.
3. SMEs need guidance on significant environmental effects methodologies.	● Review available methods and provide the best in a published form.
4. SMEs were uncertain about the role of the accredited environmental verifier to challenge the contents of a site's environmental effects register.	● Provide clear guidance on the role of the verifier.

EMAS and National Standards: Key Findings and Recommendations

Key finding	Recommendation
1. The links between national standards such as BS 7750 and EMAS are not clear to SMEs.	● Provide clear guidance on how national standards may be used with EMAS. ● Address SMEs' fears that certification plus verification will not increase cost and duplicate effort.
2. A number of enterprises fear ISO 14001 will make EMAS redundant.	● Clarify relationship between ISO 14001, national standards and EMAS.
3. A high proportion of SMEs in the UK will use BS 7750 as a route to achieving EMAS registration.	● Use those enterprises certified to BS 7750 as case studies.
4. An SME is more likely to use a national standard as a route to achieve EMAS if it is already certified to the quality standard ISO 9000.	● Develop case studies showing the links between EMAS and ISO 9000.
5. Enterprises not certified to ISO 9000 are discouraged by the close association of BS 7750 to ISO 9000 principles, especially the emphasis on documentation.	● Develop tailored promulgation messages to suit different enterprises.

Integration of Systems: Key Findings and Recommendations

Key finding	Recommendation
1. EMAS can be successfully integrated with existing management systems such as quality and health and safety. Most SMEs have integration as their ultimate aim.	● Promote the positive integration of systems. ● Develop case studies to show how integration occurs.
2. The development of stand-alone EMSs is thought preferable before integration to prevent destabilising of exiting systems.	● Promote a number of routes to achieving EMAS.

Cost and Resources: Key Findings and Recommendations

Key finding	Recommendation
1. The quantity of resources devoted to EMSs' implementation varies considerably from enterprise to enterprise.	● Provide examples of the range of costs as a percentage of turnover. ● Evaluate typical costs per stage of EMS implementation.
2. The most scarce SME resource is management time, not finance.	● Provide external advisory support time. ● Grant-aid labour-intensive stages of EMAS, e.g. initial environmental review. ● Grant-aid support organisations to provide self-help groups.
3. Initial stages of EMAS, e.g. the initial environmental review and establishing the register of legislation and environmental effects, are the most resource intensive.	● Any grant-aided assistance to enterprises should support the initial stages of EMS implementation.
4. Varying resources in SMEs means implementation will often be a staged, stop/start approach.	● Allow for recognition of achievement to the key stages of an EMS. ● Support staged approach to EMAS by staggered grants regimes and promotional material that reflects EMAS stages.

19) Towards Innovative, More Eco-Efficient Product Design Strategies

Jacqueline M. Cramer

ACHIEVING sustainable development presents an enormous challenge to society. It means that, within just a few decades, we must learn to deal much more efficiently with energy and raw materials. According to environmental experts, within the next fifty years the burden on the environment will have to be reduced to an average of one-tenth of the current levels.[1] In other words, average eco-efficiency will have to increase by a factor of ten by 2040.

Steps have already been taken within industry to increase the average eco-efficiency of products. Most of these efforts focus on incremental environmental improvements of existing working methods, products and services at operational level. Various techniques and methodologies have been developed to analyse and assess the environmental merit of such product improvements.

However, in order to reach the target of a tenfold increase in the average eco-efficiency mentioned above, more far-reaching, innovative improvements of current production techniques are required. No methodology is available yet for searching for such product changes. Some companies (among them Philips Sound & Vision/Business Electronics) have started to elaborate such a methodology. When methods and instruments are available to identify and select promising eco-efficient product design strategies, the next question is how to embed environmental issues within the business organisation in a structural way.

This chapter deals with this question on the basis of the experiences gained within Philips Sound & Vision. Before focusing on this particular case, the author will first discuss how the efforts to integrate environmental issues into product design strategies relate to current developments in environmental management systems.

■ Standardisation of Environmental Management Systems

Environmental management systems are still in the process of development. Depending on the country and the specific sector of industry, the experiences regarding the introduction of these systems may differ. Until now, however, all environmental management systems have had a common primary focus on procedures to reduce the emissions of individual plants through *process* improvements. The integration of environmental issues into *product* design strategies has received limited attention.

With the introduction of the international standard on environmental management systems (ISO 14001) and related items in the ISO 14000 series, this situation may change. More than most national and regional standards on environmental management systems, the ISO 14000 series stresses the importance of product-oriented objectives and supportive tools (e.g. lifecycle assessment and environmental labelling tools). It may therefore be expected that the implementation of the ISO 14000 series provides a good framework for environmental product improvements.

A first positive aspect of the ISO 14000 series is its international character, avoiding the proliferation of conflicting and inconsistent national and regional standards on environmental management systems. Moreover, it provides a single, more cost-effective system for multinationals to implement wherever they operate.[2] A second advantage of the ISO 14000 series is the positive support from industry for its implementation. In fact, it was industry itself that promoted the idea of an international environmental management standard.

A third advantage is that an ISO standard on environmental management can influence the behaviour of companies with respect to the environment. It leads to a standardised method of working to which companies should conform when they export to global markets. More often, customers or clients require verification that their suppliers are taking their environmental responsibility seriously. Therefore, preparing for ISO 14001 is not an option—it is a matter of survival.[3] This point can be illustrated by the response of various large, export-oriented South African companies that the author recently visited. While the influence of the national environmental legislation was (still) limited (except for water pollution standards), the acceptance of their product on the international market turned out to be crucial. These companies were therefore very interested in developments concerning ISO environmental management standards.

Thus, the importance of the ISO 14000 series in enhancing the attention to environmental issues within companies cannot be denied. However, adoption of the standard will not in itself guarantee optimal environmental outcomes. As stated in the Introduction to the ISO standard:

it should be noted that this International Standard does not establish absolute requirements for environmental performance beyond commitment, in the policy, to compliance with applicable legislation and regulations and to continual improvement. Thus, two organisations carrying out similar activities but having different environmental performance may both comply with its requirements.

■ Types of Environmental Strategies

Thus, the ISO 14000 series does not intend to set absolute targets for the environmental performance of companies. What ambitions a company has in this respect depends on its particular environmental strategy.[4] Some companies, particularly among newcomers on the market, clearly profile themselves as ecologically (and sometimes also as socially/ethically) responsible enterprises (e.g. The Body Shop; Ben & Jerry's). Such companies will continuously strive for environmental improvements of their products in order to keep or increase their market share. The incentive to improve the environmental performance has been embedded in their business strategy.

For the majority of companies, the environmental issue plays a less crucial role. This group of companies usually takes environmental items into account, when they are forced to do so or see a direct cost or quality advantage. Within this group of companies, some are gradually evolving from reactive, compliance-driven environmental management toward integrated business and environmental management focused on gaining competitive advantage. Through the adoption of this more proactive approach they aim to increase their market share and improve their public image (e.g. through becoming number one in the consumer tests on environmental aspects and/or improving the product quality). In this way, they attempt to use the environmental issue as a marketing tool, albeit in a modest way.

The management of every company makes implicit or explicit strategic choices as to the way in which it wants to position itself on the market. These choices reflect the actual content of the environmental policy that the company adopts with respect to its products. The ISO 14000 series provides the appropriate framework, while the environmental objectives set by the company determine the level of environmental ambition. The scope of this ambition can vary greatly. Roughly speaking, companies can bring about three types of environmental improvements within the product chain. They can focus on the following.

☐ Incremental, Step-by-Step Changes in Existing Products (Timescale: 1–3 Years)

The Dutch Eco-design Programme provides examples of such changes: for instance, the design of an office chair made of less material, which is recyclable and consists of fewer toxic substances; the design of re-usable plant trays for the flower auctions; the design of a face mask made of recyclable plastic which is qualitatively better than the original face mask. The environmental merits of incremental changes can be high if benefits can be reaped easily. Most companies that have already made efforts to improve their environmental performance can no longer obtain high environmental benefits from incremental improvements.

☐ More Far-Reaching Changes in Existing Products (Timescale: 2–5 Years or even beyond)

In the field of packaging, various examples can be given of such changes: for instance, new distribution techniques leading to less and re-usable transport packaging; adaptation of packaging materials into mono-materials; new display techniques in shops using less packaging material. Through more fundamental changes in the present product, high environmental merits can be achieved. This usually requires investment in R&D and initial product costs but may lead in the end to lower costs.

☐ Radical Changes in the Function of Products (Timescale: 20–50 Years)

Examples of this type of product improvement are provided by the Dutch 'DTO (Sustainable Technological Development) Programme'. This R&D programme aims to set up illustration processes to show innovative alternatives for present unsustainable developments. Examples are the substitution of meat through biotechnological techniques (so-called novel protein foods) and the development of sustainable building methods. The DTO programme aims to achieve an increase in eco-efficiency by a factor of 10–20.

The social and organisational changes needed to realise the three types of changes vary tremendously. Incremental changes in existing products require the involvement of various actors within the company itself and sometimes of its suppliers and customers. However, more far-reaching changes in existing products even call for communication and co-operation between all actors in the product chain. Radical changes in the function of the product cannot be made without fundamental changes in production and consumption. Similarly, the initial cost of implementing the three types of changes increases rapidly as the changes become more radical. Existing structures (investments) have to be readjusted or sometimes even be 'destroyed' before more radical alternatives can be implemented. Take, for example, the high financial costs of building new infrastructures or re-organising the agricultural sector in a radical way through the production of novel protein foods. However, after initial costs have been incurred, the new but radical changes may lead to substantial cost reductions.

Redesigning products in a more fundamental way requires the integration of environmental issues at a very early stage of the product creation process. When the first ideas about the new product generations are developed, the environmental issues should be taken into account. In later phases of the product creation process, the architecture of the new product has already been determined, which implies that from then on only minor changes can be made to the product.

The integration of environmental issues in the first phases of product development takes place at the company's strategic level. Senior management should decide here on the strategic environmental issues to be addressed, taking into account both environmental and market considerations. In general, proactive companies are more interested in such a strategic approach than those that make a defensive stand. Instead of waiting until environmental regulations are forced on them, the proactive companies will choose to anticipate future environmental opportunities and threats.

We can illustrate this latter point on the basis of the experiences of implementing environmental product development within the Sound & Vision division of Philips.

■ Environmental Product Development within Philips Sound & Vision

In the 1970s and 1980s, the emphasis in the environmental policy of the Philips Sound & Vision division was on incremental improvements, especially in its production processes. One of the major driving forces behind this was legislation, and the associated rules concerning licensing. Since the early 1990s, the focus has widened to encompass improvements in the consumer electronics products themselves. An initial driving force for this was the corporate environmental policy formulated by the CEO of Philips, Mr Timmer. Another reason was the growing public pressure to find socially-responsible ways of disposing of used consumer electronics goods. An additional factor was the regulation concerning the use of certain chemical substances.

In recent years, the company has initiated numerous activities to improve its products from an environmental perspective. For instance, a manual on environment-oriented product development ('eco-design') has been produced for designers. The manual includes compulsory (environmental) rules for design. These place the emphasis on avoiding banned chemical substances and instead choose less environmentally harmful ones for consumer electronics products. The following issues are addressed in the manual.[5]

☐ Material Use

The central objective is weight reduction and a reduction in the amount of potential end-of-life waste. Information is provided on material data and on criteria for material use in relation to the functionality required.

☐ Hazardous Substances or Materials

An inventory is given of hazardous substances in components and materials according to the EACEM (European Association of Consumer Electronics Manufacturers) list of relevant substances. These substances are classified in three categories of environmental release status:

- Released (according to proactive environmental standards)
- Temporarily released (better alternatives are investigated)
- Forbidden or blocked

☐ Industrial Processes

It is recommended to eliminate and/or reduce the following auxiliary materials and chemicals used in industrial processes:

- Elimination of CFC and HCFC
- Soldering
- Glues
- Metal coatings, chromates
- Lacquers, paints

☐ End-of-Life

Information is provided about the following issues:

- Cost and yield table of materials and material fractions produced at end-of-life processing
- Rules for disassembly-friendly construction
- Plastic compatibility rules
- Packaging and packaging aspects

☐ Energy Use

Information and directives are provided about:

- Energy use in operational mode
- Energy use in standby mode

☐ Environmental Design Evaluation

Design evaluation is based on:

- Environmental weight
- End-of-life costs
- The environmental release criteria to be used

In addition, the company is actively seeking the best ways of designing consumer electronics products so that they can be processed as environmentally soundly as possible at the waste stage. Apart from the products to be re-used, four material streams will essentially result.[6]

Material Stream I. Material that can be recycled on a commercial basis (metals, engineering plastics)

Material Stream II. Material for which the price of the corresponding virgin material is lower than the cost of the secondary one (some engineering plastics, glass). Under specific conditions such materials can also be recycled.

Material stream III. Material mixtures that have a recycling potential for certain functions, but that have an overall negative yield due to presence of other material types (e.g. copper from printed wiring boards, copper from wires and cables).

Material Stream IV. Material or mixtures of material that have to be sent directly to incineration. Waste fractions resulting from the processing of streams I, II and III will have to be added to this stream.

With better designs and end-of-life technology, the amount of material in category I will grow steadily in the years to come; the amounts in categories II, III and IV will shrink correspondingly. It is expected that, at the present rate of improvement a weight ratio between the four categories of approximately 1:1:1:1 for the total amount of consumer electronics products will be reached in ten years from now. By stepping up the R&D efforts, this time-span could be reduced substantially.

■ Towards More Far-Reaching Eco-Efficiency Improvements

Initially, the above product improvements were mainly incremental in nature. Gradually, the attention is now turning to more complex solutions, aimed at more far-reaching eco-optimisation of existing products. To make recycling of consumer electronics products possible on a wider scale, for instance, the company is now working on the next stages of eco-design. In that context, it has developed the concept of the 'green television', which incorporates all the accumulated environmental know-how. This concept will be used as a reference for the future generations of the product. To what extent the concept will be taken into account in the actual product planning requires strategic decisions from management. Answers are needed to questions such as: what benefits can be expected when the green TV, or certain aspects of it, are promoted and what marketing implications does this have?

To broaden the scope, the Environmental Competence Centre of Philips Sound & Vision has developed a methodology to assess which proactive environmental measures the company should now take in order to be prepared for the future. This has led to a list of more far-reaching environmental improvements that may be implemented in 1–5 years. The current situation is that the various business groups have selected various items for further investigation. These items relate, for example, to the following partly complementary, technological options:

- Minimising energy consumption and the use of raw materials and toxic substances
- Further increasing material recycling
- Optimising the life of the product (e.g. by recycling of the product or components; or by technical upgrading)

The implementation of the above options needs the commitment and support of senior management. A strategic choice should be made on how the company will deal with the particular item from an environmental and economic point of view in an early phase of product development. The marketing strategy resulting from this choice can then be translated into product requirements. Whether these requirements can be fulfilled should be checked first with R&D, and after that can be implemented in the later phases of the product creation process. This whole process of integrating environmental issues into the product creation process should be well attuned to the particular product planning procedures of the company. Within Philips Sound & Vision, a specific procedure (including milestones and release criteria) exists, in which the environmental issues are currently integrated from the first to the last phase of the product creation process.

This strategic approach to environmental product development requires co-operation between senior management, the strategists of the Philips Sound & Vision division, the product developers, the marketing experts, the researchers and the environmental specialists. Integrating environmental aspects with other aspects of business is essential. Internally, Philips is currently engaged in building up this collaboration.

■ *Three Examples*

By including environmental aspects at an early stage in the product development, Philips Sound & Vision can act proactively rather than taking defensive, corrective measures. This form of integration of environmental aspects into product development can offer opportunities for far-reaching improvements of products. Which of the potential options will be put into practice, is currently being investigated. Let us illustrate this point on the basis of the following three examples.[7]

- The reduction in the energy intensity of consumer electronics products
- The reduction of the material intensity of consumer electronics products
- The development of potential strategies to enhance the durability of products

With respect to the item 'reduction in the energy intensity', an intensive brainstorming session was held in the business groups concerned with TV, audio and VCR in order to generate and select more far-reaching environmental improvements in the energy consumption during use and standby. As improvements could be made in various parts of the product (e.g. in the components or in the printed circuit board), experts from various backgrounds were present at these workshops. The options that these experts proposed are currently being elaborated in a technical, economic and marketing sense.

Secondly, the 'reduction of the material intensity of consumer electronics products' was also elaborated in a specific tailor-made way. In order to generate options for the reduction of material intensity, close co-operation was established between Philips and one of its suppliers of materials. Various brainstorming sessions were held to identify promising alternative materials that are lighter, but at the same time have the appropriate functionality for fulfilling the demands on the product. The results of these brainstorming sessions are currently being elaborated in R&D projects.

The project related to 'the development of potential strategies to enhance the durability of products' was elaborated in a slightly different way. First, a summary of the potential options for optimising the life of products was made on the basis of a literature survey. Next, the capability of Philips Sound & Vision in meeting these options as a way to achieve further optimisation of the life of its products was assessed. At this stage, it was found important to gauge the view of the outside world on this matter. To this end, Philips Sound & Vision's Environmental Competence Centre in the Netherlands organised a brainstorming session with external stakeholders which was attended by fifteen representatives from environmental, consumer and women's groups, from the Ministry of Housing, Physical Planning and the Environment and the Ministry of Economic Affairs, from relevant research institutes and from Philips.

The participants at this session were asked which five (not more) activities they thought Philips Sound & Vision should give the highest priority in the context of the theme of 'optimising product life'. The reactions of the participants suggested a clear prioritisation. Particular attention was given to the following topics:

- Making more robust constructions
- Designing modular constructions
- Selling the use of products/leasing

These results were presented in brainstorming sessions with the business groups concerned with Audio and VCR. Establishing which additional methods stand a good chance of success in the future of Philips Sound & Vision is currently part of further internal consultation and investigation. Initial results show that products usually break down due to thermal problems (too high temperature) or defective components or joints. Only after more information has been gathered on the various advantages and disadvantages of improving the durability of the products will Philips take concrete action.

The three examples clearly show that it usually takes a number of brainstorming sessions and specific R&D initiatives before a final assessment is made of the most promising environmental opportunities to be implemented. Through these sessions and specific projects, learning experiences are built up that are used to reduce the present uncertainties about environmental opportunities and market perspectives. When the company has learned more about these more far-reaching environmental improvements, it becomes easier to integrate these endeavours into the regular product development process.

■ Conclusions

The ISO 14000 series can be a good vehicle for increasing the environmental awareness within companies. The international standard can enhance continual environmental improvements of both processes and products. The ISO standard does not intend to establish absolute environmental targets. The degree of ambition in that respect is left to individual companies. They determine the level of the playing field within the boundaries set by the environmental regulations of the Government. It depends largely on the strategic choices made by these companies which environmental improvements will be made. Of course, these choices are not made in a vacuum. Companies choose certain environmental strategies in response to societal demands and to the business environment in which they operate.

It is understandable that most companies are reluctant to move forward from incremental to more far-reaching environmental improvements. Uncertainties about the benefits restrain them from being too far ahead of others in terms of environmental performance. The first efforts, like those taken by Philips Sound & Vision, are still more the exception than the rule. Those companies that follow the latter approach, integrate environmental issues in the product planning process as early as possible. This guarantees that environmental issues become an integral part of the general business strategy. Only then can product innovation and environmental improvements go hand in hand.[8]

■ Notes

1. See R.A.P.M. Weterings and J.B. Opschoor, 'Towards Environmental Performance Indicators, Based on the Notion of Environmental Space', in Advisory Council for Research on Nature and Environment (RMNO), *Publication RMNO*, No. 96 (Rijswijk, The Netherlands, 1994).

2. See G.G. Crognale, 'Environmental Management: What ISO 14000 Brings to the Table', in *Total Quality Environmental Management*, Vol. 4 No. 4 (1995), pp. 5-15.
3. See L. Johannson, 'Tuning to Station WIIFY on ISO 14000: What's in it for You?', in *Total Quality Environmental Management*, Vol. 5 No. 2 (1995/96), pp. 107-17.
4. See M. Starik *et al.*, 'Growing an Environmental Strategy', in *Business Strategy and the Environment*, Vol. 5 No. 1 (1996), pp. 12-21.
5. See J.M. Cramer and A.L.N. Stevels, 'A Model for the Take-Back of Discarded Consumer Electronics Products', in *Engineering Science and Education Journal*, Vol. 5 No. 4 (August 1996), pp. 165-69.
6. *Ibid.*
7. See J.M. Cramer and A.L.N. Stevels, 'Experiences with Strategic Environmental Product Planning within Philips Sound & Vision', paper presented at *The Greening of Industry Network Conference*, Heidelberg, Germany, November 1996.
8. An abbreviated version of this chapter appears in J.M. Cramer, *Conference Proceedings of the 1996 Business Strategy and the Environment Conference*, held at the University of Leeds, UK, 19–20 September 1996, pp. 59-63.

20 Establishing Workable Environmental Objectives

Alison Bird

AS A PRODUCT of discussions between practising environmental managers in industry and commerce, the Institute of Environmental Management's Best Practice Programme is designed to help individuals put the theory of environmental management into practice. The programme aims to identify ways in which managers are successfully tackling specific challenges in environmental management so that these approaches can be communicated to the wider membership and built on over time.

That the workshops should focus on policy, objectives and targets was a logical step following on from the previous series looking at defining significant environmental impacts. What was not expected, however, was the eventual focus of the discussions. While initially objectives and targets were dealt with together, it soon became clear that the real challenge that managers were facing was in the progression from the identification of significant environmental impacts and the organisation's environmental policy to the development of effective and workable objectives.

Targets are milestones on the way to achieving objectives. As such, sensible targets can only be developed once achievable objectives are in place. Consequently, this chapter focuses on the specific challenges faced by managers in the workplace as part of the development of achievable objectives.

■ Theory versus Practice

Many managers are having very real problems, not so much in developing theoretically achievable environmental objectives, but in forging objectives that are achievable in practice—objectives that can stand up to challenges posed by competing business priorities and lack of commitment from key personnel.

The theoretical route to the establishment of objectives appears relatively straightforward. It involves determining the organisation's environmental impacts and evaluating which of these are significant. Objectives are then established in the light of these impacts, an action plan developed, and targets for improvement set. In practice, however, establishing workable objectives can be a challenging task. It is one thing applying 'empirical' analysis to determine what must, should, and could be done within an organisation in terms of environmental improvement, but quite another to determine whether or not it would be possible to implement changes in the context of the organisation's overall dynamics.

■ Objectives: Testing the Sincerity of the Environmental Policy

ISO 14001 describes the environmental policy as 'a statement by the organisation of its intentions and principles in relation to its overall environmental performance which provides a framework for action and for the setting of its environmental objectives and targets' (Clause 3.9). The environmental policy is generally a set of broad statements of intent, expressed in polished phraseology.

Development of policy was not greatly discussed during the workshops, primarily because this issue did not appear to pose particular problems. Rather, the principal challenges seem to lie in the process of stepping from the relatively generic policy statement to the more specific goals and commitments contained in objectives and targets.

If objectives are the acid test of the sincerity of the policy, it is *crucial* **that they are** *implementable.* Fine-sounding words and overambitious objectives rarely benefit anybody; they raise expectations that cannot be met and ultimately defeat the purpose for which they are set. Therefore, objectives do need to be both practical and achievable if they are to act as an incentive for action and not as a source of demoralisation.

This chapter focuses on how environmental managers might set environmental objectives that both reflect their organisation's environmental policy and significant environmental impacts, and provide goals that are deliverable in terms of the overall organisational dynamics.

■ Significant Impacts: A Starting Point

In a previous series of workshops,[1] a methodology was outlined for identifying significant environmental impacts. Put simply, the evaluation of the significance of impacts was linked to the importance with which various stakeholders viewed them; i.e. if a stakeholder believes that an impact is important, then it is significant.

Effective environmental management is about the management and minimisation of significant environmental impacts. The systematic evaluation of significant impacts therefore provides the foundation upon which relevant environmental objectives for the organisation are developed. These objectives will in turn reflect environmental action designed to achieve them, and targets on the way. A coherent environmental management system will contain clear and demonstrable links between significant impacts, objectives and targets and environmental action. It is likely that those certifying (or verifying) to the various environmental management system specifications will expect that such links are apparent. Consider developing a table that provides key information on the links between impacts, relevant objectives, the policy statement and other information (see Figs. 1 and 2).

■ Significant Impacts, Objectives and Continual Improvement

If an environmental impact is identified as being significant, then it is clear that it should be managed. Consequently, much debate took place during the workshops regarding whether or not every significant impact needed an associated objective. There was a degree of confusion as to whether an objective automatically required performance improvement, and therefore whether

Issue	Emissions of VOCs
Nature of impact	Photochemical smog; nuisance to neighbours
Activity	Coating; cleaning
Location	Paintshop; workshop
Significance	Legislation; public complaints; customers
Relation to policy	Commitment continually to reduce emissions to atmosphere
Associated objectives	Maintenance of legislative compliance; improved solvent management; monitoring for low VOC/VOC-free coatings; training

Figure 1: *Example of Table Linking Impacts and Objectives*

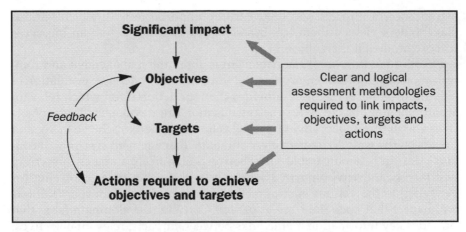

Figure 2: *Linking Impacts, Objectives and Actions*

all significant impacts needed to be linked to an improvement programme. However, having to make simultaneous improvements with respect to all impacts is clearly not practical, given the wide range of impacts generated by many organisations. There was broad agreement, therefore, that the existence of an 'objective' did not automatically imply improvement with respect to an associated impact.

ISO 14001 states that an environmental policy should include 'a commitment to continual improvement' (Clause 4.2b), and that objectives 'and targets shall be consistent with the environmental policy, including the commitment to prevention of pollution' (Clause 4.3.3). However, the Annex to ISO 14001 states that 'objectives should be specific and targets should be measurable wherever practicable' (A.3.3); the Annex to BS 7750, however, was a little more descriptive in this area and stated that 'the objectives should include a commitment to continual, year-on-year improvement in overall environmental performance, but *not necessarily in all areas of activity or at all times*. Improvement will always be possible in some areas, but may be impracticable in others, at certain times' (A.5; emphasis added). The BS 7750 Annex also states that 'areas targeted for improvement should include those where improvements are most necessary to reduce risks (to environment and organization) and liabilities' (A.5).

BS 7750 was clear, therefore, that not all objectives need to be linked to improved performance with respect to impacts. For example, it might be concluded that, if an impact was under control, an appropriate objective would be to ensure no failure in that control. Alternatively, it might be decided that although an impact was significant, appropriate control techniques are not currently available, or that financial constraints mean that action is not currently feasible. In such cases, an objective might relate to researching new techniques or to systematically monitoring available options with a view to identifying feasible options for the longer term.

By accepting that not all objectives need to be linked to improvement, it is realistically possible to link all environmental impacts to objectives, with only some of these committing the organisation to improvement (see Fig. 3).

■ *Improving, Monitoring and Managing*

In this context, three broad categories of objective can be identified: objectives aimed at monitoring and researching; objectives aimed at continuing effective management; and objectives aimed at improvement.

☐ *Monitoring Objectives*

Monitoring objectives may be set where the management of a significant impact could potentially be improved, but where financial, technological or other resource constraints mean that action is not currently feasible. Such objectives state a commitment to monitor or research specific issues likely to bring about a change in circumstance and which will allow for improvement objectives to be set. For example, a monitoring objective might be to aim to 'Keep abreast of the development of new coatings that have a reduced

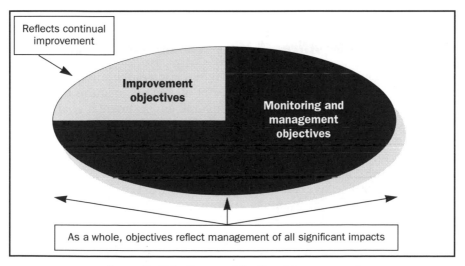

Figure 3: *Objectives and Continual Improvement*

impact on the environment in order to identify substitutes to those currently being used.'

☐ Management Objectives

A management objective may be appropriate where an impact is already being successfully managed or where improved performance is not currently feasible. Management objectives relate to the need to ensure that all controls relating to a particular impact are systematically applied. An impact might be linked to more than one management objective and may also be associated with a monitoring objective. For example, a management objective might be to aim to 'Ensure that all personnel undergo ongoing training so that they understand their specific roles in the management of the organisation's environmental impacts.'

☐ Improvement Objectives

Improvement objectives are clearly associated with an improvement programme, and may apply to areas where the organisation is required, or has decided, to improve its management of one or more impacts. Such objectives demonstrate that environmental performance is being improved and will always be necessary within a system to demonstrate continual improvement. For example, an improvement objective might state that 'The organisation aims to cut the volume of waste destined for disposal by 50% by the year 2000.'

■ Impacts and Objectives: A Many-to-Many Relationship

An organisation will generally have more environmental impacts than it is desirable to have objectives. However, not all impacts require individual objectives. An objective is, after all, an 'overall environmental goal, arising from the environmental policy, that an organization sets itself to achieve and which is quantified where practicable' (ISO 14001, Clause 3.7), and as such may apply to a number of impacts. What is important is evidence of a relationship between an impact and at least one objective.

Equally, since objectives are broad goals, one impact could have a number of related objectives. For example, an impact might be related both to an objective to monitor developments in techniques designed to improve its control, and to a number of management objectives relating to, for example, training, monitoring, water and energy use as well as waste generation (see Fig. 4).

■ *Developing Objectives: Part of a Bigger Picture*

Developing objectives is not a stand-alone process, but should be an integral part of a broader evaluation of options for tackling significant environmental impacts and the consequent development of a workable and implementable action plan. It is important to bear the notion of 'achievability' in mind when assessing what must, should and could be done and in setting objectives that reflect the outcome of this assessment.

■ *Developing an Assessment Methodology*

There is no fixed methodology for determining where environmental improvement should be concentrated nor for identifying what type of objective might be established. Deciding what objective to set will involve: balancing the extent to which the impact is an issue to stakeholders; the influence of those stakeholders; the extent to which the impact is already being effectively managed; and resource availability. In many cases, the need for improvement may be built into legislative requirements. In such cases, the resources required to make such improvements may be in themselves as much as an organisation can handle.

However, the following general rules might be applied to link the answers to these questions with the development of specific types of objective:

- If an environmental impact is deemed significant, it deserves one or more related objectives.
- If an impact is significant, but is already being effectively managed in a way that is acceptable to stakeholders now and in the foreseeable future, then linking it with one or more management objectives may be most appropriate.
- If an impact needs to be controlled (e.g. through legislation) and it is not yet adequately controlled, then it will need to be linked to an improvement objective.
- If an impact ought in principle to be better controlled, then either an improvement objective or a monitoring objective may be set depending on the ability to control it and available resources; management objectives may also apply.
- If an impact is resulting in unnecessary costs (e.g. waste production that could be minimised), then either an improvement objective or a monitoring objective may be set depending on available resources and other priorities; management objectives may also apply.

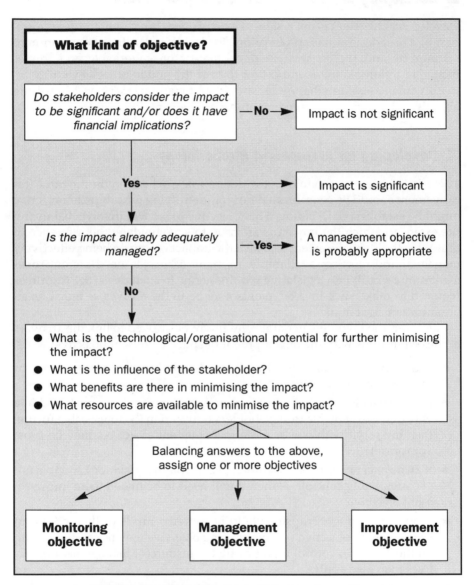

Figure 4: *Deciding Appropriate Objectives for an Impact*

Some managers may wish to develop formal assessment methodologies. There is no formally recognised approach to this; however, a number of methodologies were proposed by environmental managers at the workshops. These methodologies involve subjecting individual environmental impacts to a number of key questions relating to the perceptions of stakeholders and effectiveness of existing controls. Figure 5 provides an example of this type of methodology. An evaluation of this type will help identify a set of improvement objectives that focus on the priority areas in the context of available resources.

■ The Need for Baseline Information

Assessment methodologies and the process of putting together relevant objectives underpinned by concrete plans for action requires a good understanding of the current situation. Time and again at the workshops, managers expressed their frustration that, without adequate resources to gather baseline information on environmental performance, it was extremely difficult for them to feel confident that they were indeed focusing their resources where they are most required. The message was clear that those who had the best understanding of their organisation's current environmental situation were those who were able to deliver environmental management that yielded most benefit.

■ Deciding when Improvement Objectives will be Set

The fact that one environmental impact may appear more significant than another and might, in theory, merit improvement does not give an indication of whether improvement is actually feasible. It was clear during the workshops that, by leaping straight from the identification of an impact's significance to setting a related improvement objective, many managers have unwittingly developed a set of commitments which they feel ought to be achieved but which, in practice, are proving very difficult.

■ Costs and Benefits

Those managers undertaking the type of analysis outlined in Figure 5 tended to focus on the techniques available to control impacts and their associated costs and benefits; however, discussion of this topic is beyond the scope of this chapter. One issue raised repeatedly was the fact that, although an objective

Example methodology for prioritising action by balancing the importance of the impact against the effectiveness of its management

Legislation:

Q(a) *Is the issue subject to legislative control?*

A(a) If yes, rank as 3 as a matter of course. If it is likely to come under regulation in the foreseeable future or if indirect controls are likely to be significant, rank between 3 and 1 depending on foreseeable impact of this arising.

Q(b) *Are legislative requirements being met?*

A(b) If no, or if there is a substantial risk of requirements not being met, then rank as 3. Rank as 1 or 2 if there is a minimal risk of failure to comply or where developments in legislation could cause a future problem with compliance and where the consequence of failure would not be of great significance to the organisation or environment.

Complaints:

Q(a) *Could the impact be the cause of complaints?*

A(a) Rank between 1 and 3 depending on the potential implications of complaints. For example, rank 3 if complaints could be of high media profile and/or widespread.

Q(b) *What is the potential for the impact to cause complaints?*

A(b) Rank between 1 and 3 depending on the nature of the complaints. For example, if the issue has been the cause of: ongoing and/or vociferous complaints—rank 3; occasional substantiated complaints—rank 2; occasional unsubstantiated complaints—rank 1.

Cost:

Q(a) *Is the impact associated with avoidable costs?*

A(a) Rank between 1 and 3 depending on the significance of the costs. For example, if the impact is the source of: very high environmental/resource costs which are rising—rank 3; significant environmental/resource costs which are rising—rank 2; significant environmental/resource costs—rank 1

Q(b) *What is the potential to reduce costs?*

A(b) Rank between 1 and 3 depending on the extent to which costs are already being effectively controlled.

Customers:

Q(a) *Is the impact likely to be of interest to customers?*

A(a) Rank between 1 and 3 depending on the extent of customer interest. For example, if the impact: is an important factor in customer decision-making—rank 3; is a modest factor in customer decision-making—rank 2; could become a factor in customer decision-making—rank 1

Q(b) *To what extent are customers' interests being met?*

A(b) Rank between 1 and 3 depending on the extent to which customer interests are already being catered for.

Improvement objectives are most likely where Q(a) yields a high score and Q(b) a low one—in other words, where there is divergence between the importance of the impact and the effectiveness of its management. In some instances, for example where regulators require action, an improvement objective will *not* be optional.

Figure 5: *Example Scoring Methodology*

and its underpinning action should be achievable in theory, in practice, many such objectives were being defeated by a lack of commitment from individuals and in particular by middle and line managers.

This is why in most cases 'empirical' approaches were being used by managers to provide general guidance only; at the end of the day, managers agreed that, if relevant objectives are to be developed, 'management judgement' and proper consideration of the overall human dynamics are every bit as important as empirical analysis. An objective that cannot be 'sold' within the organisation has very little prospect of being achieved (see Fig. 6).

■ Selling Objectives

The issue of motivating individuals throughout the organisation to take environmental objectives seriously is clearly key. All too often, environmental management is still seen by managers as an externality to the mainstream business agenda and as an additional burden, competing with an already busy set of responsibilities and priorities. Discussions at the workshops frequently focused on two key questions:

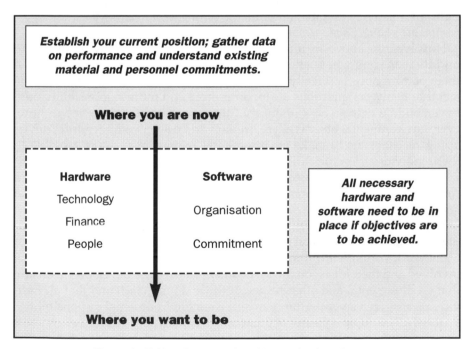

Figure 6: *Taking Organisational Dynamics into Account*

- How do you balance the importance of environmental objectives against the organisation's business objectives?
- How do you motivate people throughout the organisation to implement environmental objectives?

It was clear that there is much work required to train and educate individuals at all levels about the relevance of environmental issues and about their integral role within the business process. Such training should be practical, drawing on case studies and examples wherever possible and linking environmental management with the overall goal of continual improvement for the organisation.

However, it was evident from the workshop discussions that training alone was not sufficient. Managers and organisations that have been most effective in promoting environmental management have developed mechanisms that directly link environmental performance to the mechanisms by which individual performance is gauged. This posed a particular challenge at the middle and line management level. It was generally felt that senior managers were in a better position to appreciate the longer-term implications of environmental issues for their business and, consequently, for them. The implications for senior managers of getting environmental management right are relatively tangible in terms of: *getting work*; *getting bonuses*; *bigger profits*; *maintaining and growing the business*.

However, the above are often perceived to be only indirectly significant to middle managers whose efforts are likely to be focused on areas in which their performance is directly measured. This process is aggravated by the fact that many organisations are loading more and more responsibility onto fewer middle managers. Without middle management commitment, however, environmental objectives are unlikely to be achieved; therefore, developing mechanisms to make environmental management more tangible to these individuals is crucial. The following types of mechanisms were identified as having been successful in motivating middle managers to take environmental management more seriously: *integration into performance appraisal*; *financial and other incentives*; *improving the efficiency of areas under their control*.

It was felt that, in general, there is an underlying willingness by individuals to take environmental management seriously. What is needed is a framework within which they can effectively do so. Individuals need to be involved and their ideas sought. Mechanisms need to be developed to ensure that environmental and other responsibilities do not clash and that all individuals are given an opportunity to take their environmental responsibilities seriously. This topic is clearly of great importance in the delivery of effective environmental management in the organisation and is an area which the

Institute's Best Practice Programme may need to address in more detail in the near future.

■ *Keeping the Objectives Focused*

Another clear message from the workshops is that you won't get it right first time! Initial objectives are also likely to change over time for a number of reasons:

- Circumstances will change and, as new pressures from stakeholders emerge, new priorities and windows of opportunity will develop.
- As new techniques are developed and as the economics of change shift, new actions will become realistic.
- Goals will change as objectives are implemented and impacts that were priorities are minimised.

Managers will need to re-assess regularly the relevance of the environmental objectives that have been set, modifying them to ensure that they reflect the best use of resources in minimising environmental impacts. As a part of this process, the nature of the objectives and whether they are management-, monitoring- or improvement-oriented will evolve (see Fig. 7).

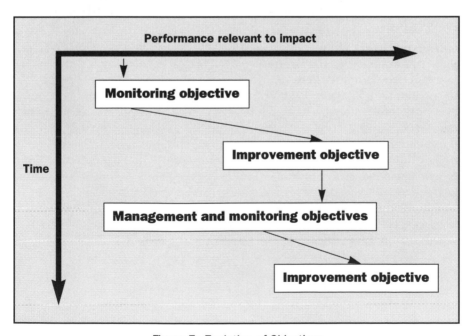

Figure 7: *Evolution of Objectives*

■ *Conclusion*

Setting workable objectives presents a greater challenge than might at first be obvious. It involves balancing the need for management and minimisation of negative environmental impacts with available resources and the dynamics and priorities of the organisation as a whole. In particular, objectives need to be 'sellable'. Not only do they need to be compatible with organisational goals, they also need to reinforce personal ones.

The human factor in establishing deliverable objectives, particularly objectives relating to improvement, cannot be underestimated. This is why formal methodologies, although they do have an important role to play, are likely to provide guidance only. All things considered, the setting of workable objectives will involve a great degree of management judgement and taking advantage of windows of opportunity as these occur. Environmental managers will not get it right first time and will need to continue to monitor the relevance of objectives so that those that are no longer appropriate can be dropped or modified and new objectives developed.

Notes

1. The outcome of the previous series of workshops was published in the *IEM Journal*, Vol. 2 No. 2.

21 Environment Risk
Assessment, Management and Prevention of Loss

David Shillito

ALTHOUGH not widely recognised as such, ISO 9000, Quality Systems, is an extremely important tool in the prevention of losses of all types. Environmental management systems (EMSs) such as BS 7750 and ISO 14001 follow logically along the same path in loss prevention.

The popularity of the original EMS standard, BS 7750, owed much to its quality systems predecessors. However, that heritage has not been completely positive. A few distinctive handicaps have emerged and are proving to be significant barriers to progress. The cynical public image of the ISO 9000 requirements for heavy bureaucratic administration and documentation is seen as a real 'turn-off' for the progressive slim-lined managements of environmentally-aware organisations. Perhaps even worse, the administration of these heavily-documented management systems is also creating a 'near-blue-collar' image for environmental managers. Unless the trend is stopped, environmental management could well become fixed with the narrow view of the ISO 9000 system, and fixed on the workplace floor rather than in the head offices and the corporate boardrooms, where it correctly belongs.

ISO 14001 was never intended to be a standard for 'effluent quality assurance' or formalisation of legal compliance management alone. This development could have been predicted if the power to achieve the 'certificate on the wall' had been more fully appreciated. The fastest and easiest way to achieve certification is certainly to base the system on pollution control and legal compliance. Once the certificate has been obtained by such a route, the potential for future development of the EMS is limited by the cultural constraints that have been created.

If the EMS, created in conformance with a standard, is to be effective across the full extent of the activities of the organisation, it must have equivalent application at all levels of management. The management activities in the boardroom are very different to those on the shopfloor. It is to be expected that the usual 'policy' undertakings to provide the necessary financial and human resources should be taken with a pinch of salt. No hard financially-oriented company would see such a commitment in terms other than being relative to broader financial policies. To most directors, compliance with regulatory requirements is not a boardroom issue unless it involves significant capital expenditure. Questions of importance to high-level environmental management usually come down to only three groups of factors:

- Unwelcome surprises: unexpected and unplanned liabilities, whether from accidents or incidents of omission
- Opportunities for improvement of operational efficiency and cost-reduction
- Threats and opportunities created by innovation in the marketplace

Issues of routine operational control, product or service quality and compliance should be capable of being handled competently within the basic direction of policy and 'management by objectives' at the level of operational middle management. If the EMS is going to be used for something more than routine operational control, it must penetrate these three areas of financial importance. This is clearly a requirement of the standard and will progressively become of greater importance.

The distinguishing feature of these three groups of financial factors is that they all involve a deeper involvement with 'risk' than is provided in the conventional assessment techniques for effects or aspects. Corporate risk management strategies are frequently well developed, but it is surprising how few of these are identified as relevant policies in operational EMS documentation. The area of difficulty is in creating the communications bridge between the corporate risk analysts and managers and the environmental assessors. The creation of such communication bridges requires much more than, as the standard naïvely states: 'establish and maintain procedures for . . .'. The main requirement is for the advisers of the board directors to think 'environment' in the widest possible terms. The proper place for risk assessment is at board level, as well as in the designers' offices or on the shopfloor.

■ *Management and Risk Assessment*

☐ *Risk Assessment*

Risk management has become almost a synonym for 'insurance', and the risk manager used to be seen as the person who 'eats lunch' with insurance brokers. In the past there was good reason for using insurance as a one-stop shop for environmental risk management: insurers have learned by experience that it is easy to underestimate environmental liabilities. As most organisations now have to carry an increasing share of their own risk liabilities, risk assessment will become increasingly more important to financial management, and thus respectable. Risk assessment has developed a poor reputation. Like many other good tools, it has been misused: too often it has been seen as being capable of being guided by politicians and public relations people. The actual language used also tends to make the subject generally inaccessible. The currently-accepted definitions of risk assessment jargon are those of the Royal Society's 1992 report _Risk: Analysis, Perception and Management_,[1] the formative reference, which updated their previous publication of 1983.

> **Hazard:** a property or situation that in particular circumstances could lead to harm.
>
> **Probability:** the probability, or frequency, of occurrence of a defined hazard in a given period, usually presented as the mathematical expression of chance (for example 0.25, equivalent to a 25% or one in four chance). Where this is not possible more qualitative descriptions are used.
>
> **Consequences:** the adverse effects or harm as the result of realising a hazard in the short or long term. (This has also been referred to as 'hazard effect'.)
>
> **Risk:** a combination of the probability, or frequency, of occurrence of a defined hazard and the magnitude of the consequences of the occurrence.

Put simply:

$$\text{Risk} = \text{Probability} \times \text{Consequence}$$

The process of 'risk assessment' for any proposal, or intention, involves four discrete stages:

1. Identifying the hazards
2. Estimating the probability of occurrence
3. Evaluating the consequences
4. Combining probability and consequence to assess risk

'Risk management' adds a further four stages by examining possibilities for:

5. Changing the hazard: either eliminating it or replacing it with a smaller one
6. Reducing the probability of occurrence

7. Diminishing the consequences
8. Developing the minimum acceptable risk strategy

In theory, the process is straightforward and can be handled at any level of complexity from heavily-quantified assessment techniques to simple rating methods.

☐ Simple Risk Assessment: An Example

The process is best demonstrated at the simplest level with an ordinary everyday example—cutting the hedge with an electric hedge-cutter. The technique here uses a three-level rating scale (low, medium or high) to evaluate probability, the severity of the consequence, and thus the risk. Hazard, probability, consequence and severity can be set out in a table (Fig. 1). Risk is assessed according to the rules in Figure 2.

Having assessed the original risks, the next stage is to see how they can be managed—stages 5, 6, 7 and 8. First, the hazard could be eliminated if the hedge were to be replaced with fencing, but this might be considered environmentally unacceptable and other hazards could be introduced. Secondly, to reduce the consequences and the probability of accidents, the height of the hedge could be reduced. The electric hedge-cutter could be replaced by environmentally-friendly hedge-cutting shears. These options might still be considered unacceptable.

The next set of management methods would be based on changing the way of working. Instead of a ladder, a good set of steps or, better still, an adjustable platform, would provide a safer and better working position. Always cutting away from the cable and not cutting in wet conditions would minimise the probability of accidents causing electric shock. The severity of the electric shock consequences could be reduced by using an earth-leak circuit-breaker plug socket—quite a cheap solution.

Also, a means of minimising the consequences of damage by flying debris or unintended contact with the cutter would be to wear heavy gloves, a hard hat, safety glasses and strong shoes: in other words to use appropriate 'personal protective equipment'. Finally, make sure your insurance policy covers such accidents.

☐ The Identification of Hazards

Hazard identification is the most important feature of risk assessment because it determines the scope of the assessment. The difficulty is in identifying all the hazards, large and small. It is equally important to have as wide a vision

Hazard	Probability	Consequence	Severity	Risk
Bodily contact with cutter	Low	Lacerations	Medium	Medium
	Low	Amputation of a finger	High	Medium
Short-circuits in wet conditions	Low	Electric shock	High	Medium
Falling off ladder (machine off)	Medium	Cuts and bruises	Low	Medium
Falling off ladder, cutter running	Medium	Serious cuts and blood loss	High	High
Flying hedge debris	High	Cuts, scratches or bruises	Low	Medium
	Medium	Damage to an eye	High	High
Cutting the cable	High	Electric shock	High	High

Figure 1: *Risk Assessment of Hypothetical Situation (Cutting a Hedge)*

Consequence class	High probability	Medium probability	Low probability
Serious consequences	High risk	High risk	Medium risk
Medium consequences	High risk	Medium risk	Medium risk
Small/temporary consequences	Medium risk	Low risk	Low risk

Figure 2: *Rules in the Assessment of Risk*

as possible of the range of problems that might occur. It has been found that people are best at identifying problems from their own fields of speciality. Using a team of people from a range of different disciplines, working together, appears to produce the most comprehensive results. It is important to be systematic or hazards can be lost. With a team exercise, it is especially important to use a systematic procedure with a chairman, a secretary to note the findings, and a programme, or the discussions will go on almost endlessly. This has been developed to a fine art in the techniques of Hazard and Operability Analysis or HAZOP,[2] which were developed in the 1970s. This technique has spread throughout the chemical and energy industries of the world.

Although hazard identification methods have been developed for conventional safety/loss prevention, the techniques are just as useful for environmental risks. Risk assessment is essential for compliance with the regulatory requirements of health and safety legislation. So often these exercises are limited to risk, resulting in consequences of harm to employees because that is all that is required by regulation. If a conventional 'health and safety' risk assessment has to be undertaken, it would seem obvious to examine the added value of extending the scope to cover other features, including significant financial loss and environmental aspects and their impacts, especially where a remediation cost can also be included. The most common reasons for restricting the risk assessment to safety are the culture of risk tolerance, lack of in-house competence and urgency.

☐ The Assessment of Probability

With safety risk assessment studies, the evaluation of probability often provides the most difficult stage. Most people tend to use an 'educated guess' about the frequencies of occurrence of incidents. To produce a significant improvement in accuracy over the educated guess, an extensive database is required. Most incident types are too infrequent to build up a sufficient number of the results to be of any statistical significance. It is necessary to use collections of data from a large number of plants with similar operations. The probabilities of the individual element operations can then be combined to provide an estimate of the frequency of occurrence. Where two things have to occur together, the numerical probabilities have to be multiplied. Where either of the two operations can occur, the probabilities are simply added.

The problem with the use of incident frequency databases is the assumption that the probability of the event occurring in your plant or office is the same as at the combination of other places that contributed information to the database—quite an assumption. We all know that the frequency of incident is determined by the way in which the plant is managed and operated. A well-managed site usually has significantly fewer incidents because poor standards of operation are not tolerated. Most plant managers will tell you that their safety record is better than the average for their industry sector, but very few will be able to demonstrate it by accident frequency rates, etc. The standard of management includes a wide variety of features, the operations to be undertaken, the quality of the working environment (lighting, etc.), physical strain, tiredness and fatigue and the way employees are motivated or demotivated, and, most importantly, the way the operators are trained, supervised and their safety-critical work checked.

One of the main lessons learned from judicial public inquiries into major disasters is that the main causes were combinations of relatively simple human failures: things that tend to go wrong in everyday conditions. Things go wrong not only on the shopfloor—perhaps the failures are more frequently found in the offices of the management. Most of these human failures could have been most easily corrected by better awareness, training, supervision and checking. These factors can be summarised by the concept that the frequency of events occurring really depends on the management's tolerance of poor or sloppy work.

Great care should be taken before assuming that the probability of an event occurring at one site will be similar to that at another location, even within the same organisation. Major differences can occur with different management cultures. Before making the assumption, compare the lost-time accident frequency rates of the two sites. This can be the best guide to whether making the assumption will be helpful. Most auditors will be able to recognise the principal symptoms of tolerance of sloppy practices.

☐ *Consequence Analysis*

In safety risk assessment, consequence analysis is usually quite straightforward because everything is contained and the elements of uncertainty are small. Very seldom do the consequences extend off-site, and then they can be quantified with acceptable accuracy. The central problem with environmental risks is that the off-site consequences are more likely to be more expensive. The consequences may not be immediate and could well remain hidden for some time. The incident may not be instantly recognised and it might well be a slow-developing, continuing affair and may not be discovered until it is too late to prevent significant damage. For example, the spillage or leakage of PCBs from an old transformer many years ago might not be discovered until the due diligence studies for the sale of the site. The cost of the clean-up when the accident happened might have been almost trivial. However, the costs of the consequences of the late discovery would also include the effect on the potential sale of the property.

In conducting environmental consequence analysis, it is helpful to consider two different types:

- **Closed consequences:** those which can be quantified (like closed questions at an audit, which can be answered with a yes/no or a number, date, etc.)
- **Open consequences:** which cannot be easily defined or quantified (the open question in an audit, where the interviewee is invited to explain and expand on the subject)

So often, open consequences involve the imponderables of providing compensation, legal claims for compensation and the costs of civil legal action, etc.

While probability of occurrence may be comparable from one location to another within one organisation, the severity of the environmental consequences will be very different because they depend on the site environment. For example, the discharge of contaminated storm water to a polluted river estuary is likely to provoke a very different reaction to that of an equivalent discharge to a high-quality salmon river in the tourist season. Consequence analysis has to be site-specific.

☐ *Combining the Probability and Consequence to Assess Risk*

The main difference between safety and environmental risk assessment comes from the difficulties in quantifying open environmental consequences. The accuracy of any risk assessment depends on the least accurate component. In 'safety', where consequences can be quantified, the probability assessment is the weak link. Greater attention has to be paid to this area. With environmental problems in consequence assessment, there is little point in producing high levels of accuracy for the probability estimates because this will not affect the overall result. The quantification of consequence costs is far more important.

The difficulty in producing accurate estimates of consequence costs has made environmental risk assessment much less quantifiable than in the safety discipline. The tendency for off-site consequences to be more complex has also contributed to the same result. The combination of these factors has resulted in most environmental risk assessments having been made using relatively simple techniques (much like the hedge-cutting case discussed earlier). This does not make them any less useful: in some respects it assists in keeping the levels of accuracy of the assessment more visible. Where highly-quantified risk assessments are made using computer software packages, there is always the tendency to believe the results and to give them more credibility than the original data actually deserved.

It is much better first to undertake an assessment using simple techniques which can be followed critically than to use more complicated methods and to lose contact with the real situation. Remember that it will be easier to see the effects of risk management options if the whole process is not lost from view in an elegant software package.

☐ *Alternative Methods for Environmental Risk Assessment*

The most important development in environmental risk assessment has come from the analysis of the databases of the insurers who have claims valuations for environmental accidents. The concept of the alternative approach is to reverse the conventional risk assessment process. Instead of identifying 'hazards', the approach is based on the identification of 'consequences'. This can be done by examining the database to establish the claims that have occurred for that sort of activity. With an extensive database, the characteristics of the environment can be included and the probability of the consequence might also be estimated giving a more complete evaluation of the risk.

Such an assessment technique would be very helpful in assessing whether insurance would be offered and the levels of the premiums that have to be charged. However, the approach would not be very helpful for risk management at the site itself. The development of this type of approach offers great possibilities to environmental risk assessors because it should allow access to much greater experience of consequence analysis. The future developments in the area of environmental risk assessment will come from the greater combination of information from insurers and operating companies. Both parties have very good reasons for being reluctant to share information. Nevertheless, it is inevitable that the common interest in risk reduction will force the development of better methods for environmental risk assessments. Eventually, low-cost packages will become available and these will allow the examination of all the risks of the organisation, not just the more significant ones. This will also contribute to overall accuracy and allow greater reliance to be placed on the results in risk management.

■ *Product or Services Liabilities*

The best place to start any form of risk assessment is in the area of greatest importance to the organisation. If senior management were asked what is most important to them, after the usual range of financial indicators it would not take long to arrive at the company's products or services, their customers, their reputation and public image, perhaps even their staff and facilities.

Risk assessments in this area would not usually be described as being 'environmental' in any way. Perhaps that is all for the good. The problem is that, in such risk assessments, the environmental consequences are often omitted. To a large extent, the same thing happens in the world of 'safety': financial risk assessments often disregard the safety implications or consequences of the intentions being considered. The effect on safety of major investments or re-organisations may well lie outside the competence of the risk assessment teams and, as mentioned earlier, this is probably the easiest way to fail in the identification of relevant hazards. Failure to recognise the consequences follows logically.

The central requirement is to be comprehensive in the examination of hazards and consequences. Rather than producing separate environmental risk assessment, it is much better to ensure that the assessment team has both safety and environmental competence. This is the best and perhaps only mechanism for integrating the financial and technical disciplines.

As suggested in the first section of this chapter, the development of an effective quality management system is the starting point for risk management for products and services. However, like all good tools, it has its limitations. It is concerned with ensuring the customer receives the goods or services specified by the contract. ISO 9001 is not overtly concerned with continual improvement. Systematic quality assurance and control are business essentials, but so is quality improvement. The approach of the ISO 9000 standard does not highlight the importance of 'criticality'. Where the 'quality' of the final product, or service, is dependent on one particular process or operation which has to be carefully controlled, that particular step can be described as 'critical' and the respective control point as a 'critical control point'. The development of techniques for identifying and assessing hazards and their respective critical process control points has been one of the major contributions of the food industry. The techniques, under the name HACCP (Hazard Analysis and Critical Control), were derived from failure mode and effects analysis. HACCP developed in the 1960s jointly by the Pilsbury Company, the US Army Laboratories and NASA in the development of foods for the US space programme. The US Food and Drug Administration have built HACCP into their regulations; the US Department of Agriculture has applied HACCP to meat and poultry inspection; and the World Health Organisation (WHO) and the British Government have encouraged the use of the system. HACCP is used to demonstrate 'due diligence' under the UK Food Safety Act (1990), which probably makes it the most widely used method of commercial risk management.

Preventing the design and generation of potentially damaging products or service is essential, but it is also necessary to be prepared for incidents when the defences fail and the unthinkable happens. Unfortunately, product recalls happen too often. The legal profession is kept busy in the civil courts resolving the difficulties of disputes between suppliers and customers. With more effective means of handling these problems, it is possible that the financial liabilities and damage to reputations could have been significantly reduced by the preparation of arrangements or emergency plans.

Most organisations develop a range of special arrangements for dealing with the problems that arise from the supply of faulty products or services. These range from the friendly customer 'help-lines', customer service departments, the involvement of the public relations and legal departments, to full crisis-management plans for use if all else fails. In organisations exposed to this type of risk, these facilities become crucial to business success. The cost of these facilities can represent a significant cost to the company, possibly substantial in comparison with the investment in quality assurance and control. Few organisations are prepared to evaluate the full cost of poor quality and even fewer are prepared to recognise the cost of the damage limitation facilities operated to maintain the organisation's reputation.

■ *Three Examples of Risk Assessment and Management*

The environmental management standards require the organisation to assess the risk of accidents, to take action to prevent them and to be prepared, through contingency planning, for accidents to happen and thus to be ready to minimise their environmental consequences.

The approach to assessment, prevention and emergency planning and protection, and the typical problems that can occur is perhaps best illustrated with three examples.

☐ *Example 1*

A company stores bulk liquids and oils in above-ground storage tanks. Following conventional practice, the tanks are bunded, but with high concrete-block bund walls to minimise ground space. The bunds look strong and secure and give everyone confidence that the risk of spillage has been adequately prevented. Deliveries are made from road tankers to discharge points adjacent to the tanks, outside the bunds. Storm water from the roadway drains to an interceptor before being directed to a local stream.

Let us examine the assessment. With these conventional industrial facilities, the most frequent accident to be expected is spillage during the delivery operations of transferring from the road tanker to the storage tank. These spillages may be trivial, from a few drips to a few grammes. Larger spills can result from badly-made connections, a failure of the hose or coupling. Such spills would normally be expected to be of a few tonnes but could also be up to about twenty tonnes—the entire content of the road tanker. Even worse, if the design of the pipes and valves is inadequate, it might even be possible to suffer an even bigger spillage through back-flow from the storage tank.

It is difficult to prevent minor spillages, so there have to be clean-up facilities available to prevent the spill reaching the storm water drains. The sorbant granules, or sausages, must be stored locally to the loading points; it is no use having them 100 metres away. The storage point must be easily visible, and the operators must know how to use them. During the accident, there will not be time to read the manual, so a training requirement is essential. Granules or sausages will not be able to handle the failure of a hose coupling, and the only safety device between the escaping liquid and the local stream is the old interceptor: when was it last cleaned out? The cleaning of such interceptors should obviously be part of the emergency prevention system.

The most frequent cause of spillage inside the bunds is overfilling the tank; for example, when the level instruments and alarms fail, or when operators ignore them. Checking the reliability of these instruments becomes part of accident prevention, as does the training of operators with respect to these alarms. The risk of a major instantaneous failure of the tank, discharging the contents quickly into the bund, is very low indeed. Tanks and their pipework and fittings simply do not suddenly split open: they usually start to leak first (this is called 'leak before failure'). This is quite a good thing, because if the tank did rupture, the out-rush of oil would probably knock down the bund wall anyway. While the bund walls may look strong, they are most unlikely to withstand the hydrostatic pressure of the liquid contents unless they are made of reinforced concrete. Bunds made out of concrete blocks and mortar do not stand much chance. Are the bund walls strong enough? The precaution of a simple hydrostatic test—filling the bund with water—might encourage more rigorous calculations to be made.

One of the problems with tall bunds is that they hide from routine observation any problems that are appearing behind them. The operators cannot see developing leaks, corrosion or even accumulations of rubbish that might have developed. In reality, the strong-looking concrete block bund may afford far worse protection than a flawed insurance policy by:

- Creating a sense of false security
- Obscuring the development of unsafe conditions, increasing risk
- Inhibiting the maintenance necessary to reduce risk

Are weekly inspections made to check these points and are the results recorded?

If it is thought necessary to provide protection against a major tank or pipework failure, rather from just the small spillages, the management should perhaps look for other solutions. It will certainly be necessary to ask if there was any action that could be taken other than trying to clean up the mess afterwards. If the answer is 'no', the emergency plans should be directed at the deployment of the facilities for cleaning up the local stream and damage limitation in the prosecution proceedings that could be expected to follow. The estimates of these costs may help to focus the minds of those involved.

It might even make other engineering solutions to keep the spillage on-site look more attractive. Possibilities include isolating the drainage system or the interceptor outlet with remotely-actuated valves. The installation of an oil/water detector at the third stage of the interceptor might also be justifiable to allow actuation of the valve isolation before the escape occurred.

The risk assessment of what could occur, the development of methods to prevent consequences, and the management actions, or emergency plans in the event of the accident, are all inter-related. It is not really possible to develop emergency plans without considering the systems that generate the emergency. This example attempts to illustrate that the risk assessment cannot really be separated from accident prevention and emergency preparedness.

☐ *Example 2*

A small manufacturing company at its works, located in a residential area, packs its products into a wide variety of similar plastic containers carrying their customers' printed labels. The customers require that the supplier should hold in stock an adequate inventory of empty containers. The result is that a large area of its warehouse facility is occupied by flammable packaging materials: not an uncommon occurrence.

This situation developed too slowly for the fire hazard to be recognised early on. The company was forced to face the problem of what would happen in the event of a warehouse fire. Discussions with the fire service identified a problem: the smoke from such a fire would necessitate the evacuation of some of the neighbouring properties unless a water-spray protection system was installed. The installation of conventional fire-fighting facilities was the logical answer. Apart from the high installation cost, this was not liked because of the potential damage to large quantities of packaging. The environmental advisers pointed out that the site drainage would not be able to cope with the fire water run-off unless the used fire water could be very carefully restricted. It appeared that the company, in its already restricted site, was being faced with investment in a new warehouse for packaging, purely for safety and environmental requirements.

On analysis, the root cause of the hazard was recognised to be the need to hold such large quantities of packaging at all. Discussions with the customers identified two possibilities:

- Redesign the packaging to use labels rather than printed containers, so that it was not necessary to store containers for each client.
- Negotiate new contracts with the packaging suppliers on a semi-'just-in-time' (JIT) basis, passing the risk on to the packaging suppliers, who

were already well equipped to deal with storage problems. (Semi-JIT maintains acceptable delivery sizes.)

Following discussions with customers and suppliers, an ideal solution was developed using a combination of both approaches. The fire risks were made manageable through investment in new package-labelling machines. Significant savings in stocking costs resulted, together with new floor space being made available. Interestingly, the revision of the packaging systems allowed the introduction of more environmentally-friendly materials.

The first action in risk management is, wherever possible, remove or minimise the hazard or replace it with more acceptable or more easily treatable options.

☐ Example 3

A small fine chemical company discovered that fume was being emitted from a scrubber handling the fume from the discharge of a road-tanker-load of acid, and that the low-water alarm on the scrubber was sounding loudly. Because the scrubber water supply had been designated as 'environmentally critical', it was fed from the water main. Further investigation showed that there was low pressure in this main: the water had been turned off. The tanker discharge operation was terminated, the plant shut down, and the emergency crews turned out in chemical suits. The emergency plan worked well, and everyone was pleased until it was discovered that, in the emergency shutdown, the surge discharges to the effluent treatment plant had caused a loss of control, resulting in a major excursion from the consent conditions for effluent quality. This also involved notification of authorities. The finding of the internal investigation revealed that:

- The water supply had failed some time before the incident with the scrubber occurred. The plant had been relying on its in-plant water storage, which was nearly empty. The scrubber was the first indication of this type of failure. Worse repercussions would have occurred if there had not been the delivery and the scrubber alarm had not sounded.
- The company had been notified about the failure of the water supply— late, but still in time to prevent serious problems. The message was confused and went to the wrong person who was out of the office. Internal communications failed.
- The design of the emergency shutdown system had not considered the effect on the effluent plant because that was not 'safety critical', only 'environmentally critical'.

- The loss of mains water pressure water had been considered but was thought to be too low a probability to be important because of the available water storage supply and promises of being promptly informed. Apparently the assessment had not considered the possibility of a failure to communicate the warning inside the company.
- The emergency shutdown system had never been tested for the entire plant at once, only unit by unit, and always with sufficient water in storage.

The recommendations were simple.

1. Improve internal communication procedures.
2. Ensure that all 'environmentally-critical' factors are considered in a new review of the design of emergency and shutdown systems.
3. Install a low-water pressure alarm on the water supply main; do not trust the supplier completely.

■ The Future: Integration of Risk Management and Environmental Economics

Risk assessment and management is a commercial tool of great importance in organisations in all sectors of the economy. In each sector, new improved techniques are being developed, tried and tested. Changes in management style and approach appear to be occurring faster than the ability of most organisations to control the results. Indeed, in some organisations, the culture to achieve change is so powerful that any suggestion that proposed changes be assessed to identify and manage any consequent risks would be seen as 'resistance to change'. This could be considered as the worst possible enemy: using 'systems' to delay and undermine the new culture and to slow down progress.

The essence of the problem is that risk assessment and the development of risk management strategies are perceived as being slow and non-productive in the short term. In any organisation where the management strategies are based primarily on short-term financial planning, the handling of low-probability events is only of secondary importance.

The argument that is used so frequently is that it is better to get on with the change and then handle any problems that may occur when more is known about the results. This 'can-do' approach has much to commend it, provided of course that there will be resources made available later on to complete all the residual requirements of the completion of the change. Unfortunately, organisational changes frequently result in the slimming down of resources. Unforeseen problems result which cascade through the system, leaving exposure to risks that had previously been managed. Organisations that depend on highly-documented systems management appear to be extremely vulnerable to this type of change. Once disrupted by loss of key people, the documentation of the systems tends to be slow to repair itself. Corporate re-engineering often results in fewer middle-management resources being available for maintenance of the old systems. The main source of the difficulties can be the 'systems culture' itself.

Where organisations are more interested in longer-term planning, the difficulties in the application of environmental risk assessment are produced more by the interdisciplinary nature of the subject. The in-plant people tend to concentrate on application in manufacture; sales and marketing people on the customer and market aspects; and strategic planners on the longer-term influences of a wide range of issues, from the plans of competitors through to real estate and expansion plans. This diversity of interests can be made to be highly productive, provided it is well controlled and integrated. Integration has also to be achieved with the conventional insurance world as well as with the other disciplines of the organisations, such as finance and safety.

Integration of the corporate environmental interests can only be achieved through ensuring an adequate awareness of the possibilities throughout senior management. The maximum benefits of environmental risk assessment are most likely to arise from the longer-term opportunities that environmental factors will bring across the marketplace. Most far-sighted companies are now addressing the opportunities for improvement in efficiency and cost-reduction within their existing operations. These could be considered to involve the internal uses of risk assessment. The companies intending to be among the market leaders in forthcoming years are having to consider the threats and opportunities created by innovation by their competitors. The use of risk assessment here can be considered to be external.

For some years now, the environmental 'gurus' have been predicting the changes brought on by the advent of 'green economics'. While it would appear that these have been somewhat slower than expected, the pressures will certainly change. It is also clearly evident that the forces of global economics could potentially counter these green trends. The balance of influence between legislation and market economics is another area containing considerable business risk.

Global factors have already entered the equation with the influence of environmental pressure groups. The influence of the Internet has been considerable in facilitating international communications between green pressure groups. It is possible that the influence of the Internet could be a major factor for better or worse in the environmental marketplace. Certainly, multinational organisations will have to learn to be more competent in assessing the impact of their international operations on their home markets.

The prospects for environmental risk assessment look exceedingly bright whether you are an industrial engineer interested in the internal affairs of manufacturing organisations, the strategic planners of multinational corporations, or governmental economists concerned with influences on world trade. Risk assessment might even provide the mechanism for international co-operation in the management of the full planetary risks, whether these be through climatic change or the higher-probability risks such as the annihilation of the human race through an asteroid collision.

■ *Notes*

1. Royal Society, *Risk: Analysis, Perception and Management* (London, UK: Royal Society, 1992).
2. Chemical Industries Association, *Hazard and Operability Studies* (London, UK: Chemical Industries Association, 1991).

Biographies

Mark Barthel, Project Manager, Environmental Management, at the British Standards Institution (BSI), has been involved in environmental issues for the last ten years. Prior to working for BSI, Mark worked for the University of Southampton in the UK as a Lecturer in Environmental Management and a Consultant in Environmental Communications. He was responsible for the ISO subcommittee on environmental management systems as well as being involved in the management of ISO/TC207. He was also the Secretary to the European standards organisation working group (CEN/PC7/WG 'EMAS'), charged with developing an European standards-based route to registration under the European Union's Eco-Management and Audit Scheme (EMAS) Regulation using ISO 14001 as a medium.

His current position within BSI is that of Product Manager of Environmental Management Systems within their Training Services division.

He received his Masters Degree in Environmental Management from the University of Southampton, and is currently working towards a PhD in Environmental Management.

BSI, 389 Chiswick High Road, London W4 4AL, UK
Tel: +44 (0)181 996 7335; Fax: +44 (0)181 996 7364; E-mail: ecobart.easynet@co.uk

---

Christopher L. Bell is a partner with the international law firm Sidley & Austin, which has offices in Chicago, London, Los Angeles, New York, Singapore, Tokyo and Washington. Mr Bell, resident in the Washington, DC office, was one of the two US delegates that negotiated the ISO 14001 EMS standard on behalf of the US ISO 14000 delegation. Mr Bell has over ten years of experience advising companies on identifying and managing a wide range of national and international environmental issues. He is currently advising a number of companies in several countries on environmental management systems matters, including whether and how to implement ISO 14001.

Sidley & Austin, 1722 Eye Street, NW, Washington DC 20006, US
Tel: +1 202 736 8000; Fax: +1 202 736 8711

Alison Bird, MPhil, is one of the founding directors of the Institute of Environmental Management (IEM). She has been involved in environmental issues since the late 1980s. After regularly covering environmental liability issues for a London-based financial publisher, she launched and edited *EnviroRisk*, a monthly publication dedicated to risk management, liability and insurance issues. After moving to Scotland, she was brought into the successful 'Centre for Environment and Business in Scotland' team to address the growing demand for professional support for the individual environmental manager. This in turn culminated in the formation of the IEM.

Institute of Environmental Management, 58–59 Timber Bush, Edinburgh EH6 6QH, UK
Tel: +44 (0)131 555 5334; Fax: +44 (0)131 555 5217

Andrew J. Blaza is an international business environmental management professional with experience in research, marketing, strategic planning and the development of innovative programmes in environmental information, training and communication. He is Chairman of the BSI Certification Advisory Council: Environment, a member of the ACCA Environmental Reporting Awards panel, the BSI QA 'Innovation in Environmental Management' Awards panel, and of the UNED-UK Executive. He has recently been appointed Senior Research Fellow in Business and the Environment at the Imperial College Centre for Environmental Technology in London. He is a co-director of PULSAR International, a UK consultancy established to help develop and promote innovative environmental communications between companies and their stakeholders, particularly final consumers.

PULSAR International, The Old Post Office, Lidgate, Newmarket CB8 9PP, UK
Tel/Fax: +44 (0)1638 500364

Nicky Chambers is an environment business professional with particular experience in environmental strategy and management, communication, education and training. She was Chair of the BS 7750 pilot programme Training Working Group and her links with BSI continue as a tutor for ISO 14001 training programmes internationally. Following a period as an independent consultant in environmental training and communication, clients of which included British Airways, BSI and Oxford Brookes University, she has teamed up with co-director, Andrew Blaza, in establishing PULSAR International to help provide the link between business action and consumer response, in the quest for sustainable consumption and production.

22 Hertford St, Oxford OX4 3AJ, UK
Tel: +44 (0)1865 247786; Fax: +44 (0)1865 794586

Jacqueline Cramer worked as an associate professor at the University of Amsterdam (1976–89) before joining the Centre for Technology and Policy Studies of the Netherlands Organisation for Applied Scientific Research (TNO). Her research relates to the question of how government, industry and social organisations can stimulate the development of cleaner production. She has been working as a senior consultant at Philips Consumer Electronics since April 1995, detached from the TNO, and has also been a part-time professor in environmental science at the University of Amsterdam since 1990. Since September 1996, she has been lecturing on environmental management at the University of Tilburg.

Her professional affiliations include membership of the Dutch Advisory Boards for Environmental Management and for Transport, Public Works and Water Amenities.

TNO Centre for Technology and Policy Studies, Laan van Westenenk 501,
PO Box 541, 7300 AM Apeldoorn, The Netherlands
Tel: +31 55 549 35 00; Fax: +31 55 542 14 58

Gabriele Crognale, PE, is an independent environmental consultant specialising in enhancing environmental management systems (EMSs) and audit programmes for industries. He helps clients develop strategic operating management systems to enhance their environmental, health and safety programmes, including the use of various management tools, and provides additional environmental management insight as part of his services. He is also an adjunct university faculty member in the Boston area where he teaches environmental continuing education courses, and provides ISO 14000 courses for Environmental Education (E3). He is currently retained by several national environmental consulting firms in the US and has consulted for several large European industrial and chemical firms.

MCG Associates, PO Box 376, Needham Heights, MA 02194-0376, USA
Tel: +1 617 9325428; Fax: +1 617 4611511

Aidan Davy is an Environmental Management Specialist, working with The World Bank's Environment Department in Washington, DC. He has recently written a World Bank discussion paper which outlines the Bank's emerging perspective on ISO 14001 and the issues for developing countries. Prior to joining The World Bank, he has worked as an Environmental, Health and Safety Manager within the Rexham (formerly Bowater) Packaging Group in the UK, where his responsibilities included developing and implementing environmental auditing and environmental management systems. He started his career at W.S. Atkins Environment, one of the UK industry leaders, where, as Senior Environmental Consultant, he worked on environmental assessments and audits of industrial and mining projects, transport projects and water resource projects for public- and private-sector clients.

Tel: +1 202 473 9131; Fax: +1 202 477 0568

Matthias Gelber is a Researcher on Environmental Management Systems at Staffordshire University, currently working on an EU-funded project implementing EMAS at airports. He is a member of the BSI Subcommittee on EMS and, having assisted companies in the UK and Germany, he is working increasingly with governments and companies in Asia on the implementation of ISO 14001.

Staffordshire University, Business School, Leek Road, Stoke on Trent, ST4 2DF, UK
Tel: +44 (0)1782 295234; Fax: +44 (0)1782 747006; E-mail: M.Gelber@staffs.ac.uk

Harris Gleckman, PhD, a Director of Benchmark Environmental Consulting, Inc., was formerly Chief of the Environment Unit at the United Nations Centre on Transnational Corporations. He has written extensively on international industry and environment, and currently works with governments, multinationals and environmental groups around the world.

Benchmark Environmental Consulting, 49 Dartmouth Street, Portland, ME 04101, US
Tel: +1 207 775 9078; Fax: +1 207 772 3539; E-mail: benchmark@interramp.com

Bernhard Hanf has been the Environmental Manager at Hipp, Germany, since 1992. He studied Agricultural Science at the University of Hohenheim from 1984 to 1990, after which he completed a one-year professional development course in preparation for becoming an environmental advisor for industry.

Hipp, Münchener Straße 58, 85276 Pfaffenhofen/Ilm, Germany
Tel: +49 8441 7570; Fax: +49 8441 757492.

Ruth Hillary is a leading business and environment researcher at Imperial College's Centre for Environmental Technology where she undertakes EU- and UK-funded research into EMSs and small and medium-sized enterprises, in particular as Project Manager of a European Commission's DG XI EMAS pilot project. She is the UK National Co-ordinator for the European Commission's DG XXIII Euromanagement-Environment pilot action. She is the founder of the Network for Environmental Management and Auditing (NEMA) and a member of the UK Government's Advisory Group on EMAS. She is the author of *The Eco-Management and Audit Scheme: A Practical Guide* and the series editor for the Business and the Environment Practitioners Series.

Centre for Environmental Technology, Imperial College of Science, Technology and Medicine,
48 Prince's Gardens, London SW7 2PE, UK
Tel: +44 (0)171 589 5111; Fax: +44 (0)171 581 0245

Dick Hortensius, currently Senior Standardisation Consultant with the Netherlands Standardisation Institute (NNI), joined the organisation in 1985. He has been involved in many standardisation projects in the field of water and soil quality and, until last year, he was the Secretary of the ISO Technical Committee on Soil Quality. His current focus is on standardisation of environmental management tools and he is responsible for the ISO subcommittee on environmental auditing as well as being involved in the management of ISO/TC 207. He is also closely involved with the activities of CEN (the European standards organisation) in implementing ISO 14001 as the European standard within the framework of the EMAS Regulation.

In his current position at NNI, he is the co-ordinator of the standards programme in the environmental field and standardisation activities for management systems.

NNI, PO Box 5059, 2600 GB Delft, Netherlands
Tel: +31 15 2690 115; Fax: +31 15 2690 190; E-mail: dick.hortensius@nni.nl

---

Jim Hutchison is a consultant in environmental management, having previously spent ten years with the British Standards Institution. He is a Research Associate and Senior Lecturer in Environmental Management at the University of Hertfordshire, where he is involved in the development of the University's new Masters Degree in Environmental Management and Business.

University of Hertfordshire, Faculty of Natural Sciences, Department of Environmental Science,
Hatfield Campus, College Lane, Hatfield, Herts AL10 9AB, UK
Tel: +44 (0)1707 284 000; Fax: +44 (0)1707 285 258

Sven Hüther is currently writing his thesis in collaboration with Hipp, analysing the energy flows in relation to environmental cost management. Previously he was involved in two research projects implementing EMAS with the IÖW (Institute for Ecological Research in Economics). He will complete his management studies, which concentrate on environmental management, at the Polytechnic in Fulda in Spring 1997.

Hipp, Münchener Straße 58, 85276 Pfaffenhofen/Ilm, Germany
Tel: +49 8441 7570; Fax: +49 8441 757492

Tomoko Kurasaka is a Co-chair of the Environmental Auditing Research Group (EARG), a non-profit-making research organisation on environmental auditing based in Japan. The EARG aims to help efforts to promote an environmentally-aware society by providing information on environmental auditing. Most of the EARG's publications are only available in Japanese, but a summary in English is available for its latest research report, *Benchmarking of Corporate Environmental Reports*. Currently, the EARG is working on a research project on the EMSs of local governments.

EARG, 3F, Sushitetsu Building, 2563 Higashi-Ikebukuro, Toshima-ku, Tokyo 170, Japan
Tel: +81 3 3443 1782; Fax: +81 3 5396 4413

Riva Krut, PhD, a Director of Benchmark Environmental Consulting, is an organisational and management consultant. Her academic research and business consulting has been with organisations and clients in Europe, the US and Africa. She has published widely on corporate and industrial global environmental management.

Benchmark Environmental Consulting, 49 Dartmouth Street, Portland, ME 04101, US
Tel: +1 207 775 9078; Fax: +1 207 772 3539; E-mail: benchmark@interramp.com

Donal O'Laoire is a Director of Environmental Management and Auditing Services Ltd, Dublin, Ireland, and a member of the Faculty of Business, Economics and Social Studies at Trinity College, Dublin. He is a member of the International Drafting Committee for the new Industrial Standard on Environmental Management, ISO 14000. He has managed large-scale projects on environmental management in Ireland and Central Europe, specifically the development of an environmental and economic legal framework for planning and development in the Czech and Slovak Republics.

Environmental Management and Auditing Services Ltd, 14 Upper Lad Lane, Dublin 2, Ireland
Tel: +353 1 6613120; Fax: +353 1 6613172

Anatoly Pichugin graduated in 1978 from the Mendeleyev Institute of Chemical Technology, Moscow, with a Degree in Chemical Engineering, and completed a PhD in Physical Chemistry in 1983. At present he is a Research Fellow in the Division of Chemical Sciences at the University of Hertfordshire, working on an EU-funded research project. He is a joint initiator and co-ordinator of an Environmental Know-How Fund Project concerned with the development of training courses in environmental management for Russian industry, local government and regulatory agencies.

University of Hertfordshire, Faculty of Natural Sciences, Department of Environmental Science,
Hatfield Campus, College Lane, Hatfield, Herts AL10 9AB, UK
Tel: +44 (0)1707 284 000; Fax: +44 (0)1707 285 258

Nigel Riglar is a Registered Environmental Auditor, and has extensive environmental management experience in both private and public sectors. He has been the LGMB's EMAS advisor since January 1995 and has advised over 200 local authorities on various aspects of introducing EMAS.

He has developed the methodology of EMAS with a particular aim to make its introduction as simple and effective as possible. This applies especially to the approach that can be taken to ensure that EMAS fits seamlessly into existing management systems and which ensures that the scheme is genuinely integrated into the operations of the local authority.

The Local Government Management Board, Layden House, 76–86 Turnmill St, London EC1M 5QU, UK
Tel: +44 (0)171 296 6596; Fax: +44 (0)171 296 6666

David Shillito is an independent specialist in environmental and safety management and in the investigation of incidents, accidents and disasters. His specialisation in accident investigation has included a number of major disasters ranging from Flixborough (1974), Abbeystead, Kings Cross Underground Station Fire, Clapham Junction Railway Crash and the Piper Alpha Platform.

He was the Institution of Chemical Engineers' representative on the BSI committee on Environmental Management Systems which wrote BS 7750. In Spring 1994 he was appointed as an Environmental Assessor to the National Accreditation Council for Certification Bodies (now the UK Accreditation Service). With UKAS he has played a major role in the environmental accreditation programmes for BS 7750, ISO 14001 and EMAS.

David Shillito Associates, Hartland Cottage, 21 Bromley Common, Bromley, Kent BR2 9LU, UK
Tel : +44 (0)171 460 8896; Fax +44 (0)171 289 0186

Ann Smith is a Principal Lecturer and Reader in Aquatic Ecology at the University of Hertfordshire. She holds a BSc (Hons) Degree in Botany from the University of Adelaide, and a PhD from the University of London (Birkbeck College) and the Marine Biological Association (Plymouth). Ann has initiated various research projects and at present is responsible for the development of a new MSc/PgDip programme in Environmental Management for Business, which began in 1996.

University of Hertfordshire, Faculty of Natural Sciences, Department of Environmental Science,
Hatfield Campus, College Lane, Hatfield, Herts AL10 9AB, UK
Tel: +44 (0)1707 284 000; Fax: +44 (0)1707 285 258

Mark Smith is currently reading for a doctorate on Eco-design innovation in small and medium sized companies with the Design Innovation Group at the Open University. This research follows completion of a Masters degree in European Environmental Policy and Regulation at Lancaster University. Other research interests include Eco-labelling and transport issues. He has been most recently involved with the publication of 'The Commercial Performance of Green Product Development'.

Design Innovation Group, Design Discipline, The Open University, Walton Hall, Milton Keynes MK7 6AA, UK
Tel: +44 (0)1908 655019; Fax: +44 (0)1908 654052; E-mail: M.T.Smith@open.ac.uk

Andrea Spencer-Cooke MA, MEM, is a freelance consultant and Council Member and Senior Associate of SustainAbility Ltd. Her work focuses strongly on corporate governance and accountability issues, including environmental management systems and company environmental reporting. Previously, she worked for the World Business Council for Sustainable Development (WBCSD) and the International Labour Organisation (ILO) in Geneva. She has a multi-disciplinary background in environmental management and social anthropology and is based in Edinburgh, UK.

SustainAbility, 49–53 Kensington High St, London W8 5ED, UK
Fax: +44 (0)171 937 7447; E-mail: info@sustainability.co.uk

Philip M. Stoesser is a graduate of Environmental Studies from the University of Waterloo, in Ontario, Canada. His present position is Senior Advisor in the Environment and Sustainable Development Division of Ontario Hydro with responsibility for the ISO 14000 standards and the development of corporate environmental management system programmes.

Ontario Hydro, 700 University Avenue, Toronto, Ontario M5G 1X6, Canada
Tel: +1 (416) 592 3966; Fax: +1 (416) 592 7097; E-mail: phil.stoesser@hydro.on.ca

Tim Sunderland has been working for the international management consultancy firm, Arthur D. Little, for six years. He works with leading national and multinational companies helping them to identify key environmental issues, develop appropriate strategies and implement effective systems of management. He has been working in the environmental management field for fifteen years.

Arthur D. Little Ltd, Science Park, Milton Road, Cambridge CB4 4DW, UK
Tel: +44 (0)1223 420024; Fax: +44 (0)1223 420021

Philip Sutton is the Director of Policy and Strategy of Green Innovations, Inc., a non-profit-making think-tank and consultancy organisation. He developed the Flora and Fauna Guarantee legislation for the Australian State of Victoria, and now works on the implementation of environmental management systems for sustainability-seeking organisations.

Green Innovations, Inc., 195 Wingrove St, Fairfield, Melbourne, VIC 3078, Australia
Fax: +1 61 3 9486 4799; E-mail: psutton@peg.pegasus.oz.au

Andy Wells is the joint Managing Director of EMSi Limited. EMSi are international specialists in the provision of environmental management training and consultancy. Andy has recently completed a research project on 'Training and Environmental Management Systems' and the opinions expressed are his own. He has been involved in environmental management for over fifteen years, has over sixteen years' experience in the training field, and has advised many organisations in the development of environmental management systems. Andy was environmental advisor to the BSI Environment Office, a training adviser to the Institute of Environmental Assessment, training verifier to the Environmental Auditors Registration Association, and member of the EARA Training Committee. His experience includes training and implementation projects for the ISO 14000 series of international standards and international environmental management and training work in many countries including Hong Kong, Malaysia, Thailand, Singapore, Taiwan, and Indonesia, as well as for many commercial and industrial sectors in the UK.

EMSi, 26 Sandringham Rd, Norwich, Norfolk NR2 3RY, UK
Tel/Fax: +44 (0)1603 663 021